Praise for Sean B. Carroll's **BRAVE GENIUS**

"This is, in short, a gripping book throughout, and Carroll deserves all praise for his double portrait of two exemplary heroes of conscience and intellect."

—*Washington Post*

"Suspenseful . . . *Brave Genius* is briskly paced and ambitiously sprawling, offering potted accounts of historical episodes large and small (the fall of France, the 1956 Hungarian crisis, Camus's famous feud with Jean-Paul Sartre, the discovery of the double helix), along with finer-grained descriptions of Camus's and Monod's work. Dr. Carroll has done some impressive archival digging, turning up fresh and often vivid details."

—*New York Times*

"Carroll beautifully encapsulates how two men seemingly so far apart in their philosophies and achievements both ended up sharing 'exceptional lives' transformed by 'exceptional events.'"

—*Scientific American*

"Carroll does a masterful job of keeping the many elements together and the story moving. . . . In 1959, C. P. Snow wrote of the "two cultures"—that gulf between science and the humanities. *Brave Genius* provides an opportunity for those on both sides of the divide to sample a potent mix of genetics, philosophy, and literature, forged in the twentieth-century tumult of war and cold war."

—*Nature*

"An exciting and impressively told tale."

—*American Scholar*

"*Brave Genius* is scintillating in its entirety, reminding us that even in a time of profound adversity, it is genius, not misery, that loves—longs for, necessitates, thrives on—company."

—*Brain Pickings*

"Wonderful . . . a great mix of wartime history, philosophy, and the tale of how molecular biology was born."

—NPR

"A remarkable book . . . Not only does it manage to avoid reducing Camus to cliché (the 'philosopher' reference notwithstanding), it brings a new perspective on Camus to light. Carroll describes the intersection of these two lives in fascinating detail, but *Brave Genius* is also an important book because, at its heart, is an argument against allowing science to be directed by political expedience."

—*Sydney Review of Books*

"Readers will learn a good deal about symbolism in Camus's fiction and biochemistry in Monod's molecular biology. But, above all, they will learn about a luminous friendship forged in dark times. A rare chronicle of valiant thinkers fighting political oppression and transcending professional boundaries."

—*Booklist* (starred review)

"Carroll deftly weaves science and history together in his account of the lives, accomplishments, and friendship of two exceptional men. . . . Spanning history, science, and philosophy, this dual biographical study of two significant twentieth-century figures will appeal to a diverse audience."

—*Library Journal*

"Skillfully combining science, biography, and history . . . an important story well-told."

—*Kirkus Reviews*

"Carroll has a winning way with words, and everything he writes about (especially difficult matters of science) sparkles with clarity."

—*Publishers Weekly*

"A brave, ambitious, unexpected book. Who knew that Sean B. Carroll, a brilliant biologist, could or would write such a work of literary, political, and scientific history? It brings many revelations, offers several heroes, but at its heart is Jacques Monod, emerging as one of the great, complete men of the twentieth century."

—David Quammen, author of *Spillover* and *The Song of the Dodo*

"Art and science are two essential components forming the very essence of what makes being human worth being human. Sean Carroll has done a yeoman's job of merging these two vital areas beautifully in this moving and carefully researched history of two great minds and two brave men. It is impossible not to be inspired by their story."

—Lawrence M. Krauss, author of *The Physics of Star Trek* and *A Universe from Nothing*

"A remarkable profile. With deep research and compelling storytelling, Sean Carroll follows these two Nobel Prize winners from the desperate depths of World War II to international fame."

—Carl Zimmer, author of *Soul Made Flesh* and *Microcosm*

"A tour de force, a gripping narrative of a pivotal time in the history of Europe and of science. Finishing *Brave Genius*, I felt inspired by the determination of the key characters in the book, by their quest for liberty in the face of great injustice, and by the power their discoveries gave to understanding the living world."

—Neil Shubin, author of *Your Inner Fish* and *The Universe Within*

"The story of two brilliant men who understood better than anyone the randomness and absurdity of life, but who fought valiantly and fiercely to make the world a better place. History, personality, and ideas come together in this amazing tale of science, philosophy, and friendship."

—Sean M. Carroll, author of *From Eternity to Here* and *The Particle at the End of the Universe*

"Probably the best book I have read in a very long time, and a timely reminder of the deep links between science and moral thinking. It is really a work of genius itself."

—K. C. Cole, author of *Something Incredibly Wonderful Happens* and *The Universe and the Teacup*

ALSO BY SEAN B. CARROLL

Remarkable Creatures

The Making of the Fittest

Endless Forms Most Beautiful

BRAVE GENIUS

A Scientist, a Philosopher, and Their Daring Adventures from the French Resistance to the Nobel Prize

SEAN B. CARROLL

B\D\W\Y
Broadway Books
New York

Published in the United States by Broadway Books, an imprint of the
Crown Publishing Group, a division of Random House LLC,
a Penguin Random House Company, New York.
www.crownpublishing.com

BROADWAY BOOKS and its logo, B \ D \ W \ Y, are trademarks of
Random House LLC.

Originally published in hardcover in the United States by Crown Publishers,
an imprint of the Crown Publishing Group, a division of Random House LLC,
New York, in 2013.

Permissions are listed on page 582, which constitutes an extension of this page.

Library of Congress Cataloging-in-Publication Data
Carroll, Sean B.
Brave genius : a scientist, a philosopher, and their daring adventures from the
French resistance to the Nobel prize / Sean B. Carroll.—First edition.
pages cm
Includes bibliographical references.
1. France—Intellectual life—20th century. 2. Nobel Prize winners—France—
Biography. 3. Camus, Albert, 1913–1960. 4. Authors, French—20th century—
Biography. 5. Authors, Algerian—20th century—Biography. 6. Monod, Jacques.
7. Molecular biologists—France—Biography. 8. World War, 1939–1945—
Underground movements—France. 9. Politics and culture—France—History—
20th century. I. Title.
D802.F8C3715 2013
572.8092—dc23
[B] 2012050707

ISBN 978-0-307-95234-9
eBook ISBN 978-0-307-95235-6

Title page spread art © Berliner Verlag/Archiv/dpa/Corbis
Book design: Lauren Dong
Cover design: Elena Giavaldi
Cover photographs: (Monod) Institut Pasteur; (Camus) Martinie/Getty

First Paperback Edition

146122990

For Olivier and Philippe Monod, Agnès Ullmann, and Geneviève Noufflard; with my deepest respect and gratitude

We live in deeds, not years;
in thoughts, not breaths;
In feelings, not in figures on a dial.
We should count time by heart-throbs. He most lives
Who thinks most, feels the noblest, acts the best.

—PHILIP JAMES BAILEY, *Festus*

Contents

IV. NOBEL THOUGHTS AND NOBLE DEEDS

EPILOGUE: FRENCH LESSONS

PROLOGUE

CHANCE, NECESSITY, AND GENIUS

Genius is present in every age, but the men carrying it within them remain benumbed unless extraordinary events occur to heat up and melt the mass so that it flows forth.

—Denis Diderot (1713–1784), "On Dramatic Poetry"

On October 16, 1957, Albert Camus was having lunch at Chez Marius in Paris's Latin Quarter when a young man approached the table and informed him that he had won the Nobel Prize for Literature.

The new laureate-to-be could not hide his anguish.

Sure, the Algerian-born French writer had been an international figure for more than a decade. He had earned great public admiration for his moral stands as well as for his novels, plays, and essays. But not yet forty-four years old, Camus was only the second youngest writer ever to receive the Nobel. He thought that the prize should honor a complete body of work, and he hoped that his was still unfinished. He dreaded that all of the fanfare surrounding the prize would distract him from his work. The demand for interviews and photographs, and the many party invitations that followed the announcement soon confirmed his fears.

Camus also worried that the prize would inspire even greater contempt on the part of his critics. Despite his public popularity, Camus had many foes on both the political right, to whom he was a dangerous radical, and the left, among them many former close comrades who had ostracized him for his clear-eyed, damning critiques of Soviet-style Communism. Both camps took the Nobel as proof that Camus's talent and influence had already peaked.

"One wonders whether Camus is not on the decline and if . . . the

Swedish Academy was not consecrating a precocious sclerosis," wrote one scornful commentator.

After the demand for interviews subsided, he paused to reply to a few well wishers. One handwritten letter was to an old friend in Paris:

> *My dear Monod.*
>
> *I have put aside for a while the noise of these recent times in order to thank you from the bottom of my heart for your warm letter. The unexpected prize has left me with more doubt than certainty. At least I have friendship to help me face it. I, who feel solidarity with many men, feel friendship with only a few. You are one of these, my dear Monod, with a constancy and sincerity that I must tell you at least once. Our work, our busy lives separate us, but we are reunited again, in one same adventure. That does not prevent us to reunite, from time to time, at least for a drink of friendship! See you soon and fraternally yours.*
>
> *Albert Camus*

Camus knew well many of the literary and artistic luminaries of his time, such as Jean-Paul Sartre, George Orwell, André Malraux, and Pablo Picasso. But the recipient of Camus's heartfelt letter was not an artist. This one of his few constant and sincere friends was Jacques Monod, a biologist. And unlike so many other of Camus's associates, he was not famous, at least not yet. However, despite his pantheon of numerous, more illustrious colleagues, Camus claimed, "I have known only one true genius: Jacques Monod."

Eight years after Camus, that genius would make his own trip to Stockholm to receive the Nobel Prize in Physiology or Medicine, along with his close colleagues François Jacob and André Lwoff.

Each of the four men's respective prizes recognized exceptional creativity, but they also marked triumphs over great odds. The adventure to which Camus referred in his letter began many years earlier, in a very dark and dangerous time. So dangerous, in fact, that the chances each of these men would have even lived to see those latter days, let alone to ascend to such heights, were remote.

This is the story of that adventure. It is a story of the transformation of ordinary lives into exceptional lives by extraordinary events—of courage in the face of overwhelming adversity, the flowering of creative genius, deep friendship, and of profound concern for and insight into the human condition.

Chance and Necessity

Several years after he won the Nobel Prize, Jacques Monod wrote a popular, philosophical perspective on the significance of modern biology for understanding humankind's place in the universe. The title he chose, *Chance and Necessity,* was taken from Democritus's dictum "Everything in the universe is the fruit of chance and of necessity." It would have been an equally apt title for Monod's autobiography, or that of any of the other three laureates. The paths their lives took, and the twists that brought them together as comrades, friends, and collaborators, were very much a product of the circumstances imposed upon them and the responses those compelled—of chance and necessity.

Many years before the honors they received in Stockholm, in the spring of 1940, the four men were living in Paris, quietly pursuing separate, ordinary lives. Camus was an aspiring but unknown twenty-six-year-old writer, working as a layout designer for the newspaper *Paris-Soir* to make ends meet while toiling on a novel in his spare time. Jacques Monod was an underachieving and, at age thirty, relatively old doctoral student in zoology at the Sorbonne. François Jacob was a nineteen-year-old second-year medical student intent on becoming a surgeon. Thirty-eight-year-old André Lwoff was the only established professional among the four; he directed the department of microbial physiology at the Pasteur Institute.

Then, in May 1940, catastrophe struck.

The German Army invaded and quickly overwhelmed France, plunging the country into chaos. This stunning event was the perverse catalyst that, as Diderot prescribed, allowed their genius to flow forth, that set the men on new paths to future greatness and into one another's lives.

Knowing of the merciless destruction wreaked upon Poland by the same army the previous fall, millions of French citizens fled the approaching Germans. Jacob was horrified at the sudden disintegration of the country but determined to carry on the fight against Hitler wherever he could. He made the agonizing decision to leave his family and France, and boarded one of the last available boats to England. There, he joined the Free French forces. He would not see his family or step foot on French soil for four years. The next time he saw Paris, it was from the stretcher of an ambulance, encased in a body cast, recovering from near-fatal wounds. The Stuka's bomb that ended his career as a surgeon was the beginning of his eventual path into science.

Monod, Camus, and Lwoff remained in France, bearing witness to the progressively harsher life under Nazi occupation. Over the next four years, occupation evolved into oppression and enslavement, accompanied by torture, deportations, and mass murder. Each man was inspired to join the Resistance against the Germans and to contribute whatever talents were useful.

For Camus, tubercular and not fit for physical action, that meant working for the underground Resistance newspaper *Combat*. During the Occupation, Camus had managed to publish his novel *The Stranger* (*L'Étranger,* 1942), along with a book-length essay *The Myth of Sisyphus* (*Le Mythe de Sisyphe,* 1942), and had even completed two plays. He was becoming known in literary circles around Paris. But the need for secrecy and anonymity in the Resistance was such that he could not reveal his identity to his comrades. Introduced under an assumed name, he simply offered his previous newspaper experience to the group. Camus began by helping select and edit articles and prepare the layout of the paper. Later, he took over as editor. Camus's voice, which in peacetime might have been limited only to the salon or the theater, found a much grander stage at the pivotal turn of the war. In his inspiring, albeit anonymous, essays and editorials, Camus exhorted *Combat* readers to take action against the German occupiers and their French collaborators: "Frenchmen, the French Resistance is issuing the only appeal you need to hear . . . Anyone who isn't with us is against us. From this moment on there are only two parties in France: the France that has always been and those who shall soon be annihilated for having attempted to annihilate it."

After being involved in the dissemination of some underground newspapers, Jacques Monod sought more direct action against the Germans. After getting his Jewish wife settled under a false identity outside of Paris, he joined the best-armed and most militant resistants, the Communist Francs-Tireurs et Partisans (FTP). He emerged as a highly capable officer and rose to become a high-ranking member of the general staff of the French Forces of the Interior (FFI), the organization that coordinated Resistance activities in the latter stage of the Occupation. Monod organized the gathering of weapons and ammunition, planned sabotage that disrupted troop movements and supplies, and helped coordinate the civilian uprisings in Paris as the Allied forces approached.

It was nerve-wracking work with deadly stakes. The threat of discovery and arrest by the Gestapo was ever-present. Capture meant either deportation to a concentration camp or execution. Several of Monod's comrades and superiors in the Resistance were arrested, deported, or shot. Monod had to go completely underground, wear a disguise, and hide out—in Lwoff's laboratory in the attic of the Pasteur Institute. Lwoff also participated in the Resistance: in his Paris apartment, he sheltered Allied airmen who had been shot down so that underground networks could smuggle them out of the country.

Many of Camus's associates were exposed. *Combat*'s printer shot himself in order not to be taken alive and to risk divulging the names of other resistants. One of the few members who knew Camus's real identity was arrested on a day she was to meet with him, and was deported to the Ravensbrück concentration camp. Camus himself was questioned by the police while carrying the layout for an issue of *Combat*. They did not find the layout, and Camus was released.

Such risks of participating in the Resistance were necessary, as Camus argued so compellingly in *Combat*. For individuals, to join was to acknowledge that the ongoing fight concerned every citizen. In taking action one person could inspire others and, Camus suggested, "at least share in the peace at heart that the best of us take with them into the prisons." Courage and sacrifice in the face of extreme danger were the only available remedies for the humiliation of military defeat and, perhaps more important, for expunging the shame felt over those French who collaborated with the Germans and heaped suffering and death upon their fellow citizens.

Resistance was also a matter of strategic importance to the Allied military effort. Supreme Allied Commander Gen. Dwight D. Eisenhower credited the FFI with greatly accelerating the advance of the Allied forces after their landings in Normandy and on the Riviera, speeding the liberation of the country, and reducing Allied losses. Eisenhower estimated the effect of the Resistance to that of fifteen divisions (approximately 150,000 regular army troops). Their losses were certainly significant: some 24,000 resistants were killed in the battles for France.

Regardless of its exact quantitative effect, the effect of the Resistance on repairing the French psyche was enormous. On the eve of Paris's liberation, Camus declared: "Four years ago, a few men rose up amid the ruins and despair and quietly proclaimed that nothing was yet lost. They said that the war must go on and that the forces of good could always triumph over the forces of evil provided the price was paid. They paid that price."

Only after the liberation of Paris did his readers learn who had actually composed such moving passages in the middle of the battle, and they loved him for it.

For his part, Camus had learned just how much words matter. He later admitted, "To risk one's life, however little, to have an article printed is a way of learning the real weight of words."

Secrets of Life

After the war, each man returned to his livelihood or, in Jacob's case, forged a new one. Like many others for whom normal life had paused during the war, and whose experiences had imparted a profound appreciation of the fragility of life and freedom, they were each imbued with a much greater sense of urgency and purpose.

Camus focused much of his writing on the moral and political renewal of the French nation. From the moment of the liberation of Paris, *Combat* enjoyed a unique prestige. The newspaper would sell out as soon as it was published. In what Claude Bourdet, a leader of *Combat*, described as one of "those accidents which condition the life of individuals, if not societies," Camus had a perfect national pulpit from which to voice his concerns and ideas. In scores of articles, Camus urged that France be rebuilt upon basic principles of equality, individual freedom, and social justice. His editorials were often the talk of Paris.

Readers also had the opportunity to discover the literary and philosophical works that Camus had written and managed to publish during the war. The terror and cruelty of the Occupation, the slaughter of tens of millions in the war (the second such war in a generation), and the horrors of the Holocaust that were coming to light had made many despair and abandon any hope for the future of humanity. Denial of any meaning or purpose in life—nihilism—was a widespread response.

But Camus vehemently rejected nihilism and took an entirely different path. In *The Myth of Sisyphus,* Camus addressed what he contended was the fundamental issue of philosophy—"judging whether life is or is not worth living." To Camus, the crux of the matter of life was the certainty of death. The practical question that certainty prompted was: How could one live a meaningful life in full knowledge of the inevitability of death?

Camus asserted that by recognizing the reality of the physical limits of one's life, one attained the clarity and freedom to make the most of life as it is. He reasoned that the logical response to the certainty of death was a revolt against death—a revolt that took the form of living life passionately and to the fullest: "Being aware of one's life, one's revolt, one's freedom, and to the maximum, is living, and to the maximum."

Camus's recipe for living life to the fullest was to do nothing in hope of an afterlife, and to rely on courage and reasoning: "The first teaches him to live without appeal [to religion] and to get along with what he has; the second informs him of his limits. Assured of his temporally limited freedom . . . and of his mortal consciousness, he lives out his adventure within the span of his lifetime."

For Camus, even Sisyphus—condemned as he was to rolling his rock uphill each day, only to have it roll back down and to begin again—was master of his own fate. Sisyphus created meaning in his own life by deciding that "the struggle towards the heights is enough to fill a man's heart." Camus concluded the essay, "One must imagine Sisyphus happy."

His reasoned optimism, born as it was in the middle of the Occupation and war, struck a chord with readers recovering from the tragedies of World War II. Camus once wrote, "In the depths of winter, I discovered that there lay within me an invincible summer." Readers in France, and then as his works were translated, millions more readers around the world, responded to that invincible summer. Camus offered a practical philosophy for living without succumbing to nihilism or appealing to religion. In the aftermath of the great calamity, Camus offered the masses a picture of a brighter future for France and the world, an alternative to the cycle of war that had darkened a half century, and that threatened to continue. He offered a choice, as he put it, "between hell and reason."

His influence was widespread and profound. One *Combat* comrade stated, "Camus taught me reasons for living." François Jacob later described his pursuit of scientific research in the most Camusian terms, as "the most elevating form of revolt against the incoherence of the universe." The author and philosopher Jean-Paul Sartre, who was a close friend for many years, described Camus as "an admirable conjunction of a person, an action, and a work."

As Camus expounded the philosophical reasons for living, Monod and Lwoff were exploring the biological secrets of life. They were joined by Jacob in 1950.

In the early 1940s, the mysteries of life were vast. Little was known, for instance, of how cells operated. At the time, physics and chemistry were the dominant sciences. While it was certain that organisms were composed of molecules, the identity of the molecules that endowed cells with the properties of life were completely unknown.

In 1944, the famous physicist and Nobel laureate Erwin Schrödinger wrote a very influential, short book entitled *What Is Life?* that examined life from a physicist's perspective. At the time of his writing, the concept of the gene was well established, but no one knew what genes were made of. Schrödinger's account of the mysteries of the matter underlying living organisms inspired many young scientists to enter biology, not the least of whom were James D. Watson and Francis Crick, who solved the structure of DNA a decade later.

It was a time for simple but fundamental questions. Monod pursued the mystery of how cells grow. He rediscovered a phenomenon in which bacteria, when given two sugars as sources of energy, used one first and then the second. Monod was asking a simple question: How did the bacteria "know" which sugar to use?

Lwoff was interested in viruses that lay dormant within bacteria. He discovered that under certain conditions these latent viruses could, in effect, come back to life. When Jacob joined the research group, they asked another question: How did the virus "know" when to become active?

The pursuit of these two apparently unrelated simple questions began in cramped and spartan laboratories in the attic of the Pasteur Institute, and led to one of the most creative, original, and influential bodies of work in modern biology. Monod and Jacob, in particular, discovered several of the major secrets of life (after DNA). Foremost among them were the first understanding of how genes are switched on and off as cells grow, and the discovery of messenger RNA, the molecule that serves as the intermediate (hence "messenger") between genes in DNA and the proteins they encode. Monod and Jacob's insights were far ahead of their time. Biologists barely had a foggy picture of what a gene was when Monod and Jacob delivered an exquisite synthesis of the general *logic* governing how genes were used. Walter Gilbert, who shared the 1980 Nobel Prize in Chemistry, described the two Frenchmen as having "made things that were utterly dark, very simple."

Their discoveries were certainly deserving of the Nobel Prize, but Monod and Jacob also displayed an extraordinary creative style that was often described in such literary terms as "taste" and "elegance." Their exceptional eloquence was coupled with bold extrapolation. The two scientists anticipated and explained the broader implications of their work for understanding one of the greatest mysteries of biology—the development of a complex creature from a single fertilized egg. It would take several decades for biologists to penetrate that mystery in depth, but Monod and Jacob provided the conceptual foundations of that effort. And their scientific impact reverberated beyond academia, for their discoveries about the inner workings of bacteria and viruses provided key tools for the birth and growth of recombinant DNA technology and genetic engineering.

Rebels

Such achievements would be admirable legacies for any scientist, but these men's concerns and talents reached far beyond the laboratory. For the former resistant Monod, the battle against totalitarian regimes was not over when the war against Nazism was won. Monod's next major clash brought him into the orbit of and friendship with Albert Camus.

Soon after the end of World War II, a new war emerged—of ideologies. It was a war between capitalism and socialism, between democracy and Communism, and the beginning of the Cold War between the United States and the Soviet Union. In France, those along the entire spectrum of political ideologies from the far left to the far right vied for power and influence. The Communist Party enjoyed strong support, particularly among the intelligentsia and workers, many of whom looked to the Soviet Union as a model of where socialism in France should be heading.

During the war, Monod had joined the Communist Party as a matter of expediency, so that he could join the FTP. But he developed reservations about the Communists' intolerance of other political views and quietly quit the Party after the war, at a time when many fellow citizens were joining. That might have been the end of Monod's involvement with Communism, were it not for bizarre developments in the sphere of Soviet science.

In the summer of 1948, Trofim Denisovich Lysenko, Joseph Stalin's anointed czar of Soviet agriculture, launched a broad attack on the science of genetics. Lysenko believed that virtually any modification could be made rapidly and permanently to any plant or animal and passed on to its offspring. His belief, while consistent with Soviet doctrine that nature and man could be shaped in any way and were unconstrained by history or heredity, flew in the face of the principles of genetics that had been established over the previous fifty years. Nevertheless, Lysenko demanded that classical genetics, and its supporters, be purged from Soviet biology.

Lysenko's outrageous statements were heralded in Communist-run newspapers in France. Monod responded with a devastating critique that ran on the front page of *Combat*. Monod exposed Lysenko's stance on genetics as antiscientific dogma and decried Lysenko's power as a demonstration of "ideological terrorism" in the Soviet Union.

The public scrutiny damaged the credibility of Soviet socialism in France. The episode thrust Monod into the public eye and made him resolve to "make his life's goal a crusade against antiscientific, religious metaphysics, whether it be from Church or State."

AT THE TIME of Monod's editorial in *Combat,* Albert Camus was having similar thoughts about the evils of the Soviet regime with its show trials and labor camps, thoughts that would eventually be articulated in his book-length essay *The Rebel* (1951).

Monod and Camus were introduced at the meeting of a human-rights group and hit it off immediately. Their attraction to each other was deep. Although the two men had nothing in common in terms of their upbringing or professions, they were kindred spirits. Francis Crick described Monod in terms that applied equally well to his new friend Camus: "Never lacking in courage, he combined a debonair manner and an impish sense of humour with a deep moral commitment to any issue he regarded as fundamental." In addition to the special bond of former resistants, Monod and Camus discovered they shared many similar concerns. Over the course of their friendship, those concerns would encompass a broad spectrum of humanitarian issues, including the state of affairs in the USSR, human rights in Eastern bloc countries, and capital punishment in France.

Monod gave Camus further ammunition for his indictment of the

Soviet Union, an indictment that terminated many of Camus's friendships with left-wing peers. Camus gave Monod access to his world of literature and philosophy.

Monod, too, was a conjunction of work and action. While Camus wrote "The Blood of the Hungarians" (1957) to arouse the world's conscience about the Soviets' crushing of the Hungarian revolution, Monod used his clandestine experience from the days of the Resistance to organize the escape of Hungarian scientists. As Monod's fame grew from his scientific achievements, he used his standing to advance many causes, including reproductive and human rights, and he was a prominent figure in the May 1968 unrest that nearly toppled the French government.

Camus had a profound influence on Monod and the philosophical ideas the biologist pursued in later years. After receiving his Nobel Prize, Monod turned to consider the implications of the discoveries of modern biology—how the answers to Schrödinger's question "What is life?" bore on the question of the meaning of life. He explained his impulse in Camusian terms: "The urge, the anguish to understand the meaning of his own existence, the demand to rationalize and justify it within some consistent framework has been, and still is, one of the most powerful motivations of the human mind." The opening epigraph of Monod's resulting, widely acclaimed, bestselling book, *Chance and Necessity,* was the closing passage from his friend's *The Myth of Sisyphus.*

THIS BOOK TELLS the story of how each man endured the most terrible episode of the twentieth century and then blossomed into an extraordinarily creative and engaged individual. It is divided accordingly—the first half is the story of how the world shaped these men, and the second half is about how they shaped the world. The dividing line is the liberation of Paris, for the preceding war and occupation were the crucible in which their characters were tested, and from which they subsequently rose to such brilliance.

Their close associates also possessed great courage and risked their lives for freedom. Two such heroines, Geneviève Noufflard and Agnes

Ullmann, have allowed their extraordinary stories to be told here largely for the first time. Indeed, this book was made possible by the discovery of and access to a great deal of previously unknown and unpublished material: letters and other exchanges between Monod and Camus, as well as eyewitness accounts of their decade-long friendship; Paris police files on Monod's initial activity in the Resistance; an unpublished wartime memoir by Noufflard, Monod's secretary in the Resistance, and original documents concerning their participation in historic events; a trove of private letters by Monod and his wife, Odette, and other family members; and a large cache of documents detailing Monod's efforts in arranging Ullmann's daring escape from Hungary.

What emerged from the many threads of Monod's and Camus's respective journeys was one story in common, the elements of which define four major episodes in their lives and form the four main sections of this book. These elements are the sudden and shocking fall of France (Part I—"The Fall"); the actions they took to fight back against the Nazis (Part II—"The Long Road to Freedom"); their initial explorations of the questions that would dominate their creative work (Part III—"Secrets of Life"); and the peak of their creative achievements and widening involvement in human affairs (Part IV—"Nobel Thoughts and Noble Deeds").

The Epilogue ("French Lessons") examines how, after Camus's death, Monod assumed part of his friend's mantle through his public commitments and writing. Both men were deeply engaged with timeless questions about finding meaningful experiences in life. They were forced to ask, by virtue of the experiences into which they were plunged, the most fundamental questions of all: What is worth dying for? And what is worth living for? Once free, they were compelled to ask: What is worth spending one's life pursuing? World War II, the Occupation, the Cold War, and the Hungarian Revolution belong to the past, but nothing has changed about the fundamental human yearning for meaning, and nothing has changed that alters the validity of their approaches. Monod argued that science had shattered traditional concepts of our purpose and place in the world. That being so, the choice remains of how to find meaning in a scientifically enlightened world.

Part One
The Fall

ALL GREAT DEEDS AND ALL GREAT THOUGHTS HAVE A RIDICULOUS BEGINNING.

—ALBERT CAMUS, *THE MYTH OF SISYPHUS*

The Arc de Triomphe in the blackout, Paris 1940.
(Photo by Brassai, *Lilliput* magazine, June 1940)

CHAPTER 1

CITY OF LIGHT

An artist . . . has no home in Europe except in Paris.
—Friedrich Nietzsche, *Ecce Homo*

Paris slipped very quietly into the New Year of 1940.

It was not the fresh blanket of snow, though one of the heaviest in fifty years, that muted the typically boisterous celebration of Le Réveillon de la Saint-Sylvestre. Nor was it the unusually cold spell that plunged Paris and much of France to well below freezing temperatures that night.

La Ville-Lumière (the City of Light) was dark and anxious. It had been so for four months.

On September 3, 1939, two days after Germany's invasion of Poland, France and her ally Great Britain had declared war. Blackouts were imposed across Paris to obscure potential targets from aerial bombing. The lights of the monuments and museums—the Louvre, the Eiffel Tower, and the Arc de Triomphe—were extinguished, the street lights along the grand boulevards and squares were veiled with a blue paint, as were automobile headlights, bicycle lights, and even handheld flashlights. Their blue beams cast an eerie, dim hue over the snow-covered city.

The cafés, clubs, cabarets, and restaurants were open on New Year's Eve, but their outside lights were off. Their windows and doors were covered to block the light from inside. The authorities extended closing time on this special occasion by three hours past the new wartime curfew, to two o'clock in the morning.

For more than two centuries, since the time of Les Lumières (the Enlightenment), when Voltaire and Diderot rethought civilization over coffee at Le Procope in the Latin Quarter, those cafés had drawn philosophers and revolutionaries from all over the world, including Benjamin Franklin and Thomas Jefferson. For the previous two decades, since the last war with Germany, the cafés and clubs of Paris had beckoned a remarkable generation of writers, artists, and musicians who made the city the artistic and intellectual center of Europe, if not the world.

In the 1920s and 1930s, the Paris literary scene drew the likes of James Joyce, Ezra Pound, F. Scott Fitzgerald, John Dos Passos, Gertrude Stein, and Samuel Beckett. Ernest Hemingway often installed himself at his favorite café, La Closerie des Lilas, in its garden of lilac trees, with notebooks, pencils, and pencil sharpener at hand. He composed some of his first novel, *The Sun Also Rises,* sitting at its marble-topped tables.

Every form of art flourished in the Montparnasse area. Salvador Dalí came to Paris from Spain and was the principal figure of the surrealists, while Russian-born Marc Chagall was a pioneer of modernism. Spanish cubist Pablo Picasso lived and worked at various times in Montmartre and Montparnasse and then settled on the rue de la Boétie, not far from the Champs-Élysées. The prolific painter was represented by Paul Rosenberg, whose well-known gallery was next door to Picasso's studio. Rosenberg would help make Picasso famous, selling his works alongside those by Monet, Degas, Matisse, van Gogh, Renoir, and Cézanne.

The music scene also thrived. Josephine Baker, Cole Porter, Coleman Hawkins, and Benny Carter came from the United States. In 1934, Belgian-born Gypsy guitarist Django Reinhardt, Parisian violinist Stéphane Grappelli, and three others formed the sensational Quintette du Hot Club de France, the most original and influential European jazz group of the era. Native legends Edith Piaf and Maurice Chevalier were immensely popular.

Paris's creative life was not exclusively the domain of artists. Science prospered as well. In 1939, Frédéric Joliot-Curie, a leading researcher on the splitting of the uranium atom, had his laboratory at the Collège de France in the Latin Quarter. Joliot-Curie, who shared the 1935 Nobel Prize in Chemistry with his wife, Irène (daughter of

Nobel laureates Pierre and Marie Curie), recognized the possibility of producing a chain reaction that would liberate massive amounts of energy. Joliot-Curie was one of the key scientists Albert Einstein cited in an August 1939 letter to President Roosevelt alerting him to the discoveries in physics that could possibly lead to the making of "extremely powerful bombs of a new type."

Fewer than two miles from the Collège de France was the crown jewel of French biology and one of the premier research institutions in the world—the Pasteur Institute. The institute was extending the many frontiers opened or advanced by Louis Pasteur (1822–1895). The primary catalyst to its formation was Pasteur's pioneering efforts in developing vaccines. It was founded specifically to treat rabies. On July 6, 1885, a nine-year-old boy named Joseph Meister was brought to Pasteur's laboratory by his desperate mother. A rabid dog had bitten Joseph fourteen times. The severe bites would surely be fatal, so Pasteur decided to try to treat Joseph with an experimental rabies vaccine that, up to that point, had only been tested on dogs. After thirteen injections over the course of eleven days, miraculously the boy survived.

After news of Meister's case spread, several children from Newark, New Jersey, were sent to Pasteur and also treated successfully with the new vaccine. The resulting acclaim led to an international fund-raising effort to establish an institute, initially under Pasteur's direction, that enabled thousands to be treated. Pasteur recruited other scientists with a similar, almost monastic, devotion to science. They would come to refer to themselves as "Pastorians." This tribe of Pasteur's associates and protégés led the world in understanding, preventing, and treating infectious diseases such as diphtheria, malaria, yellow fever, bubonic plague, typhus, and tuberculosis, and garnered four Nobel Prizes in Medicine or Physiology in just the first few decades after the initiation of the Prize. Those whose lives were touched, or saved, by Pasteur felt deep gratitude. Almost fifty-five years after being the first person treated, sixty-three-year-old Joseph Meister was working as a caretaker of the Institute.

When war was declared, the populace was placed immediately on high alert. Gas masks were issued and air-raid sirens sounded frequently. Thousands of children were moved out of the capital and into the countryside, as were most of Paris's most treasured works of art. Over the next four months, the Louvre was almost completely emptied. More than two hundred truckloads of paintings and sculptures, including the *Mona Lisa* and the *Venus de Milo,* were crated and shipped from the museum and stored in chateaux for safekeeping.

The threat of war prompted some artists and performers to leave the city or France altogether, but not Maurice Chevalier. In late 1939, he recorded "*Paris sera toujours Paris*" (Paris will always be Paris), a love song to his hometown that captured her new look under the blackouts and was a boost to her defiant, resilient spirit:

Par précaution on a beau mettre
Des croisillons à nos fenêtres
Passer au bleu nos devantures
Et jusqu'aux pneus de nos voitures
Désentoiler tous nos musées
Chambouler les Champs-Élysées
Emmailloter de terre battue
Toutes les beautés de nos statues
Voiler le soir les réverbères
Plonger dans le noir la ville lumière

Even if one puts for precaution
Latticework on our windows
Blue on our storefronts
And up to the tires of our cars
Removes the paintings from our museums
Turns the Champs-Élysées upside down
Swaths the beauty of our statues in clay
Veils the streetlamps in the evening
Plunges the city of light into darkness

Paris sera toujours Paris,
La plus belle ville du monde,
Malgré l'obscurité profonde,
Son éclat ne pert être assombri

Paris will always be Paris,
The most beautiful city in the world,
Despite the profound darkness,
Her luster cannot be dimmed

—ʍ—

WITH THE EXPERIENCE of World War I, during which France lost 1.4 million lives, still fresh in their memories, political and military leaders and civilians alike had been hoping there would not be another war. But Hitler's actions over the previous two years had convinced many of the likelihood, and some even the necessity, of battle with Germany. Nonetheless, even after the war declarations, there was still some glimmer of hope that an all-out conflict could be averted. Indeed, France and Britain had retreated from the very brink of clashing with Germany just one year earlier, in September 1938.

The path toward war began with Hitler flouting the terms of the 1918 armistice by rebuilding Germany's armed forces, and then expanding the Reich's territory through a series of military threats and political maneuvers. The Führer's moves were guided by his perception of the will, or lack thereof, of France and Britain to oppose him. He assumed that both nations wanted to avoid another bloody conflict at almost any price, even if that meant ceding the control of much of central and eastern Europe to Germany. Hitler tested the Allies' resolve at every turn.

In March 1938, Austria was intimidated by the specter of an armed invasion and manipulated politically into accepting annexation by Germany. France and Britain made no significant objections or gestures. Encouraged by his swift, bloodless takeover, and the Allies' reticence, Hitler then set his sights on Czechoslovakia, which would push all of Europe to the brink.

Hitler's aims were to absorb the Sudetenland and to conquer what remained of Czechoslovakia. How to do so without arousing France, which was bound by treaty to come to Czechoslovakia's aid if it were attacked, and Britain, which was committed to aid France if she were attacked, was a tricky proposition. But Hitler assumed that neither France nor Britain would risk a European war over Czechoslovakia, so he plotted a takeover.

After months of military, political, and diplomatic maneuvering, the situation reached a crisis in September 1938. With Hitler threatening an invasion of Czechoslovakia that would trigger their obligations, Britain and France sought some resolution that would appease Hitler and relieve them of their respective commitments. The balance of considerations was delicate. On the one hand, Britain and France could not appear too reluctant for war, or the Führer would take that as a sign of weakness to be exploited. On the other hand, they could not take too aggressive a stance, as that might provoke the belligerent dictator into a war that might escalate quickly, with unknowable consequences. There was also the matter of honor, a commitment to an ally that, if broken, would undermine the reliability of all commitments and the security of the Continent. And finally, there was public opinion, which shifted unpredictably as events unfolded. Governments that ignored this last variable did so at great risk to their longevity.

Britain's prime minister, Neville Chamberlain, initiated face-to-face negotiations with Hitler on September 15, 1938. It soon became clear to Chamberlain that the price of peace would be Czechoslovakia, or at least the Sudetenland. France's premier, Édouard Daladier, supported Chamberlain's efforts to avoid war. Chamberlain and Daladier pressured the Czech leadership to concede to Hitler's demands for the Sudetenland in order to keep the peace. The Czechs rejected the demands. France then upped the pressure by asserting that in rejecting their proposal, Czechoslovakia assumed responsibility for military action by Germany, and informed the Czech government that France would now not act if Germany invaded. The Czech government was cornered and had no choice but to bow to the demands; it could not resist Germany on its own.

Chamberlain brought the Czech concession back to Hitler on September 22. Although only a week had passed, Hitler now rejected the Czechs' capitulation as insufficient and increased his demands, which included the immediate military occupation of the Sudetenland. Chamberlain was surprised and exasperated at the Führer's change in posture and returned to London crestfallen. His cabinet rejected Hitler's new demands, as did the French and the Czechs.

In the meantime, the Czechs mobilized their armed forces and the French followed suit. The white posters plastered all over France on the morning of September 24 announced the immediate call-up of nearly a million men. French armed divisions were moved to the border with Germany.

To the general populations of all countries involved, war now appeared inevitable and imminent.

Daladier conferred with Chamberlain in London, who decided to attempt one last diplomatic effort to dissuade Hitler. Britain and France made an about-face from their previous abandonment of the

Czechs a week earlier and informed Hitler on September 27 that they would stand by Czechoslovakia if Germany attacked.

Hitler was apoplectic. He replied by vowing to destroy Czechoslovakia and to be at war with France and Britain within a week.

But, aware that the Czechs and French were mobilizing, that their combined armies were double that of the German forces, and that Britain was also readying for battle, Hitler shortly reconsidered. He wrote to Chamberlain that he was now prepared to "give a formal guarantee for the remainder of Czechoslovakia."

Chamberlain seized on the reopening of dialogue. While the citizens of each country braced for war, with many fleeing the cities via traffic-choked roads, a last-ditch campaign unfolded. Chamberlain proposed a conference to Hitler, and asked Italian prime minister Benito Mussolini to do the same. Mussolini complied; Hitler agreed and proceeded to invite Chamberlain, Daladier, and Mussolini to a summit of the four powers in Munich (the Czechs were not invited).

At the news of the invitation, Britain's war-anxious House of Commons erupted with cheers. Paris was equally relieved and hopeful. The heads of state went to Munich determined to secure the peace—the price of which was Czechoslovakia, which was to be partitioned along lines that would satisfy the Führer. The four powers promptly signed the accords on September 30. The Czechs were left with no option; as their official communiqué stated, they had been "abandoned."

Chamberlain and Daladier were greeted at home by cheering throngs. Daladier addressed the nation: "I return with the profound conviction that this accord is indispensable to the peace of Europe. We achieved it thanks to a spirit of mutual concessions and a close collaboration."

The Paris newspapers gushed with praise and relief. Former premier Léon Blum said in *Le Populaire*: "There is not a woman or man in France who would refuse MM. Neville Chamberlain and Édouard Daladier their just tribute of gratitude. War is spared us. The calamity recedes. Life can become natural again. One can resume one's work and sleep again. One can enjoy the beauty of an autumn sun."

Privately, however, Daladier had learned the lessons of dealing with

Herr Hitler. "We can never deal with Germany except with force," he told two of his generals just days after the Munich pact.

When Daladier became premier, he maintained his position as minister of war and national defense. Shortly after the Munich pact, he committed 40 billion francs, nearly 85 percent of France's tax revenue for 1939, to rearmament, as well as an additional 2.5 billion francs in a secret deal to acquire one thousand aircraft from the United States.

Daladier's mistrust of Hitler was validated in March 1939 when the Führer, mocking the assurances given in Munich just six months earlier, engineered a full Nazi takeover of what remained of Czechoslovakia. Britain and France could merely protest what was an overnight fait accompli. Poland was then surrounded on three sides by the Reich, and surely would be its next quarry.

Daladier told his cabinet: "There is nothing more to do than prepare for war." In order to increase the standing army, he increased the length of service for military reservists. With respect to future spending, he insisted to a cabinet committee, "We should not devote a single dollar of our reserves to nonmilitary purposes. It is indeed necessary to go further: the dollars and gold of which we dispose should be devoted entirely to the purchase of airplanes in the United States . . . With that sum, we will be able to create a powerful air fleet, thanks to which we will crush the Ruhr [an industrial center in Germany] under a deluge of fire, which will lead Germany to capitulate . . . it is the only means of finishing the war. I do not see another."

The time for appeasement had passed. The military leadership prepared war plans. By July 1939, Gen. Maxime Weygand claimed that, due to the rearmament initiatives, France had the best-equipped army in the world and that there was no doubt of victory. The general's confidence was bolstered by the facts that the French Army, which had reached more than 2.4 million men by late August 1939, was comparable in size to the German Army, and that the French held an advantage in the number and quality of tanks. With the additional security of the heavy fortifications of the Maginot Line, which ran the length of the French-German border, the French leadership firmly believed that it would be folly for Germany to attack, but if they did so, together with the help of France's Allies (mainly Britain), Germany would be defeated, as in the previous war.

Public opinion had also shifted as Hitler's territorial thirst appeared ever more insatiable. France and Britain had commitments to aid Poland if it was attacked, and by late August Poland was clearly in Hitler's gun sights. This time, however, last-minute diplomatic heroics similar to those during the Czech crisis of the previous year failed. Hitler gave the order to attack, and on the morning of September 1, the Wehrmacht and Luftwaffe began to pummel Poland.

Daladier ordered a general mobilization on September 2.

While split along many political and ideological lines before September, the diverse press quickly adopted the same themes. *L'Intransigeant* said, "War has been imposed on France and she has no other choice but to fight; France and her Allies are fighting a Nazi-created religion of hatred, brutality, and lies."

In *Le Populaire,* former prime minister Léon Blum wrote, "The Nazis have compelled the most peaceful of nations to go to war for the defense of her liberty, existence, and honor."

Across the political spectrum, from Socialists to conservatives, the war was unwanted but had become a necessity. It was to be a "just war," according to the Catholic daily *La Croix,* to decapitate "the modern Attila," a "struggle between civilization and barbarity."

If the last war was to be any guide, this spirit and unity were going to be essential for the struggle ahead.

The French military leaders, however, did nothing to save Poland in the first crucial moments of the war. After a week's delay, they finally launched an invasion of the German Saarland. The offensive was given enthusiastic coverage in the press, one newspaper calling it a "brilliant attack." But in reality, the Army moved just five miles into Germany, into territory where villages had already been evacuated.

Despite overwhelming superiority on the western front—with some eighty-five well-armed French divisions facing thirty-four largely reserve German divisions—the French did not attack and thus did not draw away any of the pressure from Poland, and neither did they threaten any vital German areas. Poland crumbled in eight days. With

Poland annihilated, the French command secretly ordered a retreat from the Saarland at the end of the month.

But Germany took no significant action against France, either. Days, then weeks, passed. Troops worked and even played in full sight and range of each other across the front. The military communiqués published on the front pages of the Paris newspapers became progressively shorter and repetitive: "Night calm on the entire front" or "Nothing to report" or "Routine patrols" were standard entries.

The war acquired new names. At first it was *la guerre d'attente* (the war of waiting). In England, it was dubbed the "Bore War," then the "Phoney War." Soon, a new moniker was offered by Henri Lémery in *Paris-Soir—"la drôle de guerre"*—the funny war.

—~—

By the New Year, four months had passed, and still nothing had happened. The long pause nourished hope that perhaps with further patience, resolve, and diplomacy, calamity might again be avoided, as it had been so narrowly in 1938.

There were many soldiers on leave in Paris restaurants, cabarets, and theaters that cold night of La Saint-Sylvestre. Neither they, nor the citizens who toasted the New Year with them, could know that the Phoney War was then half over, that there would be four more months of waiting. Nor could they imagine that this would be the last such celebration in a free Paris for a very long time.

The last verse of Chevalier's song played on radios and phonographs that night:

Même quand au loin le canon gronde	Even when the cannon is roaring in the distance
Sa tenue est encore plus jolie . . .	Her dress is even prettier
Paris sera toujours Paris!	Paris will always be Paris
On peut limiter ses dépenses,	One can limit what she spends
Sa distinction son élégance	Her distinction, her elegance

N'en ont alors que plus de prix	Are only all the more priceless
Paris sera toujours Paris!	Paris will always be Paris

as Parisians pondered what the New Year would bring. *Le Figaro* assured that: "Throughout this night, on each floor, deep in everyone's heart, the same burning hope arose: 'that 1940 will be the year of victory.'"

CHAPTER 2

PLANS

Since France, the deadly enemy of our people, is pitilessly choking us and depriving us of power, we must not shrink from any sacrifice on our part that will contribute to the destruction of France as the master of Europe.

—ADOLF HITLER, *Mein Kampf* (1926)
(banned in France)

ACROSS THE BORDER IN BERLIN, THE NEW YEAR'S MOOD WAS DEcidedly different. In a thirty-minute radio address, Joseph Goebbels, the Reich minister for public enlightenment and propaganda, reviewed the past year and looked ahead to 1940:

> The year 1939 was so dramatic and filled with historical splendors that one could fill a library writing about them. One hardly knows where to begin . . . our people began to restore its national life in 1939, beginning a great effort to throw off the chains of constraint and slavery and to once again take our place as a great power after our deep fall.

He justified the takeover of what remained of Czechoslovakia after annexation of the Sudetenland in late 1938. Then he rationalized the invasion of Poland, blaming the "London warmongering clique," a series of purported incidents against Germans, and provocations by Poland (such as mobilizing its reserves) as warranting action by Germany. "The Führer had no alternative but to answer force with force," he claimed. Goebbels exonerated Germany of any blame for the current climate in Europe and railed against the French and British governments:

> On September 2, London and Paris gave Germany an ultimatum, and declared war against the Reich soon after . . . The war of the Western powers against the Reich had begun . . . No one can doubt that the warmongering cliques in London and Paris want to stifle Germany, to destroy the German people . . . We 90 million in the Reich stand in the way of their brutal plans for world domination . . . They have forced us into a struggle for life and death.

Finally, Goebbels looked into the future:

> It would be a mistake to predict what will happen in the New Year. That is all in the future. One thing is clear: It will be a hard year, and we must be ready for it . . . As we raise our hearts in grateful thanks to the Almighty, we ask his gracious protection in the coming year.

THE SPEECH WAS reported and quoted prominently in the Paris papers. *Le Matin* described it as a "harangue."

Goebbels's speech did not tip the Führer's hand. Nowhere in the bombast was any specific hint about the plans for 1940. Would the Germans attack and, if so, when and where? Allied intelligence and the High Command had to weigh scenarios and plan accordingly.

The one possibility that was thought to be least likely was a direct invasion from across the border with Germany, along which France had constructed the Maginot Line. The cornerstone of France's strategy for the defense of the homeland, the Line was an extensive system of fortifications built along the entire frontier with Germany. The Line was born out of the costly experiences of the French military in World War I, when the French leadership was caught by surprise by the German invasion in August 1914.

French forces suffered very heavy losses in the opening months of the war. In just a few weeks, the German Army had reached the Marne River, only forty-three miles from Paris. There was great fear that the capital would be captured. However, the deep thrust of the German armies had left gaps and weaknesses in their lines that French commanders identified and exploited. French and British forces counterattacked, pushing the Germans away from Paris and into defensive positions that began four years of stalemate on the western front. More than 2 million soldiers fought in the battle of the Marne, with the two sides suffering more than 500,000 casualties. The French alone lost more than 80,000 men. One key aim of the Maginot Line was to hold up any surprise invasion long enough for sufficient forces to be mobilized that could thwart the assault before it advanced deep into France.

A second inspiration for the construction of the Maginot Line was the epic battle of Verdun in the late winter and early spring of 1916. The city was surrounded by eighteen large underground forts that had been constructed around the turn of the century; these would save its French defenders. Flanked on three sides by German forces determined to take the stronghold, the French forces led by Gen. Philippe Pétain endured massive artillery bombardments, including poison gas shells, while resisting the German offensive. Although they took several forts, the Germans were not able to sustain the attack and take the city. For his leadership, Pétain was hailed as a hero—the "Savior of Verdun"—and was named marshal of France after the war.

Pétain's experience in defending the Verdun forts and his great stature made him a key proponent for the building of the Maginot fortifications. Pétain believed very strongly that the Line would provide multiple strategic advantages. One of the foremost concerns of the French leadership was the conservation of manpower. During World War I, France lost 27 percent of all able-bodied men between the ages of eighteen and twenty-seven. This greatly reduced the birthrate in France, which was already considerably smaller in population (39 million) than Germany (59 million) at the end the war. Fortifications could be manned with fewer troops than open field formations. Furthermore, the Line would not only deter enemy forces from directly attacking France but force them to take routes through Belgium or Switzerland and therefore perhaps divert major battles from French soil.

The Line stretched across the length of the German frontier, from Switzerland to Luxembourg, at a depth of ten to fifteen miles from the border to the French interior. It was comprised of a series of large fortresses and smaller forts that were spaced about ten miles apart in order to allow for mutual artillery support during battle. Each fort contained underground structures for housing, feeding, and arming the crews, as well as extensive networks of connecting tunnels, telephone lines, and supplies intended to last at least three months. The forts were connected by rows of antitank obstacles and dense barbed wire all along the front. The formidable challenges presented to attackers instilled great confidence in both the leadership and the public that the country was well protected.

As early as 1935, however, some dissenting voices were making themselves heard. Col. Charles de Gaulle, who served under Pétain in World War I and was taken prisoner in the battle for Verdun, was concerned that the focus on defense compromised the opportunity to take the offensive. De Gaulle thought France needed armored forces capable of rapid offensive movement. That notion was rebuffed by the then minister of war, who stated, "How can anyone believe that we are still thinking of the offensive when we have spent so many billions to establish a fortified frontier!"

De Gaulle continued to press the case for armored motorized divisions, to the point where his relentless advocacy earned him the nickname "Le Colonel Motor."

The protection of France's northern frontier with Belgium, however, posed a different set of considerations. For centuries, the favored invasion route into France was through the Belgian plain, as it had been in 1914. Pétain, French minister of war in 1934, insisted that to meet the threat of an invasion, French forces "must go into Belgium!"

Entering Belgium raised some sensitive issues. Because Belgium was a sovereign country, France was reluctant to enter preemptively without a request or permission from the king. The difficulty that waiting for permission posed was that Belgium might either decline to ask for assistance or refuse entry to Allied troops, as it did in the first days of World War I. Any delay on Belgium's part could handicap the Allies' war plans and increase France's vulnerability.

Some parts of Belgium were well protected. The Maginot Line connected to the Belgian fortification system, the strong point of which was Fort Eben-Emael, a fortress between Liège and Maastricht (the

Netherlands) that protected key bridgeheads into Belgium from Germany. There were also some natural obstacles protecting Belgium's frontier with Germany, including the Meuse River and the hilly Ardennes Forest. At the same Senate commission hearing in 1934, Pétain was asked about the possibility of an invasion through the Ardennes, which lay immediately northeast of Sedan, France. Pétain replied, "It is impenetrable, if one makes some special dispositions there. We consider it a zone of destruction . . . the enemy could not commit himself there. If he does, we will pinch him off as he comes out of the forest. This sector is not dangerous."

In accord with Pétain's analysis, only very sparse, light fortifications were constructed on the Franco-Belgian frontier. French war planning focused on countering an attempted invasion that was expected to cut through the Belgian plain. The centerpiece of the plan was for Belgian forces to delay the German advance while French and British troops rushed into Belgium to form a defensive line as far east as possible.

In January 1940, Allied intelligence and war planners were analyzing and debating invasion scenarios when the German plans fell into their laps.

On the foggy morning of January 10, German major Erich Hoenmanns was flying another major, Helmuth Reinberger, in his Messerschmitt to Cologne when he lost power. Thinking he was over Germany, he attempted an emergency landing. He crash-landed in Mechelen-sur-Meuse, Belgium, a few miles from the border. Reinberger happened to be carrying detailed plans for an attack on Belgium and the Netherlands. Once the two majors realized where they were, Reinberger tried to burn the documents. But Belgian border guards arrived on the scene and salvaged some of the papers from the fire.

Belgian intelligence officers were first concerned that the papers were a plant intended to throw them off the actual German plans. But after interrogating the prisoners and examining the documents, they concluded that the papers were probably authentic. The bits they could read indicated the Germans were planning to invade the Netherlands and Belgium, and to do so very soon (though not written on the documents, the planned date was January 17). The plan also included a diversionary attack on the Maginot Line.

The German command learned of the plane crash and was deeply worried that its plans had been compromised. Hitler was furious. Rather than delay, he ordered that the attack go ahead as planned, before the Belgians, the Dutch, and their Allies could respond.

The Belgians passed on to the French, British, and Dutch the information gleaned from the charred documents, as well as additional intelligence warnings pointing to an imminent invasion. Dutch and Belgian troops were put on alert, and France began to mass formations on the Belgian border in preparation for entering.

The Germans, however, got wind of the alerts and realized that they had lost the element of surprise. Then the weather deteriorated. Chief of Staff Gen. Alfred Jodl explained to Hitler that for the attack to succeed, they would need at least eight days of good weather; he suggested that the attack be postponed until spring. Hitler agreed, telling Jodl that "the whole operation would have to be built on a new basis in order to secure secrecy and surprise."

The planned route of the German invasion was, in fact, exactly the one expected by the French command. They were satisfied that they understood the German command's reasoning and methods.

COLONEL DE GAULLE, however, was deeply concerned that the Allied plan to form a defensive line did not take into account the new capabilities of mechanized warfare.

On January 26, just days after the crisis of the Belgian invasion had passed, he made one more attempt to alert the High Command of the need for greater mobility. He sent a memo to eighty high officials, an unusually brazen gesture for a lower-ranking officer.

With the specter of Poland's destruction still fresh, he warned:

> The enemy would take the offensive with a very powerful mechanized force both on land and in the air; . . . because of this our front could at any moment be broken; . . . if we ourselves had no equivalent force with which to reply there would be a grave risk of

> our being destroyed . . . The French people must not at any price fall into the illusion that the present military immobility conforms to the character of this war. On the contrary, the motor gives to the means of modern destruction a power, a speed, a range of action, such that the present conflict will be marked by movements . . . [the] speed of which will infinitely surpass the most amazing events of the past. Let us not fool ourselves! The conflict which has began [*sic*] can well be the most widespread, the most complex, the most violent, of all those which have ravaged the earth.

The generals had long before heard enough from de Gaulle, and ignored his pleas.

With hard evidence that Hitler had aimed to invade Belgium and the Netherlands—two neutral countries—and would in all likelihood try again, Premier Daladier took to the airwaves on January 29 to deliver a scathing assessment of the Nazis' intentions. In a radio address to the French people entitled "The Nazis' Aim Is Slavery," he left no doubt about the nature of Hitler's regime:

> At the end of five months of war one thing has become more and more clear. It is that Germany seeks to establish a domination over the world completely different from any known in history.
>
> The domination at which the Nazis aim is not limited to the displacement of the balance of power and the imposition of supremacy of one nation. It seeks the systematic and total destruction of those conquered by Hitler, and it does not treaty with the nations which he has subdued. He destroys them. He takes from them their whole political and economic existence and seeks even to deprive them of their history and their culture. He wishes to consider them only as vital space and a vacant territory over which he has every right.
>
> The human beings who constitute these nations are for him only cattle. He orders their massacre or their migration. He compels them to make room for their conquerors. He does not even take the trouble to impose any war tribute on them. He just takes

> all their wealth, and, to prevent any revolt, he wipes out their leaders and scientifically seeks the physical and moral degradation of those whose independence he has taken away.
>
> Under this domination, in thousands of towns and villages in Europe there are millions of human beings now living in misery which, some months ago, they could never have imagined. Austria, Bohemia, Slovakia and Poland are only lands of despair. Their whole peoples have been deprived of the means of moral and material happiness. Subdued by treachery or brutal violence, they have no other recourse than to work for their executioners who grant them scarcely enough to assure the most miserable existence.
>
> There is being created a world of masters and slaves in the image of Germany herself . . .
>
> For us there is more to do than merely win the war. We shall win it, but we must also win a victory far greater than that of arms. In this world of masters and slaves, which those madmen who rule at Berlin are seeking to forge, we must also save liberty and human dignity.

The conflict between France and Germany had remained, however, largely a war of words. The frontier with Germany was quiet throughout a bitterly cold January, the third coldest on record. Many parts of Northern Europe were experiencing the coldest winter in a century. In mid-January, temperatures plunged to below zero in Paris, Amsterdam, and Berlin. Another severe cold wave struck in mid-February.

By all conventional wisdom, the extreme cold and snow were deterrents to any potential overland assault. However, the six months of tension and inactivity since the declaration of war were wearing on the morale of the French troops. It was increasingly difficult for commanders to maintain discipline and a heightened sense of alert. Separated from their families, businesses, and farms, soldiers were granted more leave than regulations and circumstances normally would have allowed.

—ɯ—

When they did take leave for Paris, soldiers found a city in which citizens were trying to go about their normal routines, and planning for their futures, almost as if there were no war.

The universities had remained open. Twenty-nine-year-old zoology doctoral student Jacques Monod had not been called up during the general mobilization. He had been excused from active military duty several years earlier due to a slight limp in his left leg caused by a bout with polio. He continued to teach his classes at the University of Paris (La Sorbonne) while pursuing his thesis research. After receiving his *licencié ès sciences* (equivalent to a bachelor's degree) in 1931, Monod had spent several years drifting among laboratories in search of a scientific problem on which to focus. He hoped to finally complete his PhD in 1940.

Handsome, an experienced sailor, and a talented musician, Monod lacked focus, not confidence. His father, Lucien, was a painter, engraver, art historian, freethinker, and scholar. His mother, Charlotte Todd MacGregor, was an American born in Milwaukee and descended from Scottish immigrants. The Monods instilled a love for literature and art in Jacques and his brother, Philippe. Lucien also greatly admired Darwin and passed that interest on to Jacques. The Monods were also passionate about music and encouraged Jacques's development as a cellist and aspiring conductor. The question in the Monod household was not whether Jacques would achieve great things, but whether he would be the next Beethoven or the next Pasteur.

He was a long way from either destiny; he was prone to many distractions. Having been raised in Cannes, he loved the sea. In the summer of 1934, he took the opportunity to sail on a natural-history expedition to Greenland on the *Pourquoi-Pas?* In 1936, he was again offered a berth on the *Pourquoi-Pas?* He elected instead to accompany one of his mentors, Boris Ephrussi, for a year at the California Institute of Technology in Pasadena, ostensibly to learn genetics in the laboratory of Thomas Hunt Morgan, the 1933 Nobel laureate for his pioneering discoveries about the nature of the gene. Morgan's lab was then the world's hub of genetics.

Monod took advantage of opportunities in California, but not so much those of the scientific kind. He formed and conducted a Bach society that gave frequent concerts. He enjoyed an active social life, hobnobbing with the Southern California elite, but he accomplished nothing scientifically. Ephrussi was very disappointed. He thought Monod was gifted but terribly undisciplined. At the end of his stay, Monod was even offered a position as conductor of a local orchestra. Even though he did not make the most of his scientific connections while in California, it was a very fortunate choice to have gone there. The *Pourquoi-Pas?* sank in a hurricane off Iceland, and all but one crew member was lost.

Monod decided not to stay in California. After returning to France, he finally settled upon a research project studying the growth of bacteria. But Monod was not close to Pasteur, either scientifically or temperamentally; he was not even at the eponymous institute. Monod was at the Sorbonne, which, despite having a storied history as an institution, was far behind the times in biology and not at all the peer of the esteemed Pasteur Institute. No one at the Sorbonne, not even his thesis supervisor, had the least interest in Monod's research. He was completely on his own. He had to do all of the work himself, from preparing the sterile media and glassware to performing his experiments.

And yet Monod did not abandon music. In 1938, he formed and directed a Bach choir, La Cantate, which performed in public to considerable critical acclaim. He did settle down, however, in his domestic life. That same year, he courted and married Odette Bruhl, an orientalist at the Musée Guimet near the Trocadero Gardens. An expert in Tibetan painting and an experienced field archeologist, Odette had a knowledge of art and music and a passion for the outdoors that were perfect complements to Jacques's passions.

The newlyweds became parents the following year when twin sons Olivier and Philippe were born on August 5, 1939, just four weeks before war was declared—a declaration that Monod did not think would happen. Only the day before Hitler's invasion of Poland, Monod had written to his father, "There will be no war. Hitler is much smarter than Wilhelm II and he knows what it would cost him. His bluff having failed, put together with the complicity of the father of the people [aka Stalin; the German-Soviet nonaggression pact had been signed a week earlier], he will try to get way without too much damage. I only regret that the English are too polite with him. They should not have bothered writing him long letters. They should have told him to piss off, without any further explanation."

Despite his skepticism about war breaking out, the new father was thinking about his family responsibilities. He shared his hopes for his children with his parents:

> *I would like to raise them as I was. I would like for them to learn naturally, effortlessly, almost without knowing it, that the love of beautiful things, critical thinking, and intellectual honesty are the three essential virtues. This way, they will like things for themselves, will judge for themselves. This way, they will be real men, as there used to be, they won't be fooled by intellectual snobs and political scoundrels. They will know how to live above and outside of a century which is only getting deeper into infamy, lies, and stupidity. I love you my dears because I know that it is because of you that I possess some of these virtues that I wish for them to have.*

Monod had not been called up during the first general mobilization, but France continued to make preparations, so he expected to be drafted in some capacity. He wanted to serve not merely as an *auxiliaire* in support of the regular armed forces, but as part of them. So, rather than waiting to be mobilized, he took the initiative to request officer training. Seeking a branch where he could use some of his scientific background, Monod hoped to join the engineers, specifically the 28th Engineering Regiment, because he learned that there was only one platoon in the military engineering group that was based at Versailles, near Paris. If he was accepted, he would be able to see Odette and the twins regularly. The training would take seven or eight months and would require Monod to study electricity, Morse code, radio, topography, and other technical subjects. Odette approved of the whole idea, as it was both a rearguard assignment and comparatively safer than other options, such as the air corps.

In February, Monod learned that he had been accepted and would have to report for initial training beginning sometime in mid-April. At the end of the month, he took a first step by applying for his heavy vehicle license, the *permis poids lourd*. Fond of riding his motorcycle around Paris, Monod would have to drive four- to thirteen-ton vehicles in the engineering regiment. He so impressed the examiner with his driving skills that he was also given a *permis transport en commun*—a license for transporting more than nine people at a time. He shared the news with Odette, who was with her mother and the twins in Dinard on the Brittany coast: "I demonstrated dizzying panache, balanced with prudent caution . . . The instructor assures me that it is a first [to obtain both licenses at the same time, without having applied for one]. As you might think, I am consumed with vanity." However proud Monod was, his most important credential still eluded him. He told his brother Philo, "The laboratory has been put on ice." Three years into his doctoral work, eight years past his bachelor's degree, and about to turn thirty, he was still not sure when he would complete his PhD.

The mobilization of men in all lines of work caused many disruptions. With much of the farm labor force mobilized, some food shortages were inevitable. Meat rationing and restrictions on the sale of alcohol were imposed. Another effect of the general mobilization was that many businesses were shorthanded. It was the possibility of landing a job that brought twenty-six-year-old Albert Camus to Paris from his native Algeria in mid-March. Camus was not called up for military duty when his fellow colonists were mobilized. He was exempted on account of having contracted tuberculosis when he was seventeen.

In the previous two years, he had worked as a reporter and editor of *Alger Républicain*, a fledgling left-wing daily. Although the war seemed far away from Algiers, its declaration spelled the end for the editorial positions that the paper, and especially Camus, had pursued.

Camus's outlook was shaped by his very humble origins. His father, Lucien, an agricultural worker, died of wounds received at the battle of the Marne when Camus was less than a year old. He was raised by his mother, Catherine, a deaf, largely mute, and illiterate cleaning woman whom Camus adored. Camus and his mother shared their gasless, sparsely furnished apartment with his older brother, a partially paralyzed uncle, and his grandmother. Despite his poverty, with no books, newspapers, or even a radio at home, Camus exhibited academic abilities in reading, writing, and speech that were noticed early.

In primary school, Camus fell under the influence of his teacher Louis Germain, a freethinker devoted to secular, democratic principles, who instilled in his students the values of honesty and sincerity, along with a love for soccer. Germain became a father figure to Camus, gave him two hours of supplementary lessons each day, and encouraged him to go on to high school—as opposed to going to work, as most children Camus's age did.

Camus won a scholarship to a lycée in Algiers, where most of his classmates came from much more privileged backgrounds. Undernourished, pale, and shabbily dressed, Camus nevertheless carried himself with pride and dignity. In time, he charmed and earned the respect of his classmates. Young Camus learned to be equally at ease with people of all classes, but he identified with those who were, like him, poor underdogs.

It was in high school that his appetite for literature and philosophy blossomed, thanks in particular to his teacher Jean Grenier, himself a writer and philosopher. Reading Nietzsche, Malraux, Gide, and Grenier's own writing stoked Camus's ambition to write. It was also during high school that Camus contracted tuberculosis in his right lung. In his Algiers neighborhood, in the period before antibiotics, the disease was often fatal. Camus was hospitalized and underwent repeated pneumothorax therapy, in which his lung was deliberately collapsed. His survival was in doubt for some time. But he did recover, and his long convalescence gave him plenty of time to reflect on his mortality.

His early brush with death gave birth to an intense sense of purpose and urgency. The precocious philosopher began to make notes on the question of how, in light of the certainty of death, one should live life. "Should one accept life as it is? . . . Should one accept the human condition?" he jotted in a notebook. "On the contrary, I think revolt is part of the human condition."

His sense of urgency spilled over into his romantic life. Before he turned twenty-one, while a philosophy student at the University of

Algiers, Camus married a beautiful young woman, Simone Hie. Unfortunately, she was a morphine addict and the couple was estranged within a year, though not officially divorced until six years later. Being married was only a technicality for Camus, who became involved with a number of women, often at the same time.

Camus was determined to become a writer. After receiving his diploma in 1936, his original plan was to become a teacher like his mentors, a member of the civil service, and to write in his spare time. His tuberculosis, however, made him unlikely to be hired. Camus had to stitch together a series of odd jobs. He performed in a traveling acting company for Radio Alger, tutored students, and cofounded the Algiers House of Culture. He also wrote almost nonstop. He published his first book, a set of essays entitled *L'Envers et L'Endroit* (The Wrong Side and the Right Side) in 1937. The first printing was just 350 copies. The next printing would not be for twenty years.

Camus began work on a novel, but he still sought a steady income. He was offered and accepted, of all things, a job as a technician at the Institut de Météorologie at the University of Algiers, collating and organizing historical weather data. He performed his job well, grateful for the salary that enabled him to write in his off-hours. While working at the institute, he managed to complete essays for a second book, *Noces* (Nuptials), published in 1939.

Camus resigned his meteorological post when a better opportunity came along, one that would allow him to put his writing skills to work every day, and to be paid for it—as a journalist for the newly founded *Alger Républicain*. At first, Camus covered the routine beats of a city reporter: local government, courts, crimes, and car accidents. He soon initiated a literary review, penning his analyses of new works by Sartre, Huxley, and many more authors.

Working full time, Camus still managed to plot out his own body of work. At the outset of 1939, he jotted down his list of projects in his literary notebook. Among them were three works in three different forms: a novel, a play, and an essay, all on the philosophical theme of the absurd—the dilemma posed by the human search for meaning and the seeming indifference of the universe to that human concern. For several years, Camus had immersed himself in the philosophers and writers who had wrestled with how to respond to the absurd condition. Many previous thinkers had taken the path to nihilism, to the denial that life had any value. Camus was determined to develop a different view, one that both embraced absurdity as an essential truth and valued life to the fullest.

The declaration of war made Camus despair, both privately and publicly. He saw the war as another unnecessary, avoidable, disastrous, absurd chapter of history that would consume the lives of those who did not make it nor wish for it. He wrote in his journal: "They have all betrayed us, those who preached resistance and those who talked of peace. There they are, all so guilty as one another. And never before has the individual stood so alone before the lie-making machine . . . The reign of beasts has begun."

The day after the invasion of Poland, Camus wrote in *Alger Républicain,* "Never have left-wing militants had so many reasons to despair . . . Perhaps after this war, trees will flower again, since the world always finally wins out over history, but on that day, I don't know how many men will be there to see it."

Despite his total opposition to the war and his tuberculosis, Camus attempted to enlist, twice. He felt that it was a matter of expressing his solidarity with those who were being drafted. He was rejected, twice.

The war put immediate strains on the newspaper: circulation and advertising dropped, newsprint became scarce, and government censorship intensified. A two-page evening paper was created, *Le Soir Républicain,* with Camus at the helm. *Alger Républicain* folded while Camus jousted with the censors, writing under a series of pen names. His antiwar editorials outraged officials and alienated readers at a time when most were calling for unity. On January 10, 1940, the government suspended publication of the paper and the police seized its remaining copies.

With no good job prospects in Algeria, Camus hoped to secure a position with a Paris newspaper—in spite of the danger posed by moving so much closer to the war front. He decided to move there alone. As he had once explained to a close friend, "For my works, I need freedom of mind, and freedom, period." In early 1940, he had two lovers on his mind: Yvonne Ducailar, a graduate student at the University of Algiers; and Francine Faure, a mathematician and talented pianist who lived in the city of Oran. Camus was unwilling to give up either woman, and equally reluctant to commit to one over the other. In his nascent essay on the absurd, Camus asked, "Why must one love so few people in order to love a lot?" Camus chose to love many, but not always solely on his terms. Before he left Oran for Paris, and under pressure from Francine's family, Camus promised to marry her once he was divorced from his first wife.

On March 16, Camus arrived in Paris as it was still thawing out from the remnants of the frigid winter. Ice chunks were floating in the Seine, and the gray, rainy skies were a depressing contrast to the sunny, warm, and fragrant Algeria he had left behind.

Camus soon obtained a position as a layout designer with *Paris-Soir,* a widely circulated daily that catered to more pedestrian tastes than *Le Figaro* or *Le Temps.* Its war reporting reassured readers of France's military prowess, with minimal concern for actual facts. The issue on Camus's day of arrival in the capital declared that France had some five hundred laboratories working on secret weapons and that the German Army, Air Force, and Navy had no idea what was waiting for them.

Camus had no reporting or writing responsibilities. Outside of his shift, he had the remainder of his day to pursue his philosophical works. As Europe stumbled toward what he believed would be certain disaster, Camus withdrew into himself and poured all of his energy into writing. In his notebook, he issued his own instructions: "Now that everything is clear-cut, wait and spare nothing. At least, work in such a way as to achieve both silence and literary creation. Everything else, everything, whatever may happen, is unimportant . . . More and more, faced by the world of men, the only possible reaction is individualism."

—w—

AS CAMUS SETTLED in at *Paris-Soir,* the war began a new chapter. The change was not on the front—the communiqués continued to state "Nothing to report"—but in the government. Daladier had taken no action when the Soviet Union had suddenly invaded Finland on November 30, 1939. The Finns fought valiantly but received no assistance. Their ordeal and their bravery were covered prominently in the papers. Some thought that Great Britain and France should have intervened against the USSR, regardless of the more proximate threat from Germany. After holding out for more than three months, Finland capitulated on March 12, 1940.

Prompted by growing dissatisfaction over Daladier's conduct of the war, a vote of confidence was taken in the Chamber of Deputies on March 20. Daladier lost. He had won a unanimous vote only a month earlier. Daladier resigned, and Paul Reynaud, Daladier's highly respected, intelligent, and decisive minister of finance, became premier.

The new Reynaud government declared that its purpose was "to arouse, reassemble, and direct all of the sources of French energy to fight and to conquer."

In late March, just one week after becoming premier, Reynaud traveled to London to confer with the British leadership about war plans. In particular, he wanted to discuss potential offensive measures that could be taken to prevent vital supplies, specifically Swedish iron ore, from reaching the Reich.

Even though Norway was neutral, Reynaud suggested that a force be sent to occupy the key port of Narvik. The premier found a like-minded ally in the Lord of the Admiralty, Winston Churchill, who had previously recommended the very same mission to Prime Minister Chamberlain. "The ironfields . . . may be the surest and shortest road to the end," Churchill wrote in late 1939. Other officials concurred that the German war effort could not last a year without the ore.

Chamberlain, however, wanted to act only with the agreement of Norway and Sweden. Both countries rejected the plan. Disappointed but resolved to take some action, Churchill suggested to Chamberlain, instead of occupation, the mining of Norwegian territorial waters. This would force German cargo ships out from the protection of neutral waters and into international waters, where they could be seized or sunk. Chamberlain agreed to Churchill's proposal and added to it the dropping of thousands of mines into German rivers and canals, as well as the bombing of the Ruhr industrial region of Germany.

Reynaud brought the proposals back to his cabinet. War Minister Daladier and Chief of Staff Gen. Maurice Gamelin rejected the mining of German waters and the bombing of the Ruhr, arguing that it would provoke Hitler to retaliate upon France.

The British went ahead with just the plan for the mining of key points along the Norwegian coast, dubbed Operation Wilfred. The date was set for the morning of April 8, with ships to begin heading for Norway on April 5.

Energized by this initiative, Chamberlain offered his assessment of the war to date to a gathering of his Conservative Party on April 4, 1940:

> When we embarked upon this war in September last, I felt that we were bound to win, but I did think of course, that we might have to undergo some very heavy trials, and perhaps, very severe losses. That may be so still. But I want to say to you now that after seven months of war I feel ten times as confident of victory as I did at the beginning.
>
> I do not base that confidence on wishful thinking, which is pleasant but dangerous . . .
>
> When war did break out German preparations were far ahead of our own, and it was natural to expect that the enemy would take advantage of his initial superiority to make an endeavour to overwhelm us and France before we had time to make good our deficiencies. Is it not a very extraordinary thing that no such attempt was made? Whatever may be the reason—whether it was that Hitler thought he might get away with what he had got without fighting for it, or whether it was that after all the preparations were not sufficiently complete—however, one thing is certain: he missed the bus.

CHAPTER 3

MISADVENTURES IN NORWAY

Castles in the air—they are so easy to take refuge in. And so easy to build too.

—Henrik Ibsen (1828–1906), *The Master Builder*

At two a.m. on April 3, 1940, three troop transports disguised as coal ships left the German port of Brunsbüttel at the mouth of the Elbe River. They were followed by several more transports on the next several nights. During the night of April 6, fourteen destroyers and a heavy cruiser left their bases at Wesermünde and Cuxhaven, followed the next morning by two battleships from Wilhelmshaven, and the following night by torpedo boats, cruisers, minesweepers, and support vessels from Helgoland, Kiel, and Wesermünde. By April 8, much of the entire German naval surface fleet, more than fifty vessels in all, were at sea.

Seven months after conquering Poland, Hitler was on the move again and taking another bold gamble. His ships were headed not to Britain or France, however, but to Norway. One crucial objective lay a thousand miles away, above the Arctic Circle: the port of Narvik. Ten destroyers were to offload two thousand troops there to seize and hold the port. W-hour, the time of invasion, was set for the morning of April 9.

"Operation Weserübung" had been planned for months. Its primary purpose was to secure the source of Swedish iron ore for German industry. Germany imported more than ten million tons annually from northern Sweden. Much of that was transported by rail to Narvik, then shipped by sea to German ports. For several months Hitler and his commanders fretted that the Allies were plotting some maneuver that would deny Germany its ore.

They had good reason to worry, for at the very time that the German fleet was dashing for Norway, the British Navy was launching Operation Wilfred to mine the Norwegian coastal waters.

Norway was thus being approached by two navies on a collision course. Neither adversary knew that the other was on the move, nor did neutral Norway know that it was about to be engulfed in a broadening war.

THE GREAT GAMBLE that Hitler had taken was endangering his navy. The German command knew very well that the British Navy was much stronger. Its members hoped that the invasion would catch both Britain and Norway by surprise and that their objectives could be secured before either country could respond.

There was also a calculation in the British plans. If Germany reacted to the mining by attempting an invasion of Norway, the Allies would then have the pretense for breaching Norway's neutrality and pursuing their own occupation of vital ports such as Narvik.

The mining took place as planned early on April 8, which put British ships in Norwegian waters as the German fleet approached. The first encounter was between the British destroyer *Glowworm* and two German destroyers and the German cruiser *Hipper*. The *Glowworm* was badly damaged but managed to ram and damage the *Hipper* before sinking.

Other German ships were also sighted as they approached Norway. A Polish submarine sank a German troop transport ship, and the German cruiser *Blücher* was sunk by fire from a coastal battery guarding the entrance to the port of Oslo. But the Germans landed most of their troops safely at Narvik, Trondheim, Bergen, and other ports. Paratroopers took control of airports and airfields.

A much smaller force landed at Copenhagen and seized Danish airfields, meeting little resistance. Under the threat of bombardment, the Danish government capitulated within six hours. The Germans captured most of their objectives as planned on April 9. Despite the many Allied sightings of, and encounters with, the German fleet, Hitler had taken the two neutral countries and the Allies by surprise.

THE FRENCH COMMAND was baffled. No sooner had Reynaud digested the welcome report of the successful British mining operation than a news flash of the German fleet's movement toward Norway reached him. He contacted Adm. Jean Darlan, who was completely unaware of any German maneuvers.

On the morning of April 9, as reports of German successes in taking ports came in, General Gamelin told Reynaud, "You are wrong to get excited. We must wait for more complete information. This is a simple incident of war. Wars are full of unexpected news."

The French and British governments pledged their full assistance to Norway. The newspapers condemned the German attacks on two neutral countries. *Le Figaro* asked, "Will the lesson be learned by other neutrals?"

AFTER THE SUCCESSFUL troop landings, the German Navy did not fare well. The next day, April 10, five British destroyers caught five German destroyers in Narvik harbor, sinking two and damaging the other three. Ten British dive-bombers sank a cruiser in Bergen. Then, on April 13, the battleship HMS *Warspite,* supported by nine destroyers, sank or crippled the eight remaining German destroyers that had offloaded troops at Narvik, as well as one U-boat. In just a few days, the German Navy had lost half of her destroyers and much of her entire surface fleet.

Le Figaro heralded the news in a bold headline, calling it an "Overwhelming Naval Victory" at Narvik. A summary of the developments in Norway concluded: "The situation is thus better in Norway than it first appeared."

The losses suffered by the German Navy boosted Reynaud's confidence. The premier outlined his government's approach to the war in a secret session of the Senate on April 18. "It would be absurd to throw ourselves head-on against the Siegfried Line [Germany's fortified western front]," Reynaud explained. Rather, he offered that the best strategy was to deprive Germany "of the supplies which are vital for her in making war—iron ore from the north [Sweden] and oil from the south . . . It is only by engaging in these distant operations against the enemy that we can employ our sole superiority over Germany—naval power." Such distant operations offered the added appeal, of course, of shifting the war away from French soil.

Reynaud summed up the war policy as "defensive on land, offensive in blockading Germany." He told the Chamber of Deputies, "It is extremely doubtful that Hitler has the means of taking the offensive."

The vast majority of Parisians believed the same. The running joke around the capital in April was prompted by the report in a newspaper of a speech in which Goebbels purportedly claimed that Hitler would be in Paris by June 15.

At the Petit Casino, a comedian asked the orchestra leader, "Have you been to the École Berlitz?"

The bandleader asked, "Why?"

"To learn German, so you can talk to Adolf when he gets here."

The crowd howled with laughter.

Goebbels would be off by eight days.

Marching and Morse Code

Just as the news from Norway signaled that the war was taking a new turn, Jacques Monod reported for duty. His mind was focused not on the war, however, but on his separation from Odette. His first posting was not to Versailles, but to Montpellier, in the far south of the country almost five hundred miles from Paris and even farther from Dinard—too far for a casual visit. Jacques bridged the distance by writing to Odette virtually every day, telling her after his first full day in the barracks, "Don't worry about me, the only things I am thinking about are the fact that our separation will not be very long, and that we know for sure that we will be very close to each other in a few weeks." Jacques reminded Odette that what mattered was not his immediate assignment, but the branch to which he would be assigned after the short course of basic training in Montpellier ended. That was supposed to be the École des Transmissions (Signaling School) of the 28th Military Engineers in Versailles. He reassured Odette: "Except for rotten luck, I will be sent there [Versailles] soon."

Jacques regaled Odette with the details of his daily routine, and stories about his new mates in the barracks. On his second full day, he reported:

> *Training started seriously today. We have to know how to march, to turn around, to salute, and so on. I assure you that it is not that boring and it is often even funny. The incredible awkwardness of a number of poor devils is absolutely hilarious. There is in particular one poor man who cannot manage to march and whose walk is simply amazing . . .*
>
> *Luckily the people in the barracks are all good sorts, although there is sometimes a certain austerity, on the account of the presence of two priests and a seminarian. As expected, I am already on a first-name basis with one of the priests, the pastor of a parish in Lyon. But I am a little afraid that he will try to save my soul.*

Jacques's letters were unfailingly upbeat. He told Odette that the food was fine, his bed was comfortable, and that he even enjoyed the long marches, which were making him fit despite the stress on his weaker left leg. He loved learning Morse code, which was required of everyone in his branch. He was getting good at transmission. The hardest part for everyone was receiving messages and learning the sounds of the dots and dashes, but Jacques found that his musical ear and sense of rhythm gave him an advantage. After receiving a package from Odette containing a blank and some sweets, which earned him the envy of the barracks, he demonstrated his proficiency at the bottom of his return letter:

· _ _ _ · _ · _ ·· _ _ · (*"Je t'aime"*)

Jacques was able to confirm the official information that courses would begin in Versailles on May 7, so he would be in Montpellier for about twenty days in total. Odette was buoyed by Jacques's news and by his spirit. She wrote to his parents and marveled at his morale, his curiosity about everything, and the energy with which he approached every task. She told them: "I hope after the war, he will not want to remain in the Army!"

In three weeks, Jacques expressed only two complaints. First, when four days went by without letters from Odette due to problems with the mail, he wrote to her, "I've been feeling very isolated and far away from you . . . If only you knew how I am waiting for them, and how those four days are centuries." And second, he was so busy and isolated that he did not know what was going on outside the base. He asked Odette, "What is happening in the world my sweetheart? It seems to me that I don't know anything anymore. I have a hard time reading the newspapers, I read them irregularly and I really don't know anymore where things stand."

The Stranger

Camus could not help but read the newspaper, and he was much less optimistic about unfolding events than his fellow Parisians, or the propagandists at *Paris-Soir.* As the Norwegian front opened, he wrote to a friend, "Events are going at such speed that the only wise and courageous attitude to have is silence. This can be used as a sort of sustained meditation which will prepare us for the future." Camus felt that his only option was to "wait and work."

He was also not at all enamored with Paris, whose pace Camus found overwhelming. "You can't live here, you can only work and vibrate here," he told a fellow writer.

When he came back from *Paris-Soir* to his room at the Hôtel Madison, on the Left Bank of the Seine and facing the historic Saint-Germain-des-Prés church, he shut out the outside world and focused on his novel. He had worked at it, on and off, and in various versions, for more than three years. Threads of the story, its characters, and its atmosphere came from the people and places in Algeria he knew so well, as well as his reporting experiences with the *Alger Républicain.* Speaking of Algiers, he told Francine, "I see the form and content around me in the poverty . . . the simple people, and their resigned indifference. They give an image of a rather frightening world without tenderness." The protagonist-narrator of the story is Meursault, a clerk living in Algiers who is indifferent to events and conventions of everyday life. Meursault exhibits no grief at his mother's death, no interest in the question of marriage to his girlfriend, no remorse over his killing of a man, and, most important to Camus's philosophical intentions, no belief or interest in God, not even when facing execution. Camus explored several different titles such as *A Happy Man, A Free Man,* and *A Man Like Any Other* before settling on *The Stranger.*

When he arrived in Paris, he thought he had already written about three-quarters of the story. Thereafter, he wrote like "a desperate man," often suffering from headaches and fevers that tested his endurance. The challenge that he had given himself was to express a philosophical idea—the absurd—and reactions to it in novel form. It was a studied effort. Camus's extensive reading and literary reviews had made him an acute observer of literary styles. In a review he wrote of Jean-Paul Sartre's *La Nausée* (*Nausea*), Camus suggested, "A novel is only philosophy put into images, and in a good novel, all the philosophy goes into the images."

To craft those images, Camus drew upon his own sense of isolation in Paris, of being a stranger. His first entry in his notebook after arriving in the city was:

> What is the meaning of this sudden awakening—in this dark room—with the sounds of a suddenly strange city? And everything is strange to me . . . What am I doing here, what is the point of these smiles and gestures? . . .
>
> Strange, confess that everything is strange to me.

HE SENT PROGRESS reports in love letters to his fiancée, Francine, as well as to Yvonne Ducailar, and to his former roommate and lover Christiane Gallindo, who typed his drafts. To Yvonne, he declared that he was working as if he were on a "tightrope, in passionate and solitary tension." To Francine, he explained, "I've never worked so much. This room is miserable. I live alone and I am weary, but I don't know if it's despite all this or because of it that I am writing all I wanted to write. Soon I will be able to judge what I am worth and decide one way or another."

At times, he sensed the story was all falling into place, telling Christiane that "at certain moments, what power and lucidity I feel in myself!" Other days, he despaired. After rereading all that he had written, he wrote to Francine that "it seemed a failure from the ground up."

The urgency with which he wrote was also spurred by the uncertainties created by the new developments in the war. As much as Camus tried to insulate himself in his room, when at *Paris-Soir* he could not avoid hearing the news and feeling the anxiety in the capital. One letter to Yvonne began, "I am writing to you from the newspaper office, amid the general hysteria created by events here. Men will die by the thousands so there is something to be excited about." Indeed, he might be one of those soldiers, as he was due to take another examination for the draft in May. He assured Yvonne, "I don't care if I am accepted. What I have to do and live through, I can do as well in the middle of battle as in the middle of Paris." He then added in Meursault-like fashion, "As for the risks of death, they are of no importance."

On May 1, he completed his first draft of *The Stranger.* He wrote immediately to Francine:

> *I am writing to you at night. I have just finished my novel and I'm too overexcited to think of sleeping. No doubt my work isn't finished. I have things to go over, others to add and rewrite. But the fact is, I've finished and I wrote the last sentence. Why do I turn immediately to you? I have the manuscript in front of me, and I think of all it cost me in effort and will—how much involvement it required—to sacrifice other thoughts, other desires to remain in its atmosphere . . . I am going to put these pages in my drawer and start work on my essay, and in two weeks I'll take it out again and rework the novel.*

Blunder

With the loss of the protection of their destroyers, and the ability to withdraw if necessary, the relatively small force of German troops in Norway was very vulnerable. Hitler was so concerned that his generals had to talk him out of abandoning Narvik.

The Allies decided to mount a counterattack to retake some ports. The original plan was to concentrate on Narvik. But Norway's King

Haakon IV requested that Trondheim also be recaptured. The British complied with the king and divided their available assault forces between the two objectives.

One force was landed both north and south of Trondheim, with the mission to attack it from the flanks. The soldiers never reached the town. The British were outflanked by the Germans, who had complete command of the air. The British were forced to retreat, then ordered to evacuate a little more than a week after landing. The Trondheim assault forces suffered more than 1,500 casualties without taking any ground.

A second force landed near Narvik. With their original strength reduced by the Trondheim mission, it was decided not to attempt an assault on Narvik right away and to wait to amass a larger, overwhelming force. By early May, nothing had yet happened.

With the crisis in Norway averted, Hitler turned his attention to other plans.

THE DEVELOPMENTS IN the Norway campaign were followed very closely in both London and Paris. The failure to take Trondheim and the delay in attacking Narvik would have dire political repercussions. The first major operation of the war under Chamberlain's direction had been bungled, no matter how rosy a picture the prime minister tried to paint to the House of Commons on May 7, of how the British troops "man for man showed themselves superior to their foes."

The opposition mocked his explanation for the "reverse" in Norway with shouts of "Hitler missed the bus"—painfully reminding Chamberlain of his boast a month earlier. Then, a succession of speakers voiced their doubts about Chamberlain's leadership.

After patiently waiting his turn, Leo Amery, a friend and fellow party member of Chamberlain's, held the floor for more than an hour: "I confess that I did not feel there was one sentence in the prime minister's speech this afternoon which suggested that the government either foresaw what Germany meant to do, or came to a clear decision when it knew what Germany had done, or acted swiftly or consistently throughout the whole of this lamentable affair." Amery continued,

"What we have lost is one of those opportunities which do not recur in war. If we could have captured and held Trondheim . . . then we might well have imposed a strain on Germany which might have made Norway to Hitler what Spain was to Napoleon."

Amery urged, "We cannot go on as we are. There must be a change . . . This is war, not peace . . . Just as our peace-time system is unsuitable for war conditions, so does it tend to breed peace-time statesmen who are not too well fitted for the conduct of war . . . Somehow or other we must get into the Government men who can match our enemies in fighting spirit, in daring, in resolution and in thirst for victory." Amery looked straight at Chamberlain and, quoting Oliver Cromwell, said, "You have sat too long for any good you have been doing. Depart, I say, and let us have done with you. In the name of God, go!"

The debate continued into the night and the next day. David Lloyd George, prime minister during World War I, continued the drumbeat of opposition, opining that "nothing which can contribute more to victory in this war than that he [Chamberlain] surrender the seals of office."

Pummeled from every quarter, Chamberlain spent the next day—Thursday, May 9—consulting with his advisers and vacillating between holding on to office and resigning.

ACROSS THE CHANNEL, Premier Reynaud was exasperated with General Gamelin. All he had heard since the first days of the Norway crisis were excuses for not anticipating the German move on Norway, for not having troops in position to deploy, and for delays in attacking German positions. Gamelin laid all blame on the British for the failure of the operations. Reynaud told his aides, "I have had enough. I would be a criminal if I left at the head of the French army that nerveless man, that philosopher!"

On May 9, he summoned his cabinet to make his case to sack Gamelin, telling them that if Gamelin was left in charge, the French were "certain to lose the war." His cabinet, however, did not share his concern. Daladier spoke in defense of Gamelin and vigorously opposed his firing. Crestfallen, Reynaud said, "As I cannot make my point of view prevail, I am no longer Head of the Government." He then asked the group to keep his resignation secret until a new government could be formed.

ENGINEER MONOD, LIKE all other Frenchmen, was oblivious to the leadership crisis in his government. He arrived in Versailles and moved into his barracks, where, as the eldest member of the 126th section, he was given the honor of being the head of the section, and the responsibility for getting the unit settled. Jacques's immediate concern, however, was to see his wife as soon as possible. He wrote to Odette in Dinard on May 9 to tell her that he would be free after noon on Saturday the eleventh until Sunday evening the twelfth. He told her, "I can only think of one thing at this moment: Will I see you Saturday or Sunday?"

HITLER ALSO HAD plans for that weekend. At five p.m. on May 9, at the very moment that the leaders of both Great Britain and France were pondering their resignations, he left the Finkenkrug railroad station outside Berlin aboard his special armored train. The Führer was accompanied by his secretaries and various members of his inner circle, including General Jodl. Just before nine p.m. the train stopped near Hanover so that Hitler could get the weather forecast for the next day. Assured of clear skies, he ordered the code word "Danzig" to be forwarded to all units on the western front.

CHAPTER 4

SPRINGTIME FOR HITLER

We have assured all our immediate neighbors of the integrity of their territory as far as Germany is concerned. That is no hollow phrase: it is our sacred will . . . The Sudetenland is the last territorial claim which I have to make in Europe.

—ADOLF HITLER, September 26, 1938

WHILE NINE GERMAN DIVISIONS (120,000 TROOPS) HAD BEEN committed to the invasion and occupation of Norway and Denmark, 136 divisions, more than 2 million men, had been amassing on the German border with Luxembourg, Belgium, the Netherlands, and France.

They had not gone unnoticed. In the Netherlands, all military leaves had been cancelled and all personnel ordered to join their units. The civilian population was asked to limit rail travel in order to facilitate the troop recall. All public buildings and installations were placed under armed guard. Holland, a neutral country, was assembling the largest army in its history.

PHONE CALLS STARTED coming into the French command just after midnight on Friday, May 10. At one a.m., General Gamelin was awakened with a message from an agent who was behind enemy lines: "Columns marching westward." Premier Reynaud received urgent word from Brussels that both the Belgian and Dutch armies noted increased activities on their front.

As dawn broke, Germany attacked across a 175-mile-long front with Holland, Belgium, and Luxembourg. First, there were artillery barrages, then waves of bombers, then armored columns and troops invaded the Low Countries. In Holland, paratroopers seized forts and airfields.

Shortly after six a.m., the Belgians and Dutch both formally requested the assistance of the Allies. Within minutes, Gamelin was notified and the orders were given for French and British forces to cross into Belgium and to speed north to Holland—the long-planned maneuvers to establish a continuous line of defense across Belgium and Holland.

Reynaud had to withdraw his resignation and to put aside his differences with Gamelin. Hitler had preempted any change in leadership.

Gamelin prowled the halls of his command post, even smiling as he learned the direction of the German attacks. He had long planned for this clash of armies. Altogether the Allies had at least as many men as the enemy, about 152 divisions in total: 104 French, 15 British, 22 Belgian, and 11 Dutch. The French had a slightly larger number of tanks than the Germans, as well as a substantial advantage in artillery pieces. The Allies were outnumbered, however, in aircraft by about two to one, and in antiaircraft guns by almost three to one.

Gamelin was, one corporal observed that morning, "absolutely confident of success." He issued the order of the day to the troops:

> The attack that we had foreseen since October was launched this morning. Germany is engaged in a fight with us to the death.
>
> The order of the day for France and all her Allies are the words: Courage, energy, confidence.

PARISIANS AWOKE TO a beautiful sunny morning, to choruses of birds, and the sound of air-raid sirens. A few planes flew over the city, but no one was sure whether they were French or German. Out of months of habit, few bothered to head to the air-raid shelters.

As bulletins came over the radio of the German attacks on the Low Countries, the collective response in France was "Finally!" After eight months, the long-anticipated battle had arrived. Civilians and soldiers alike appeared relieved that the tension had been broken, and they looked forward to the fight. "The Boches have business with somebody their own size now!" is what A. J. Liebling, Paris correspondent for the *New Yorker,* heard on the street. "They will see we are not Poles or Norwegians," said many. A corporal had recently written, "The real roughhouse is about to begin. So much the better! It will be like bursting an abscess!" Liebling's friend Captain de Cholet phoned that morning to say he was returning to the front. "It's good that it's starting at last. We can beat the Boches and have it over by autumn," the captain added.

At the offices of *Paris-Soir,* a military specialist told staff members, "That's it, Hitler has made his mistake."

That evening, Premier Reynaud addressed the country over the radio:

> Three free countries, Holland, Belgium, and Luxembourg, were invaded this past night, by the German army.
>
> They called to their aid the Allied armies.
>
> This morning, between seven and eight o'clock, our soldiers, the soldiers of freedom, crossed the frontier.
>
> This centuries-old battlefield of Flanders our people know well!
>
> Opposite us, hurling himself at us, is the centuries-old invader.
>
> Everywhere in the world, every free man and every woman watches and holds their breath before the drama that is about to play out . . .
>
> The French army has drawn its sword, France gathers itself.

The official military communiqués for the day reported forty-four enemy planes downed over France, the Dutch claimed to have shot down seventy German planes and blown up four armored trains, and the Belgians assured that the German attack was contained at all points. The Dutch Army command at the Hague was "satisfied that they have the situation in hand." Allied losses were not reported.

Chamberlain, however, was one casualty of the momentous day. He resigned and Winston Churchill became prime minister.

MONOD TRIED TO intercept Odette. He had not yet received a reply to his letter of May 9, and he hoped that Odette had not left Dinard for their rendezvous. He hurriedly composed a short letter from his bed, where he was recuperating after receiving a shot of the typhoid/tetanus/diphtheria vaccine. Writing on graph paper with a shaky hand, Jacques said: "If I have not heard from you by tomorrow morning, I will think that you did not leave Dinard, and I will feel quite relieved. In any case my dear angel, if this letter comes to you in time, I beg you to postpone your departure until the situation and the events are clarified a bit. I beg you not to risk leaving the kids and undertaking such a trip in the current conditions."

Despite the German attack, Odette had in fact left Dinard for Paris. That morning, she had run into a friend in Dinard who was leaving for Paris in the afternoon and offered to take her along. She arrived at Jacques's barracks at eight in the evening, and after some waiting she was delighted to finally see Jacques. Despite ongoing air-raid alerts and the battle that was unfolding, he was allowed to spend the next day at home with Odette, where his brother, Philo, came to join them. Jacques briefed them on all of his activities at the Signaling School—the maneuvers and exercises, the courses in physics and radio, and the reports he was obliged to write as head of his section. He knew nothing more about the conduct of war than what was in the newspapers.

THE NEWSPAPERS ON May 11 carried passionate appeals to patriotism and unity. In *Le Figaro,* Wladimir d'Ormesson penned a tirade against the enemy "assassins" and claimed, "We have unlimited confidence in the leaders of our armies. Behind the lines, let us be worthy of them." In *Le Matin,* Jean Fabry urged, "Let us have confidence in our soldiers and their commanders."

Those commanders were dealing with some of the surprises and setbacks of the first hours of the campaign. At Eben-Emael, the massive fortress defending key bridges over the Albert Canal, German gliders landed on the roof, where there were no defenses. The specially trained troops promptly disabled many of the fort's large guns. By the second day, the fort had surrendered. The Germans took more than a thousand prisoners at a cost of just six men killed.

The fort and the canal had been expected to hold up the German assault while the advancing Allied troops maneuvered into position. In one operation, the Germans had eliminated a vital element of the Belgian fortification system and secured bridges across the canal. They began flooding into Belgium. The quick loss of the fort was a setback; nevertheless, the Germans were proceeding on the course that Gamelin and his commanders had expected.

The military communiqués stated, and newspapers dutifully reported, that on May 11, Allied forces were advancing rapidly and thirty-six enemy planes were downed and, on May 12, Allied forces were in place on Belgian and Dutch soil, thirty enemy planes were downed, and the pressure on the Dutch had been ameliorated by the actions of the Royal Air Force. On May 13, a combined British and French force even landed in Norway, finally, and captured Bjerkvik, just north of Narvik. This was the first opposed amphibious landing of the war.

All seemed, or at least was reported to be, in control. The Allies' movement into Belgium and Holland was, according to *The Times* of London, "brilliantly prepared and executed."

Unfortunately, the armies were right where the Germans wanted them to be.

Le Choc

The French military communiqué of May 12 also mentioned that the enemy was making an important effort in the Ardennes region in Belgium. Pétain had deemed this rugged forest with narrow, winding roads "impenetrable"; it would at least be folly to attack through it.

An enormous concentration of German armor was in the process of mocking that doctrine. Seven panzer divisions with more than 1,200 tanks and 134,000 troops cut right through the forest. Lead elements reached the critical natural barrier of the Meuse River on May 12, beyond which lay northeastern France. The region was lightly defended, mainly by second-line troops—older soldiers who had been called up from the reserves, as they were not expected to receive the brunt of the German attack.

The Germans had changed their plans. After the capture of Majors Hoenmanns and Reinberger in Belgium in January, along with the latter's copy of the then-imminent plan for an attack on Belgium and Holland, Hitler and the German command worried that the original battle plan had been compromised. Moreover, Hitler was concerned that the attack would bog down among the heavily fortified defenses of the Allies. The solution, then, was to go where there were not any fortifications, where they were not expected and could achieve surprise.

Hitler and his staff conceived a two-part plan. First, in order to set a trap for the Allies, they would begin the invasion by the expected routes into Belgium and Holland. This would, and did, draw the Allied armies north. Then, in a second offensive, the Germans would cut through the Ardennes south of the Allied armies, cross the Meuse River around Sedan, race west to cut off the northern armies from their supply lines and reinforcements, and destroy them. This daring plan to make a "sickle cut" across Northern France all the way to the sea would rely on overwhelming force and unprecedented speed.

Beginning on the morning of May 13, the French defenders on the Meuse were pounded by an intense air bombardment. For eight hours, more than a thousand planes struck in waves, shattering the nerves of the troops and pinning them in their bunkers. Assault teams then crossed the river at several places and took key vantage points.

Preoccupied with the offensive farther north, which in fact was going moderately well for the Allies, Gamelin and other commanders did not perceive the threat from the Ardennes. Little air support was provided either to bomb the armored columns or to counter the German dive-bombers. German forces that crossed near Sedan were right at the "hinge" between two French armies, thus piercing the Allied front.

When word reached various headquarters that the Meuse had been breached, there was shock, confusion, and, finally, concern. Gen. Gaston Billotte recognized the danger and ordered counterattacks to bomb the bridges the Germans had laid and to repel the incursion. Billotte told the Air Force, "Victory or defeat hangs on those bridges." The scene at Gen. Alphonse Georges's command post was merely pathetic. As the general exclaimed to another, "Our front has been pushed in at Sedan!" he and other officers broke into sobs.

The bombing of the bridges was ineffective. Worse yet, the Allies,

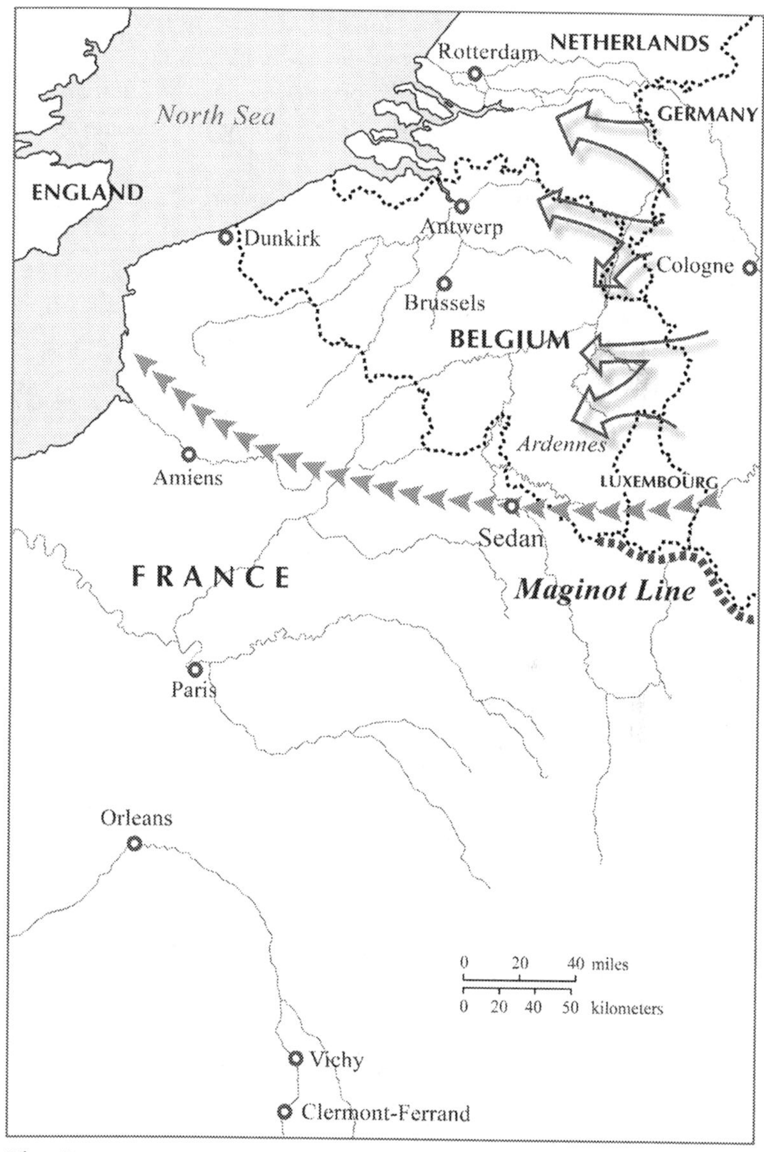

The German invasion. Open arrows indicate the line of attack launched on May 10, 1940, against France, Belgium, and the Netherlands. The closed arrowheads indicate the path of the "sickle cut" taken across France to cut off the French forces that moved north to repel the invasion of Belgium. (Map by Leanne Olds)

already at a significant disadvantage in the air, suffered terrible losses. The British RAF alone lost thirty of their seventy-one bombers. The ground counterattack was repeatedly postponed, and the German advance continued. Confronted with overwhelming force, some units panicked and fled.

The public was largely in the dark about the gravity of the rapidly developing crisis. Communiqués offered reassuring but completely vague statements about "hurling counterattacks" and "the enemy making an important effort in spite of increased losses." The press was operating under stringent military censorship, which prevented the reporting of much negative news. Despite being in the offices of *Paris-Soir,* Camus learned very little. The paper sought to maintain the public's faith by inflating claims and offering up pithy slogans. The pages Camus laid out reported that the Germans were losing a hundred planes a day and that "France has many trump cards in its hand and does not need to bluff."

More important, the government was also dangerously unaware. Early on in the battle of the Meuse, Premier Reynaud was being reassured that threats were being met. Finally, on the evening of May 14, he heard from Gamelin himself about the Army's precarious position. Reynaud was so alarmed that he sent an urgent message to Churchill to request more fighter planes: "If we are to win this battle which might be decisive for the whole war, it is necessary that you send at once ten more squadrons."

Churchill and his command were concerned about the RAF's losses and had to consider the need to defend England should Hitler continue to advance. And there was more startling news that evening—the Dutch Army was surrendering.

Before Churchill could even reply to Reynaud, the premier called the prime minister directly at seven thirty the next morning, Wednesday the fifteenth. "We have been defeated," he blurted in English. "We are beaten; we have lost the battle."

"Surely it can't have happened so soon?" Churchill replied.

"The front is broken near Sedan; they are pouring through in great numbers with tanks and armored cars," Reynaud explained.

Churchill said he would fly over to meet with Reynaud. When he landed on the sixteenth, the German front had already penetrated sixty miles beyond Sedan and was a little more than seventy miles from Paris. Churchill was told that the Germans were expected in Paris within a few days. Churchill met with Reynaud, Daladier, and Gamelin in an elegant room at the Quai d'Orsay. He saw "utter dejection" on every face. Gamelin went over the current war map, indicating how far the Germans had progressed. Churchill had been through such anxious moments in World War I. In 1914, the Germans' initial thrust carried them to the Marne, where they were finally halted and pushed back by counterattacks. He sought to quell the panic.

"Where is the strategic reserve?" he asked Gamelin. Churchill asked again in French, *"Où est la masse de manoeuvre?"*

Gamelin turned to Churchill, shook his head, shrugged, and said, *"Aucune."* None.

Churchill was stunned. The French had assumed that they could maintain a five-hundred-mile-long defensive front and had made no provision to meet any breakthrough. Churchill looked out the window to see clouds of smoke from bonfires in the courtyard onto which officials were dumping wheelbarrows full of documents.

Churchill needed to rouse the French leadership's failing spirit. He cabled London to ask for the ten squadrons that Reynaud wanted. His cabinet agreed, in spite of its concern for the defense of England. At a minimum, for the sake of history, England could say it gave the French the assistance that was requested.

French communiqués continued in their vagueness. On Wednesday night, the fifteenth, it was reported that in the area of Sedan "where the enemy had made some progress, counterattacks were ongoing with tanks and bombers."

THE NEXT DAY'S edition of *Le Figaro* stretched the truth of the battlefield situation:

> For the moment, the principal battle is taking place on the crossings of the Meuse . . . For the enemy, the concern is to break the vast hinge at the elbow of the Meuse between the Allied armies

> of Lorraine [to the south] and Belgium [to the north]. In spite of the terrible shock, where all possible means were employed, this attempt has only partially succeeded. The surprise of the suddenness and intensity of the attack did not rattle the gallantry of the defenders. This valor is receiving incessant support . . .
>
> It will be, without a doubt, several days before the outcome of this battle appears more clear.

The morning communiqué for the sixteenth declared, "It is in the best interests of the conduct of operations not to furnish precise information on the actions in progress."

It was also in the best interest of civil order. German forces were dashing for the sea in order to cut off the northern armies. The rout was under way.

ODETTE HAD STAYED in Paris, hoping to spend more time with Jacques, but he and the rest of his unit were confined to the barracks. They were only able to see each other very briefly in what Odette described as a "dark hovel" that served as a visitors' room. Odette decided to return to Dinard to be with the twins and her family. It was a difficult parting. Jacques felt great anguish at seeing Odette go, alone, to make a potentially hazardous journey. Odette did not know when she would see Jacques again, as the war seemed to be moving very quickly.

Relieved when he heard that she'd arrived safely in Dinard, Jacques told her not to worry if there were aerial bombardments, as he was safe. Moreover, he added, "In any case, I do not believe, I cannot believe, that the situation is as serious as some seem to think it is. It was much worse, on many occasions, in '14." Two days later he wrote, "It seems to me that after the first shock, everyone is regaining their strength and hope. At least that is what is happening here, but we hear only so much from the outside . . . My dear, I am maintaining the hope, the certainty that this nightmare will pass, that we will remain free, that we will be together, and we will love each other more and better than ever, if that is possible." In closing his letter the next day, he suggested,

"Do as I do, don't listen to the radio too much, think about me as I think about you, be hopeful. It is a necessity and a duty."

Prayers

The Germans' thrust west meant they were bypassing Paris, at least for the time being. That brought some relief to the French government and kindled hope that the lengthening German spearhead might be vulnerable.

"Le Colonel Motor" de Gaulle had been given command of an armored division that, at the outbreak of the battle, still had no tanks. He was assigned the task of trying to halt the enemy in the region of Laon, about seventy-five miles northeast of Paris. De Gaulle was under no illusions as to the dire state of the Army. On May 16, during reconnaissance, he came across streams of refugees fleeing Belgium and, worse, many soldiers who had lost their weapons and their units: "At the sight of those bewildered people and of those soldiers in rout . . . I felt myself borne up by a limitless fury. Ah! It's too stupid! The war is beginning as badly as it could."

He resolved, "Therefore it must go on. For that, the world is wide. If I live, I will fight, wherever I must, as long as I must, until the enemy is defeated and the national stain washed clean."

De Gaulle received several battalions of tanks, about 150 in all, and assembled all of the forces he could muster in the vicinity. He attacked on May 17. Despite heavy pressure from Stuka dive-bombers, and the complete absence of his own air support, de Gaulle's division managed to inflict heavy casualties on the enemy. He attacked again farther west on the nineteenth, until he encountered overwhelming resistance.

Meanwhile, Premier Reynaud was thinking how the course of events would have been different if he had someone else as commander in chief instead of Gamelin. If the leadership was to be any better from then on, he would first have to get Daladier, Gamelin's chief backer, out of the War Ministry. On May 18, he moved Daladier to the Foreign Ministry and took over the War Ministry himself. He also wanted to restore the waning morale in the armed forces. The same day he brought France's greatest hero from World War I, eighty-four-year-old Marshal Pétain, into his government as his vice premier.

The newspapers gushed. The unanimous view was that this living embodiment of French pride and honor, who had served "everywhere and always" with brilliance, would prove of inestimable value at this critical time, just as he had in the last war.

ON SUNDAY, MAY 19, Reynaud, members of his cabinet, deputies, and senators attended a special Mass at Notre-Dame. A huge crowd gathered outside the eight-centuries-old cathedral, which was enshrouded in sandbags and had its stained-glass windows removed as a precaution against bombs. Inside, Msgr. Roger Beaussart led the audacious prayers: "What have we come to ask of God? Victory . . . We ask Him for this because we are fighting a war not for money or worldly power, but to preserve spiritual values without which there is no reason to live . . . We face a barbarism and a perverse cruelty of which there is no other example in history . . . We have in our hearts an indelible faith in final success. Saints of France, protect us, grant triumph to France and the Allies."

The immense throng exhorted in unison:

Our Lady of Paris, we have confidence in you!
Saint Michael, defend us in combat!
Saint Denis, defend France!
Saint Louis, protect those who govern us! . . .
Saint Geneviève, protect Paris and France!
Saint Joan of Arc, fight with our soldiers and lead us to victory . . .
Saints of France, we have confidence in you!

After a final blessing, the organist played "La Marseillaise." More than a few dignitaries had tears in their eyes.

But those prayers would go unanswered. In this war, there would be no miracle on the Marne. The Savior of Verdun would save nothing.

WHILE THE GERMANS had bypassed Paris, at least temporarily, anxiety in the capital was spreading. Camus wrote to Francine, "As the days go on and the dangers become clearer, Paris becomes more anguished." Some packed what belongings they could and left, but many stayed for their families, their jobs, or school.

Some of the staff of *Paris-Soir* left for Nantes, but Camus remained, as he wrote to Yvonne, "in the middle of an almost deserted newspaper; on double duty." He continued: "Paris is dead and there is a latent threat. We go home and wait for the alarm or for anything at all. I'm stopped regularly in the streets for verification of identity: the atmosphere is charming."

With the military situation deteriorating, Camus thought again about enlisting. He wrote the Army to offer his services. He explained to his fiancée why he would take such a risk: "This war has not stopped being absurd, but one cannot retire from a game when the game becomes deadly . . . If I am accepted, I'll finish my work in the middle of the brawl, I'm sure, just as I would have done in the silence and solitude of Paris." Camus was still waiting to hear back from the authorities in late May.

Monod heard Reynaud's address to the Senate on May 21 over the radio. The premier opened by stating bluntly, "The homeland is in danger," and went on to speak of the "disaster" that had befallen certain armies. Jacques had spent his days digging trenches, and while he had jauntily reported to Odette, "I think that I missed my vocation of earthwork contractor," it was time to face the situation. He wrote Odette after Reynaud's speech:

My dear angel,

I am back from the cafeteria, where I heard Reynaud's speech at the Senate. His pure frankness and brutality let me think that there is still a chance to pull through. But even without believing the worst, we must still think about our kids, and what we would do if the total catastrophe were to materialize. We don't have to look far away, and I see only one solution. In case every hope was to be lost, you must try to find a way to England. Once there, you can seek asylum with [distant cousins] the Glehns and the Marshes.

After giving Odette their addresses, Jacques continued:

My darling, I am writing you all of these things coldly, without believing that they are real or possible . . . I am asking you to do and to organize everything as if it had to be carried out. I have total confidence in you and in your courage, my darling. I know that when the time comes you will do everything so that our children live free. As far as I am concerned, I will never believe in the total and final victory of those people, even if it would appear to be as total and final as possible. Trust me, my darling. I will find a way to rejoin you if there is nothing more that can be done here.

All of this being said, my darling, I should add that I don't believe a word of it. My courage and my trust rest on you. Not a minute, not a second, are you away from me, my darling. I hold you against me, my darling. I love you more than anything in the world.

Retreat

General Gamelin was finally relieved of duty. He was replaced by another World War I icon, the highly decorated, seventy-three-year-old Gen. Maxime Weygand. But it was too late for him or any other French commander to reverse the tide. The only hope for the isolated northern armies to survive was to make for the Channel coast and disembark there for England.

The operation was a British initiative. Lord Gort, commander of the British Expeditionary Force (BEF), had alerted London that evacuation might become necessary. The Admiralty thought that at best 45,000 troops or so might be rescued, which would still leave the bulk of the BEF and all of the French Army stranded. By May 26, Gort's forces had withdrawn to the Dunkirk bridgehead.

Any British doubts about the necessity of evacuation were erased on the evening of May 27, when King Leopold of Belgium, without giving any notice to Britain or France or even his ministers, asked Germany for an armistice and surrendered unconditionally. The retreating British forces were now fully exposed on one flank; the Germans would have free passage through formerly defended territory straight to the sea.

To protect Dunkirk, the French First Army dug in at Lille and was ordered to fight to the last man in order to delay oncoming German divisions from reaching the bridgehead. Under constant bombardment from the Luftwaffe, a flotilla of all sizes of boats managed to evacuate an astounding 338,226 troops (198,229 British) over the span of nine days. More than 200 boats were sunk, the RAF lost 474 planes, casualties were heavy, and all equipment had to be left behind, but the scale of the evacuation was seen, particularly in Britain, as a miracle.

As always, the French authorities tried to cast the operation in the most favorable light. In the middle of the evacuation, the public was merely told that Allied forces were pursuing "with vigor, in the middle of constant combat, and in good order, the execution of movements decided by the command."

Churchill, however, reminded the relieved British Parliament, "We must be very careful not to assign to this deliverance the attributes of a victory. Wars are not won by evacuations."

One Last Stand

After the capitulation of the Dutch and the Belgians, and the Dunkirk evacuation, the French stood largely alone, with just sixty divisions up against more than twice that many German divisions, which were better armed and more mobile. With the RAF removed to guard the British homeland, the Germans also had complete air superiority.

The sickle cut had carved a line across northern France, so the Germans controlled all territory north of the line. Ninety percent of France, including Paris, was still controlled by the French, whose forces held a four-hundred-mile-long front from the Channel all the way to Switzerland. But the options facing the French leaders were grim. They could fight on against a numerically superior force, they could try to evacuate more troops to Britain or to colonies in Africa and continue the fight from abroad, or they could seek an armistice.

Reynaud was determined to fight on. He told his generals, "Since I am convinced that no peace and no armistice will be acceptable, the armies must fight on as long as possible, and the government must be ready, if necessary, to leave the soil of France."

Defeatism was infecting Reynaud's cabinet, spread in no small part by Marshal Pétain. From the moment the marshal entered the government, he believed that the war was lost and that France should seek peace terms. Moreover, he was convinced that he, as an eminent soldier, would be able to secure better terms with the enemy.

To help resist this faction, Reynaud again shuffled his cabinet. On June 5, he dismissed Daladier and made Charles de Gaulle, who had been promoted to brigadier general only two weeks earlier, undersecretary of the Ministry of National Defense. De Gaulle's appointment infuriated Weygand and Pétain, both of whom detested their brash forty-nine-year-old former protégé and saw him as an arrogant upstart.

De Gaulle was candid with Reynaud. "The disproportion between our forces and the Germans is so great that, barring a miracle, we have no longer any chance of winning in Metropolitan France, or even holding there," he told the premier. "If the war of '40 is lost, we can win another. Without giving up the fight as long as it is possible, we must decide on and prepare for the continuation of the struggle in the Empire."

Reynaud agreed and asked de Gaulle to fly to London to "convince the English that we will hold out, whatever happens, even overseas if necessary. You will see Mr. Churchill . . . the reshuffling of my Cabinet and your presence by my side are signs of our resolution."

ON THAT MORNING of June 5, the Germans launched a new offensive to crack the French lines. General Weygand's Order of the Day declared the stakes: "The Battle of France has begun. The order is to defend our positions without thought of retreat . . . The fate of our country, the safeguarding of her liberties, the future of our children, depends on your tenacity."

And they did fight. Some units, such as the 14th French Infantry, led by the young general Jean de Lattre de Tassigny, fought valiantly. During these desperate days of early June, even when the government and the command knew that the battle for France was hopeless, French forces inflicted nearly five thousand German casualties per day, almost double the rate of the first three weeks of the war.

Day by day, however, they were pushed back. By June 9, the Germans were just forty miles from Paris. Each hour, as the battles raged with the Germans, the battle inside Reynaud's government intensified. Weygand told Reynaud, "We are at the end of our reserves." He advised the premier that it would be wise for the government to evacuate the capital.

Pétain pressed the premier further by reading a formal note, which stated: "The necessity of asking for an armistice to stop hostilities if, of course, the conditions of the armistice, though hard, are acceptable. The salvation and future of the country demand that we proceed in this way with courage."

Reynaud rebuffed Pétain, insisting that "no honorable armistice with Hitler" was possible.

That morning, Weygand had issued another appeal to his troops:

> The safety of the nation demands of you not only your courage but all the persistence, all the initiative, all the fighting spirit of which I know you are capable.
>
> The enemy has suffered heavy losses. He will soon reach the limits of his effort.
>
> We have reached the final quarter hour. Hold fast.

The cabinet resolved to leave Paris for Bordeaux the very next day. For the members of the government, June 10 would be long, momentous, and in the words of de Gaulle, "a day of agony."

Exodus

Everywhere there was retreat. The French armies were in full retreat from the Germans. The government was about to retreat from Paris. Civilians were abandoning their towns and villages as well as the capital. In Norway, the collapse of the Allied armies in France forced their evacuation from Narvik. On the morning of June 10, Norway formally surrendered.

At the daily briefing, Reynaud asked Weygand how long it might be before the Germans were in Paris. "In 24 hours, if the Germans know how weak we are," Weygand answered. "But it will probably be a little longer. They will probably encircle Paris first rather than attack directly."

Later that afternoon, as the Germans surrounded Paris on three sides and approached even nearer to the capital, and as the ministries hurriedly prepared to relocate, Mussolini added to France's miseries—Italy declared war on France. Reynaud addressed the nation by radio that evening:

> We are in the sixth day of the greatest battle in history.
>
> For six days and five nights, our soldiers, flyers, and the Royal Air Force have faced an enemy with force superior in number and in armaments. In this war, which is no longer a war of fronts but of deep points of support, our armies have maneuvered in retreat.
>
> They have only abandoned each point of support after having inflicted severe losses. The kilometers gained by the enemy are littered with destroyed tanks and downed planes . . .
>
> The trials that await us are heavy. We are ready. Our heads do not bow.
>
> It is at this precise moment, while France, wounded but valiant and still standing, struggles against German conquest, when she fights for the independence of all people as well as for herself, is the moment when Mr. Mussolini chooses to declare war. How to judge this act? France has nothing to say. The world that watches us will judge . . .
>
> In the course of its long, glorious history, France has met many difficult trials.
>
> It is then that she has astonished the world. France cannot die.

In his last act before leaving Paris, Reynaud cabled a message to President Roosevelt. Reynaud knew that Roosevelt's hands were tied by the US Congress, which limited how far the president could go in committing America to aiding the Allies. Nonetheless, Reynaud felt that a declaration by the United States that indicated greater future involvement was the one action that could prolong the fight. At this stage, he had nothing to lose by asking.

Mr. President:

I wish first to express to you my gratitude for the generous aid that you have decided to give us in aviation and armament.

For six days and six nights our divisions have been fighting without one hour of rest against an army which has a crushing superiority in numbers and material.

Today the enemy is almost at the gates of Paris.

We shall fight in front of Paris; we shall fight behind Paris; we shall close ourselves in one of our provinces to fight and if we should be driven out of it we shall establish ourselves in North Africa to continue the fight and if necessary in our American possessions.

A portion of the government has already left Paris. I am making ready to leave for the front. That will be to intensify the struggle with all the forces which we still have and not to abandon the struggle.

May I ask you, Mr. President, to explain all this yourself to your people to all the citizens of the United States saying to them that we are determined to sacrifice ourselves in the struggle that we are carrying on for all free men.

This very hour another dictatorship has stabbed France in the back. Another frontier is threatened. A naval war will begin.

You have replied generously to the appeal which I made to you a few days ago across the Atlantic. Today this 10th of June 1940 it is my duty to ask you for new and even larger assistance.

At the same time that you explain this situation to the men and women of America, I beseech you to declare publicly that the United States will give the Allies aid and material support by all means, "short of an expeditionary force." I beseech you to do this before it is too late. I know the gravity of such a gesture. Its very gravity demands that it should not be made too late.

You said to us yourself on the 5th of October 1937: "I am compelled and you are compelled to look ahead. The peace, the freedom and the security of 90% of the population of the world is being jeopardized by the remaining 10% who are threatening a breakdown of all international order and law."

Surely the 90% who want to live in peace under law and in accordance with moral standards that have received almost trusty acceptance through the centuries, can and must find some way to make their will prevail.

The hour has now come for these.

Paul Reynaud

AROUND MIDNIGHT, REYNAUD and de Gaulle climbed into a car together and headed south for Orléans.

One crucial issue that Reynaud did not address before leaving was whether Paris would be defended. Weygand had informed him that day that Paris was to be an open city, and that it was his intention not to defend it. Unfortunately, this decision was not passed on to the military commander of Paris or to the public. Instead, the French radio declared, "Should the Germans reach Paris, we shall defend every stone, every clod of earth, every lamp-post, every building, for we would rather have our city razed to the ground than fall into the hands of Germans."

Up until the time when the government departed, the population had been draining out of Paris. Now the stream of refugees turned into a flood. Thanks to the radio announcements and the appearance of handbills declaring *"Citoyens! Aux armes!"* (Citizens! To arms!), the widespread expectations were that Paris would be bombed and that there would be house-to-house fighting as the Army and civilian volunteers made a last stand to defend the capital. While about one-third of the city's population of nearly three million left during the first month of the war, nearly that many would leave in just the next several days. Overall, 70 percent of the metropolitan area's prewar population of five million would take to the roads, joining the columns of civilian refugees from Holland, Belgium, and northern and eastern France, as well as soldiers heading to or away from the front. The swarms of refugees appeared to Antoine de Saint-Exupéry, aviator and future author of *The Little Prince,* as though "somewhere in the north of France a boot had scattered an ant-hill, and the ants were on the march."

With the government leaving and danger imminent, it was also time for the newspapers to shut down or to relocate. Almost all newspapers ceased publication on June 10 or June 11. The owner of *Paris-Soir,* Jean Prouvost, was as well informed as anyone about the situation, for he was also Reynaud's minister for information. He had begun preparations well in advance for moving the newspaper's operations to the south. He gave the order on June 9 for the remaining staff to clear out of Paris and to head for Clermont-Ferrand, 230 miles away, where arrangements had been made with a publisher who had offered the use of printing facilities there. That publisher was the former prime minister Pierre Laval.

Camus was asked to drive one of the paper's executives. Accompanied by another staff member, they drove all night on roads choked with refugees and abandoned automobiles. The executive chatted with Camus to make sure that he would stay alert. When they reached the center of Clermont-Ferrand, the car was out of gas, oil, and water. As the radiator steamed, Camus suddenly realized that in the rush to leave, he might have left some manuscripts back in his room in Paris. He jumped out of the car and threw open the trunk, and was relieved to find in his valise the complete text of *The Stranger.*

Abandoned

On June 11, the sun did not rise.

Parisians awoke to a dark, eerie sky unlike any they had seen before. Clouds of dense smoke filled the air, blotting out the morning light. A fine black soot fell across the city. In a capital already besieged by wild rumors—the two most optimistic being that the French Army had turned back the Germans and that America had entered the war—the black fog was reported to be everything from German smokescreens to a sign of the apocalypse.

The actual source of the heavy smoke was the deliberate destruction of oil depots on the outskirts of Paris to keep the supplies out of the hands of the enemy. But Parisians could not learn this, as there were no longer any newspapers to investigate or report the cause.

Under the dark sky, the parade of people hurriedly leaving the capital continued. Those Parisians who remained watched the exodus from their doorways or window ledges. They saw an endless procession of cars and trucks overloaded with passengers and furniture, pushcarts piled with family treasures, horse-drawn wagons, bicycles, and streams of people on foot.

There were no taxis or buses to be found. All of the train stations but two, Gare de Lyon and Gare d'Austerlitz, were closed. Around the latter were mobs of people of all sorts of nationalities—Belgian, Dutch, British, as well as French—desperate to go anywhere away from the approaching Germans. In the chaos, many children became separated from their parents, some by accident, some when parents chose to put them in the limited seats available on the few trains that were running. By midday, the wind had blown the black fog away and the day had become sunny and very hot, adding to the misery of those crammed into the stations or sitting in railcars.

After several days of confusion, Monod's unit finally received orders to leave. Until then, it had not been clear whether they would retreat or be assigned to the defense of Paris. The order came as he was digging antitank trenches around Versailles. They left in the middle of the night on a train headed south, away from the Germans, their destination unknown. Monod tried to keep Odette updated on his whereabouts. Before leaving, he wrote, "I don't know where we are going nor how . . . Take good care of yourself and our little ones. You are my only reason to be. I love you. May God help you." The next day he wrote her from a cattle car in the Argenton-sur-Creuse train station, 180 miles south of Paris. The train was not strafed or bombed, but derailed on its way to Périgueux, another seventy-five miles southwest of Argenton, in the Dordogne region. Monod would reach the ancient town on the thirteenth of June.

Desperate to contact Odette, he sent cables to Dinard but received no reply. He feared that Brittany had been cut off by the Germans, and that Odette had left and joined the swarms of refugees. He was able to get a letter off to his parents in Cannes, asking them to contact him in Périgueux if they heard from Odette. He told them, "The only things that matter in the current cataclysm are Odette, the kids, you, Philo . . . In the face of the horrible tragedy that is threatening us with devastation, we must look deep inside ourselves for the last drop of hope and energy." With Cannes just forty miles from the Italian border, he had to worry, too, about his parents. He added, "I think that the despicable bastards next to you are going to leave you alone. The newspapers seem to say so. I love you, my dear parents, but I don't have the courage to write more."

By Wednesday, June 12, many Parisians who had intended to stay left. With no customers, workers, or supplies, shops and restaurants closed. With no guests, hotels closed. Some stores were abandoned without the owners even bothering to lock the doors. Neighborhoods were empty.

The sky was eerily quiet for the first time since the battles began: No Allied or German planes were overhead. But the distant rumble of artillery could be heard.

At 12:15 p.m. General Weygand at last informed Gen. Pierre Héring, the military governor, that Paris was to be declared an open city. No defense would take place around or within the city. Notices were posted the next day that urged the population "to abstain from all hostile acts and . . . to maintain the dignity and composure required by these circumstances."

By the evening of June 13, the city was virtually deserted. There were no streetlights. No lights shone whatsoever. The Place de la Concorde and the Champs-Élysées were empty and so quiet that one could hear the echo of individual footsteps.

AT DAWN THE morning of June 14, two motorcycles roared down the Champs-Élysées, followed by two command cars flying the swastika flag. By eight a.m., a motorized unit entered the city and raced to City Hall. Wave after wave of trucks, tanks, and motorcycles arrived, along with columns of gray-uniformed soldiers. The invaders fanned out across the city, replacing the French tricolor with the German swastika and taking up positions at the monuments and along the great boulevards. Submachine-gun-bearing sentinels were posted around the Arc de Triomphe, and cannons were pointed up each of the boulevards leading to the monument. After two generals paid their respects at the Tomb of the Unknown Soldier, the parading began and went on all day.

Parisians, if they could bear to watch, looked on with a combination of disbelief and dread.

German troops parade down the Champs-Élysées in front of the Arc de Triomphe on June 14, 1940. (AP Images)

CHAPTER 5

DEFEATED AND DIVIDED

A house divided against itself cannot stand.

ABRAHAM LINCOLN, June 16, 1858

ANOTHER CATALYST TO THE MASS EXODUS WAS THE RECOMMENDation, issued by radio, that all men aged eighteen to fifty leave Paris immediately. The concern was that these men would be conscripted by the Germans, or at least taken into custody.

Nineteen-year-old medical student François Jacob was in the middle of second-year exams when it became clear that the Germans would soon enter the capital. He and three of his classmates secured a black Citroën 11 and took off south, just a few days ahead of the invaders.

It took Jacob and his comrades several days to cover the 250 miles to Vichy, through and around the masses of refugees crowding the roads—"a river of torment and fear" that flowed south. Six to eight million refugees were displaced from their homes and farms, overwhelming every city and village they reached. There was nowhere for the dazed and exhausted castaways to sleep, little to eat, and not enough water. Jacob and his fellow travelers saw for themselves the disarray of the military as they shared the roads with defeated soldiers in retreat.

The grandson of the first Jewish four-star general, Jacob had been raised to believe in the "indestructible framework of the country." In just a few days on the road, Jacob had witnessed "a whole nation disintegrate." Staring quietly out the car window at the disaster around him, he thought about the seemingly secure world that had once surrounded him—"the country, the Republic with its institutions and its laws, its army and its justice. And suddenly the whole edifice has caved in."

Jacob stopped in Vichy to see his father, Simon, who had taken refuge there with his two grandmothers. But he and his friends were determined to continue southwest, to get as far away from the invaders as quickly as possible. On the morning of June 17, Jacob's twentieth birthday, the very day he would become eligible for military service, he left behind his father, who was on the verge of tears as he slowly waved good-bye to his son, unsure when or if they would see each other again.

Crawling along the road clogged with cars, trucks, bicycles, and carts, Jacob was overwhelmed by the magnitude and the swiftness of the collapse: "Everything I believed in, everything that I thought I'd believe for life, everything that seemed the very basis of our existence, forming our protective armature, that seemed to shape our view of the world: all this crumbled in an instant. In an instant, the country has foundered. In an instant, despite its great men and its great schools, its generals and its institutions, its teachers and its senate, it has collapsed, body and soul."

Talk in the car turned toward whom to blame. Foremost, there was the madman Hitler, who, "foaming at the mouth and screaming maledictions, had coolly decided to put the world to the test of fire and blood." But, Jacob thought, guilt also fell on the "governments of bunglers, manipulated in the shadows by degenerates, [that] did not know how, or did not dare, to stop him."

The government had first taken flight for Tours and then, when that city too was about to be taken by the Germans, it would move quickly on to Bordeaux, a port city on the Garonne River in southwestern France. The bungling and manipulation had not yet reached its lowest point.

To Surrender or Fight?

Over the course of three long days in Tours, Premier Reynaud's cabinet and military commanders had wrangled over the two viable choices before them—to continue the fight from abroad or to seek an armistice. General Weygand and Marshal Pétain led the latter contingent, while Reynaud and de Gaulle searched for any possible means of carrying on the fight. Reynaud was supported by Churchill, who flew over to meet with the command. Churchill wanted the French to hold out at all costs, for that would buy the British more time to prepare, and hopefully would weaken what Hitler could throw at England. In March, France had pledged to Britain that it would not seek a separate peace agreement with Germany; Reynaud reiterated that pledge.

The division within the government, however, was becoming wider, and the pressure on Reynaud to surrender was increasing as the military situation deteriorated. As the ministers looked out the windows from their meeting room in the Château de Cangé at the pathetic sight of refugees streaming by, General Weygand urged, "If an armistice is not demanded immediately, disorder will spread to the armies as it already has to the population."

Reynaud continued to refuse. Weygand took another approach, arguing that a government established outside of France itself would have no recognized authority and could not regain what had been lost.

Marshal Pétain lent Weygand his support, stating, "The duty of the government, regardless of what happens, is to remain in the country or else it will not be recognized as such." He concluded, "The armistice, in my opinion, is the necessary condition for the perpetuity of an eternal France."

ONCE THEY REACHED Bordeaux, de Gaulle sought to bolster Reynaud's resolve. He told the premier, "For the last three days we have been speedily rolling toward capitulation . . . I myself refuse to submit to an armistice. If you remain here you are going to be submerged by the defeat. You must go to Algeria at once. Are you—yes or no—decided on that?"

"Yes!" Reynaud declared.

"In that case," de Gaulle said, "I should go tomorrow to London to arrange for British shipping to help us. Where will I find you again?"

"You will find me in Algeria," Reynaud answered.

REYNAUD ASKED THE cabinet to consider moving the government to North Africa. Georges Mandel, minister of the interior, backed the proposal. He reminded the cabinet of the pledge Reynaud had made to

President Roosevelt that France would continue the fight from North Africa, if necessary. Pétain and Weygand maintained their opposition to any relocation.

For the next two days, June 15 and 16, the impasse between Reynaud and Weygand and the division within the cabinet continued. Reynaud secured the agreement of the heads of the Senate and Chamber of Deputies to move the government to North Africa. Before a critical cabinet meeting on the afternoon of the sixteenth, when the ministers were likely to arrive at their ultimate decision—armistice or going abroad—de Gaulle phoned Reynaud from London with a stunning development: the British government was proposing the formation of a union between Great Britain and France. The language was as stirring as the idea itself:

> At this most fateful moment in the history of the modern world, the Governments of the United Kingdom and the French Republic make this declaration of indissoluble union and unyielding resolution in their common defence of justice and freedom . . .
>
> The two Governments declare that France and Great Britain shall no longer be two nations but one Franco-British Union.
>
> The constitution of the Union will provide for joint organs of defence, foreign, financial, and economic policies.
>
> Every citizen of France will enjoy immediately citizenship of Great Britain; every British subject will become a citizen of France.

Reynaud was astounded and delighted. He had not been able to solicit any concrete commitment from the United States, but now their cross-Channel ally was offering a bold, historic gesture. The proposal had been forged in just two days by French diplomats. After some personal persuasion by de Gaulle, it had been endorsed by Churchill.

It arrived not a moment too soon. Reynaud's ministers had been slipping down the slope toward an armistice. Reynaud believed that the British proposal would turn the tide.

But he was sold out. Someone leaked the proposal to the cabinet before he could announce it. Weygand and others rallied opposition to the idea. When Reynaud told the cabinet he was planning to meet

Churchill the next day to announce the union, he was greeted with silence. Several opponents then spoke out against the union, expressing indignation at the proposal that "would have placed France in a state of vassalage."

These ministers apparently gave less consideration to what an armistice with Germany would produce. Mandel warned them, "You imagine that by capitulating you will sleep in peace and pick up the routine of an easy, comfortable life. But the war will continue over your heads . . . the war will begin again on our soil . . . By surrendering, you fancy you will win rest and quiet. Instead you will reap only the contempt of the world and ultimately of yourselves." He concluded that there were two types of men in attendance, "the brave and the cowards" or "those who wish to fight, and those who do not."

Some ministers believed that Britain would soon be defeated and that there was no point to joining a doomed partner. Reynaud was shattered. He would later recall the moment as "the greatest disappointment of my life." The discussion turned back to an armistice.

Exhausted and despondent, Reynaud feared that the cabinet would force him to call for an armistice. Resolute that he could not do so, Reynaud tendered his resignation, telling President Albert Lebrun that if he wanted to follow such a policy, he should "go and ask Marshal Pétain!" Reynaud added, "I'm told he has his cabinet list in his pocket."

Lebrun, who was also opposed to the armistice, tried to talk Reynaud out of resigning. Failing to do so, and under the impression that the majority of the cabinet favored an armistice, Lebrun asked Pétain to form a government. The marshal promptly opened his briefcase and produced a list. "There is my government," he told Lebrun.

Usually, it took three or four days for a new premier to assemble a cabinet. The ready-made list indicated that Pétain and his supporters had been plotting a new regime for some time. The new cabinet included many who had pressed for an armistice and excluded those who had vowed to continue the war. Weygand was named minister of national defense. Mandel and de Gaulle were out.

Within two hours, just after midnight on June 17, the cabinet met. They agreed unanimously to ask the German government for the conditions of an armistice. The inquiry was submitted through a Spanish intermediary.

Before there was any reply from the Germans, Marshal Pétain addressed the nation by radio at 12:30 p.m. on the seventeenth:

> Frenchmen! On the appeal of the President of the Republic, I have assumed today the direction of the Government of France.
>
> I am, in heart and thoughts, with our admirable army who, with every heroism and without precedent, have continued a glorious military tradition against an enemy superior in numbers and in arms, our army which has known, by its magnificent resistance, how to fulfill its duty toward our Allies.
>
> Certain of the help of the ex-Service men whom I have the honour to command, and assured of the confidence of the whole people, I give myself to France to help her in her hour of misfortune.
>
> In these painful hours I am thinking of our unfortunate refugees, and all their extreme distress. I express to them my compassion and my solicitude.
>
> It is with a heavy heart I say we must cease the fight. I have applied to our opponent to ask him if he is ready to sign with us, as between soldiers after the fight and in honour, a means to put an end to hostilities.
>
> Let all Frenchmen group themselves around the Government over which I preside during this painful trial, and affirm once more their faith in the destiny of our country.

François Jacob and his friends had stopped for lunch at a little inn in a village in the Auvergne. From an open window, they heard the "quavery voice" of Marshal Pétain. As they took off again for the coast, the conversation in the car was animated. Jacob's best friend, Roger Dreyfus, took the lead by snarling about Pétain, the military's failures, and the incompetent government. He cursed "the traitors, the crooks, the dirty bastards of every stripe" who had led the country to ruin. Then he rallied his mates, saying, "Even so, we're not going to let ourselves be had. We're not going to wait here for the SS to arrive so we can smile prettily for them. When they are here, that will be the end. That doddering old fart Pétain won't keep them from doing what they want. You don't talk it over with the Nazis. You bash their faces in. There's only one thing to do: go on with the fight. And to fight, you have to get out of France. Go where you can."

Jacob agreed. Having been raised in a military family, in which not just his illustrious grandfather but his father and uncles all served, Jacob decided that it was his turn. "We are not going to shrink before a threat worse than any ever seen," he told his companions. "You don't negotiate with a Hitler. Either you destroy him or he'll destroy you. I, too, am for getting out of France, for fighting wherever we can."

Before taking that leap, before leaving France, Jacob wanted the approval of his family. When the car reached Arcachon, on the Atlantic coast west of Bordeaux, he sought out his uncle Henri, who had arrived there two days earlier. Amid the chaos and anxiety, the sight of his uncle was a great comfort.

They walked on the beach after dark. Uncle Henri was also the personal physician of three-time prime minister and current president of the Chamber of Deputies, Édouard Herriot, who was in Bordeaux with the government. Uncle Henri knew all about the intrigues among the political and military leaders wrestling over France's fate.

He told François about the struggle over whether to continue the war, then turned and said, "If you can, go to England rather than Africa."

Au Revoir, La France

Jacob and his comrades were among many thousands facing the decision of whether to remain in France or, if they wanted to leave, how to get out and where to go.

Charles de Gaulle had already made his decision. When the government changed, he knew that surrender was imminent, which he could not accept. Fearing interference or arrest by Weygand, he slipped out of Bordeaux on the morning of June 17 and flew in a British plane to an airport outside London. He arrived just as Pétain was asking for the fighting to stop.

De Gaulle went to see Churchill for permission to use the radio to address the French people. The next day, from the BBC Broadcasting House, he spoke as though he were "concentrating all his power in one single moment." In a loud, commanding voice he declared:

> The leaders who, for many years past, have been at the head of the French armed forces, have set up a government.
>
> Alleging the defeat of our armies, this government has entered into negotiations with the enemy with a view to bringing about a cessation of hostilities. It is quite true that we were, and still are, overwhelmed by enemy mechanized forces, both on the ground and in the air . . .
>
> But has the last word been said? Must we abandon all hope? Is our defeat final and irremediable? To those questions, I answer—No!
>
> Speaking in full knowledge of the facts, I ask you to believe me when I say that the cause of France is not lost. The very factors that brought about our defeat may one day lead us to victory.
>
> For, remember this, France does not stand alone. She is not isolated! Behind her is a vast Empire, and she can make common cause with the British Empire, which commands the sea and is continuing the struggle. Like England, she can draw unreservedly on the immense industrial resources of the United States.
>
> This war is not limited to our unfortunate country. The outcome has not been decided by the Battle of France. This war is a world war. Mistakes have been made, there have been delays and untold suffering, but the fact remains that there still exists in the world everything we need to crush our enemies one day . . . The destiny of the world is at stake.
>
> I, General de Gaulle, now in London, call on all French officers and men who are at present on British soil, or may be in the future, with or without their arms . . . to get in touch with me.
>
> Whatever happens, the flame of the French resistance must not and shall not die.

With the electricity out across much of France and so many refugees on the roads, few people actually heard the original speech. But de Gaulle's appeal was reproduced in some newspapers and spread by word of mouth.

Camus did not hear de Gaulle. He was bogged down on the road to Bordeaux among the refugees and retreating soldiers. When the Germans approached Clermont-Ferrand, Jean Prouvost issued the order for the *Paris-Soir* staff to move once again and follow the government to Bordeaux. When Camus finally reached the town, he thought of leaving to go fight abroad. But he missed the last boat from Bordeaux and was stuck.

MANY PARISIANS OR other citizens from the north who had family in the west or south of the country had retreated to relatives' homes to escape the battle. They suddenly had to decide whether to flee the country altogether. It was an agonizing decision. Since most families had at least one member in the military whose fate (prisoner, casualty, or refugee) was typically unknown, leaving meant risking being cut off completely from loved ones. Odette was in that dilemma.

While Jacques was garrisoned in Versailles and then evacuated to Périgueux, Odette and the twin boys had in fact remained in Dinard, along with Odette's mother, her three sisters, and their children. But on June 17, as the Germans were advancing and the people were fleeing in front of them, the whole family planned to make their way south to Biarritz, on the coast just eleven miles from the Spanish border. As Odette and her sisters—Suzanne, Madeleine, and Lise—packed suitcases for the car trip, Odette's mother watched over the twins. For some reason, Grandmother Bruhl was holding a tube of the sleeping tablet Gardenal and left it next to Olivier when she left the room. The ten-month-old opened the tube and swallowed the tablets. Panic ensued as the sisters went to find a doctor amid the chaos of all of the refugees in Dinard. Once the doctor was located, it took most of the afternoon for Olivier to be treated. The family abandoned packing and the trip. The next day, as the sound of artillery drew near, Suzanne, Madeleine, and their children went to Saint-Malo, succeeded in getting space on a military tug and on a cargo ship, and made it to England. Odette and the twins stayed in Dinard with her mother and Lise.

With German armored columns approaching quickly, Jacob and Dreyfus decided to find a boat leaving as soon as possible; the other two passengers in the Citroën got cold feet and elected to stay in France. Jacob and Dreyfus first asked around Bordeaux. No luck. Then in Bayonne, they found a note on the British consulate door stating: "For all military inquiries concerning the French, apply to Saint-Jean-de-Luz." They headed to the small resort town just a few miles north of the Spanish border.

The evacuation in progress was dubbed Operation Aerial. Like Dunkirk, it was a British operation to get soldiers out of the reach of the Germans. British warships provided cover while several troop ships, including the Polish liners the SS *Batory* and SS *Sobieski*, were loaded. Two Polish divisions had participated in the defense of France. Their commander in chief and leader of the government in exile, Gen. Władysław Sikorski, refused to capitulate to the Germans. Churchill pledged to evacuate all of the Polish troops that could be loaded.

Jacob and Dreyfus wandered around Saint-Jean-de-Luz looking for passage. A cavalry lieutenant in a blue képi told them to come to the port around five o'clock. The official fighting was supposed to be over, so the officer added, "Don't make yourselves too conspicuous before then."

When they returned, fishing boats were ferrying the last of the Polish troops to the large ships. The French police had formed a line to prevent French civilians from boarding the boats. In front of Jacob, a civilian tried to pass. When a policeman stopped him, he pretended not to understand French and blurted out, "Swastika," the most Polish-sounding word he could think of. The policeman let him pass. Jacob and Dreyfus took advantage of the distraction to slip past and sneak aboard the small boat that shuttled them out.

Soon they were aboard the SS *Batory*. No one was sure whether the ship was going to England or North Africa. The man who had bluffed his way on board turned to Jacob and asked, "Ever hear of de Gaulle?

He's a general, I heard him on the radio. He said he is going on with the war, in England. He said that, sooner or later, we will beat them."

Revenge. That was the one motivation that would have to sustain Jacob in the uncertain days ahead—that and the hope of someday, sooner or later, returning to France. As his ship left for England on June 21 under the cover of darkness, it was impossible to guess when, or if, he would see France again.

Divided

Many members of the government were torn over whether to remain or to go abroad, and were running out of time to decide. With the terms of the armistice unknown, and likely to be severe and perhaps unacceptable, the presidents of the Senate, the Chamber of Deputies, and the Republic urged Pétain to accept a division of the government. On June 18, Pétain agreed to a plan in which he would remain in France with several ministers while President Lebrun, the heads of Parliament, and those members of Parliament who wished to would go to North Africa. The cabinet concurred, and plans were set in motion to depart on June 19 and 20.

De Gaulle took to the airwaves again on the BBC on the nineteenth, reiterating his call to resistance:

> It is the bounden duty of all Frenchmen who still bear arms to continue the struggle. For them to lay down their arms, to evacuate any position of military importance, or agree to hand over any part of French territory, however small, to enemy control, would be a crime against our country. For the moment I refer particularly to French North Africa—to the integrity of French North Africa . . .
>
> Soldiers of France, wherever you may be, arise!

Weygand and Laval tried to block the departures to North Africa. On June 21, Laval told President Lebrun, "You will not leave, you must not leave . . . When it is known that you chose as the hour of departure the very hour in which our country was sunk in the greatest distress, there will be but one word on every lip: desertion . . . Perhaps an even graver word: treason." Despite the plan agreed to days earlier, Pétain then declared that if Lebrun insisted on leaving, he would have him arrested.

That same day, the *Massilia* sailed for North Africa with twenty-seven members of Parliament. That evening, the French armistice delegation was led to the Forest of Compiègne, the site of the 1918 armistice that the victorious Allies had imposed on a defeated Germany. The Germans had even retrieved from a museum the very railroad car in which the 1918 terms were signed. Hitler appeared briefly the next morning to savor his triumph.

The French delegation was instructed to break off negotiations if Germany demanded the French naval fleet, any overseas territory, or occupation of all of France. The twenty-four terms were severe but came up short of these three breakpoints. The fleet was a complex and sensitive issue. France had promised Britain and the United States that its navy would not fall into German hands. The French had to finesse an agreement by which the fleet would be partially demobilized and disarmed under German supervision, and trust in a promise that the Germans did not "intend" to use the fleet for their own purposes. Churchill would say of the German promise, "What is the value of that? Ask half a dozen countries."

Three-fifths of France, including the Atlantic coast, Paris, and the north were to be occupied, while the south and southeast (the *zone libre*) were to be administered by a French government. Another article forbade French nationals from fighting against Germany for other states. Those who did would be treated as *francs-tireurs* (guerrilla fighters) and shot upon capture—a sober warning to those in Britain or North Africa who wanted to continue fighting. There was nothing to be negotiated. France signed the terms on June 22.

Once de Gaulle learned of the terms of the armistice, he took to the airwaves again:

> The French government, after having asked for an armistice, now knows the conditions dictated by the enemy.
>
> The result of these conditions would be the complete demobili-

> sation of the French land, sea, and air forces, the surrender of our weapons and the total occupation of French territory.
>
> It may therefore be said that this armistice would not only be a capitulation, but that it would also reduce the country to slavery . . .
>
> I call upon all Frenchmen who want to remain free to listen to my voice and follow me.

Churchill added his own condemnation, stating that his government did not feel "that such, or similar terms, could have been submitted to by any French government which possessed freedom, independence, and constitutional authority." He continued: "A British victory is the only hope for the restoration of the greatness of France and the freedom of her people." Churchill then added a call to "all Frenchmen, wherever they may be, to aid to the utmost of their strength the forces of liberation."

Pétain was outraged and broadcast a reproach to Churchill: "The French government and people heard the statement of Mr. Churchill yesterday with grief and amazement . . . The French can but protest the lessons given by a foreign minister . . . The French will be saved by their own efforts, Mr. Churchill should know this . . . He should know that the French are showing more grandeur by admitting defeat than in trying to avoid it by vain and illusory efforts."

The new regime also stepped up pressure on its opponents. General Weygand ordered de Gaulle to return home to his post, stripped him of his rank, and initiated his court-martial. De Gaulle was not intimidated. He continued with his radio broadcasts from London. On June 26, in reply to a broadcast by Pétain in which the marshal claimed that the French government "is still free," de Gaulle directly addressed his former commander:

> M. le Maréchal, in these hours of shame and anger for *la patrie*, one voice must answer you. This evening that voice will be mine . . .
>
> You were led to believe, M. le Maréchal, that this Armistice, sought from soldiers by a great soldier such as yourself, would be

honorable for France. I think you now know where you stand. This Armistice is dishonorable. Two-thirds of our territory handed over to the occupation of the enemy—and what an enemy!—our whole Army completely demobilised, our officers and soldiers, who are now prisoners, kept in captivity. Our Navy, our aircraft, our tanks, our armies are to be handed over intact, so that the enemy may use them against our own Allies. Our country, our Government, you yourself reduced to servitude. Ah! In order to obtain and accept such an act of enslavement there was no need for you, M. le Maréchal, there was no need for the victor of Verdun—anyone would have done.

FRANCE HAD FALLEN from confident to conquered in just six weeks. But France, or what remained of her, was not merely defeated but divided. She was divided spiritually between the minority who wanted to continue the fight, and those who accepted defeat. She was divided and estranged from her former close ally, not only unable to receive Britain's aid but perhaps about to contribute to its downfall. She was divided physically, with part of the population forced to live under occupation by their conquerors, and the remainder to live under a separate government. And she was divided between leaders: one, an elderly hero who thought that the war was over, that Britain would soon be defeated, and that a new Fascist-dominated European order would be established; the other, a rogue general in exile who thought that the war had just begun and would never be over until France and liberty had been restored.

In the last days of June, as the nation and the world struggled to comprehend what had befallen France, her future was unknowable. Would she, as the Pétain government pledged, "turn over this dark page in history" and be a "country which will continue to live with soul uplifted and free"? Or would she, as de Gaulle predicted, be enslaved "beneath the German jackboot"?

For individual citizens, whose lives had been plunged into chaos and who despaired for their families, country, and freedom, the fundamental question was how to reclaim their future. How could they regain their liberty? Whom should they follow? Should they put their faith in the marshal, who promised to restore order to the nation, and who offered that only through "composure and labor" would France revive. Or should they heed the general who, though without any army, vowed to reconquer what had been lost?

SOME, LIKE JACOB, had already chosen their path. In the course of time, Monod and Camus would find theirs. Some, however, saw no future. Joseph Meister, Louis Pasteur's first patient, veteran of the First Battle of the Marne, and caretaker of the Pasteur Institute, retreated to his apartment, closed the windows, and turned on the gas on his stove.

Part Two
The Long Road to Freedom

Noble souls, through dust and heat, rise from disaster and defeat the stronger.

—Henry Wadsworth Longfellow, *The Sifting of Peter*

CHAPTER 6

REGROUPING

Faced with crisis, the man of character falls back on himself. He imposes his own stamp of action, takes responsibility for it, makes it his own.

—CHARLES DE GAULLE

ALL ACROSS FRANCE, THE ANTS THAT HAD BEEN SPILLED BY THE invasion were trying to return to their nests. With the armistice signed, some six million refugees who had fled the German advance, mostly by escaping to the south and west, now had to make their way back to their homes, if they still had them. Amid the chaos and uncertainty, most thought first about finding and reuniting with loved ones. With the rush to escape the Germans, and the collapse of all public services, most families did not know the fate of their members in uniform. In six weeks, France had suffered more than 90,000 soldiers killed, 200,000 wounded, and 1.8 million taken prisoner by the Germans. And many of the roughly 3 million soldiers who were not captured or casualties, like Jacques Monod, in turn did not know where their families had ended up.

Jacques did not stay put in Périgueux. After learning of the armistice, his unit headed farther south, with Jacques at the wheel of a bus carrying twenty-five soldiers. He stuck to secondary roads running along the edges of the granite plateaus of the Massif Central. After a twenty-four-hour journey, he reached an abandoned seminary in Saint-Sulpice-la-Pointe, north of Toulouse. The unit decided to move in and await further orders.

Jacques had not heard from Odette for more than two weeks—since he was in Versailles. He knew that the Germans must have reached Dinard and that it was part of the occupied zone. He was desperate to know that she was safe but did not even know where she was. In order to try to contact her, and to let his family know where he had ended up, he sent three letters on June 26 from Saint-Sulpice—to Odette in Dinard, to Odette at their apartment in Paris, and to his parents in Cannes. In the first, he wrote:

> *Mon amour,*
>
> *I don't have any hope of reaching you this way. But I want to try everything. Nothing is more important to me than to find you. I will only start living again when I have found you. I don't have the courage to write more in this letter, which is only a sounding.*
>
> *I adore you.*

To his parents he wrote:

> *My Dears,*
>
> *Will this letter ever reach you? I am just hoping. If it does reach you, my dears, telegraph me, and write to me right away. I know nothing of Odette, nothing of you, nothing of Philo. Nothing matters to me more than to find you again, to find my kids and my wife. It is impossible to think of anything else, to understand anything else. I adore you.*

A week later, he was comforted to receive a telegram and a letter from his parents saying that they were fine, and it included one hundred francs, for he had had no money for more than two weeks. But by July 3, he had still not heard from Odette. He wrote back to his parents to say that he was in "terrible anguish" over not knowing where his family was. He was thinking of placing a notice in newspapers throughout the occupied zone to try to locate her.

On July 7, a telegram arrived from his mother—Odette was safe in Dinard! Odette had received Jacques's "sounding" on the fourth but could not send a telegram right away because the service was not yet working again. She immediately wrote a letter to Jacques and one to his parents in Cannes, but the latter arrived first. After much hesitation she had decided to stay in Dinard rather than join the massive streams of refugees on the roads.

After Odette's letters finally reached him, Jacques poured out his relief:

My dear angel,

So it is true, you are over there. I found you again, I could not believe it despite Maman's letters and cables. It took your two letters, from the 4th and the 6th, received today, to convince me that I could breathe and live again. I could not bear much longer this burden of anxiety. And now, I wish to have a God to thank . . . You were absolutely right not to leave. All of my anxiety was coming from the fact that I thought you had left, and were then stopped somewhere in the dreadful rabble. God only knows where you would be had you tried that.

After telling her of his adventures, he raised the question of where they could meet once he was demobilized. Article IV of the armistice stated: "French armed forces on land, on the sea, and in the air are to be demobilized and disarmed in a period still to be set." Jacques hoped that might occur relatively soon, but everything was vague at the time. They would have to wait, apart, until the situation was clearer, perhaps a few weeks. He asked Odette, "Hold my little ones against your heart. What a joy for us, my darling, to have in them all of the reasons to hope."

Vichy

Camus was one of the millions of displaced vagabonds forced to improvise from day to day. After being stranded among refugees in Bordeaux, he made his way back to Clermont-Ferrand, in the non-occupied zone. He thought that the *Paris-Soir* staff would head back to Paris and continue publishing the paper there amid the Occupation. But Prouvost, his publisher, was named high commissioner for propaganda by the new Pétain government that was headquartered for a short while in Clermont, and then settled less than thirty miles away in the spa town of Vichy. Camus thus had a ringside seat to the machinations of the new regime.

The shame of defeat and the humiliation of having to bow to the Germans were about to be exploited politically, and as a result, the anguish of those who yearned for France to regain her freedom was about to be compounded.

The master architect of the new Vichy government was Pierre Laval, the publisher who had made his Clermont printing facilities available to the *Paris-Soir* newspaper team. Laval had twice been premier but had been excluded from power since 1936, and he was deeply resentful. While Pétain had earned his reputation at Verdun, Laval had been a pacifist during World War I, a position that garnered him a place on the Carnet B (a list of potential subversives) and the nickname "Pierre Loin-de-Front" (Pierre Far-from-the-Front). A chain-smoker who wore a trademark white cravat, Laval had gained a fortune between the wars, but little respect. He was widely seen as a shady character with dubious motives. He wanted back into the government, and he saw his opportunity in the wake of France's collapse. His strategy was to serve and manipulate Pétain, whom Laval saw as only a figurehead—"a vase on the mantelpiece." His first coup was to maneuver into the post of deputy prime minister in late June.

Laval believed that France would secure the best peace terms with Germany and Italy only by emulating the victors' dictatorships. He thus sought to dissolve Parliament and to equip Pétain with sweeping powers. He helped draft an audacious measure, without precedent in France's history, for the National Assembly to pass in short order. The initial text read:

> The National Assembly gives all powers to the Government of the Republic, under the signature and authority of Marshal Pétain, President of the Council, to promulgate by one or several Acts the new Constitution of the French State.
>
> This Constitution will guarantee the rights of work, of family, and of the country. It will be ratified by the Assembly which it will set up.

To achieve his aims, Laval exploited both Pétain's own vanity and the hero status the old warrior enjoyed among many members of Parliament. Their opponents would have been those who wanted to continue to prosecute the war from abroad, except the most effective members were stranded or detained in North Africa and unable to counter Laval's maneuvers.

To Camus's disgust, *Paris-Soir* joined the chorus of worship that heralded Pétain as France's savior. The newspaper ran an article proclaiming that the eighty-five-year-old had "the arteries of a man of forty." Camus saw right through the dealings in Vichy, writing to Yvonne Ducailar to lament "the cowardice that surrounds me." He told her, "What we are going to experience now is unbearable to think of, and I am sure that for a free man, there is no other future, apart from exile and useless revolt. Now the only moral value is courage, which is useful here for judging the puppets and chatterboxes who pretend to speak in the name of the people." Camus wanted to flee for Algeria, "the last French soil that is still free (without that ignoble thing called Occupation)," but there was no transportation available. He was trapped, at least for the time being.

In the first days of July, Laval finessed Pétain and worked on the members of Parliament. He did not have to push the marshal too far to his own views, as Pétain blamed the British, the Parliament, and even French schoolteachers for the debacle. Pétain viewed the latter as unpatriotic Socialists who had failed to instill a sense of duty in the young. Pétain expected Britain to fall within weeks and to be completely destroyed by Germany. He believed that France, under his leadership, would survive relatively intact, albeit functioning like a province of Germany. The themes of his new order were "Work, Family, and Homeland."

Laval cajoled and pushed the Chamber of Deputies to accept his sweeping proposal, warning some of the deputies of a possible military coup if they failed to act, bribing others with promises of positions in the future government. He told them, "Parliamentary democracy lost the war. It must give way to a new regime: audacious, authoritarian, social, and national." Even more boldly, he declared, "We have only one road to follow, and that is a loyal collaboration with Germany and Italy. We must practice it with honor and dignity. And I am not embarrassed to say so. I urged it during the days of peace."

A humiliating defeat was to morph into "loyal collaboration"? The specter sickened Camus. "Cowardice and senility, that's all we are being offered," he wrote to Francine in Oran. Only two weeks after the armistice and days before the National Assembly acted, Camus perceived very clearly what the new regime was about. He foretold what the consequences would be for France:

> *Pro-German policies, a Constitution like those of totalitarian regimes, horrible fear of a revolution that will not happen, all this as an excuse for sweet-talking enemies who will crush us anyway, and to preserve privileges which are not threatened. Terrible days are in store: famine and general unemployment along with the hate they bring, which won't be prevented by an old geezer's speeches . . . We must also be aware of anti-British propaganda, which hides the worst motivations . . . I don't need to tell you that we have lost the battle, and that those who led us in spite of ourselves to this disastrous war have further strengthened their power after this unnamable defeat.*

Pétain's government included Admiral Darlan and General Weygand. Both leaders of the houses of Parliament, each of whom had long been staunch Republicans, gushed with praise for Pétain and supported Laval's proposal. Speaker Édouard Herriot said, "Around the Marshal, in the veneration which his name inspires in us all, our nation has rallied in its distress. Let us be careful not to trouble the accord which has been established under his authority." On July 9, the Chamber voted 395 to 3, and the Senate voted 229 to 1 to affirm that the Constitution needed revision. The next day, the National Assembly voted 569 to 80 to dissolve itself and, in effect, to turn the country over to Pétain and Laval.

For weeks at Saint-Sulpice, Monod spent most of his days playing bridge and chess, hoping for another letter from Odette and waiting for his demobilization instructions. The mail was slow and irregular. It took seven to ten days for a letter to reach its destination, so information about where they might reunite or about work prospects in Paris were out of date before one or the other received a response. There was some good news: Philo was demobilized and made it home to Cannes, and Odette's sisters Suzanne and Madeleine made it safely to England. But Jacques and Odette's relief was tempered by their anguish at learning that Odette's brother Étienne and her brother-in-law Ado had both been taken prisoner. With just Odette's sister Lise and mother remaining in France, Jacques and the twins would be even more the center of Odette's world.

Jacques wanted Odette to leave Dinard as soon as possible, because the German formations on the coast were targets for the RAF, but not before he knew where he was going and when. Finally, he was officially demobilized on July 29, and the family was reunited in August in Paris. They returned to their apartment on rue Monsieur-le-Prince, near the Sorbonne, and tried to pick up the threads of their former life. Jacques had contacted the rector of the Sorbonne while he was in Saint-Sulpice. He was desperate to return to his studies on bacterial growth and finish his doctorate. He resumed his research in the very old laboratory that opened onto a gallery of stuffed monkeys.

CHAPTER 7

ILL WINDS

I embark today on the path of collaboration.

—MARSHAL PHILIPPE PÉTAIN, October 30, 1940

LIKE OTHERS WHO RETURNED TO PARIS OR OTHER NORTHERN parts of the country, the Monods had the added challenge of not just reuniting their family and regaining their livelihood but of facing everyday life in the presence of a foreign army. The changes to the capital were difficult to stomach.

Across the city, it seemed as though the red-and-black swastika flags flew from nearly every window. The Germans had taken over most government buildings and had requisitioned many choice hotels and private homes as offices and residences. The Majestic Hotel had become military headquarters, the Ritz the home of the Luftwaffe, and the Chamber of Deputies was converted into offices for the Kommandant of Greater Paris. Parisians who needed permits or had other business with the Germans had to go to the large Kommandantur Building, identified in huge white letters, that dominated the Place de l'Opéra. All of the official facilities were heavily guarded by armed sentries.

At every intersection, there were signposts with black and white gothic lettering in German. There was no French traffic; virtually all of the vehicles on the streets belonged to the German military, or were sleek black staff cars that ferried officers around the capital.

And then there were the soldiers. The Monods could not avoid them when they took the twins out for a stroll in their double pram. Blue-gray and dark-green uniforms were everywhere—in the cafés, on the grand boulevards, and at the monuments. Some restaurants had been converted to soldiers' canteens.

While the German officers and bureaucrats indulged in the spoils of war, and their soldiers enjoyed the Parisian sights, the costs of their victory were being paid by the French. Article XVIII of the armistice agreement stated simply: "The French Government will bear the costs of maintenance of German occupation troops on French soil." In August, those costs were set at 20 million reichsmarks *per day*, retroactive to June. The value of the mark was then fixed at 20 francs, an overvaluation by at least 50 percent. The 400 million francs per day that the French were obliged to pay represented about 60 percent of national income, and was enough to support an army of 18 million men. As a result, Germans in Paris had enormous spending power and soon cleared out store shelves of all kinds of items—perfume, jewelry, clothes, wine, chocolate—that had not been available in Germany for years.

As Camus had predicted, hunger quickly became one price of defeat. While some rationing had been in place since the beginning of the Phoney War, food shortages were made much more severe by damage to crops, the reduction of imports from French territories, the displacement of labor that left stores and farms shorthanded, and massive German requisitions. Much more stringent rationing was imposed in Paris in September, and throughout France by October 1. People could obtain ration cards at local town halls after standing in long, slow lines. They were allotted different amounts of food according to their age and occupation. Adults over twenty-one but under seventy years old were entitled to 350 grams of bread per day, 350 grams of meat per week, and 500 grams of sugar, 300 grams of coffee, and 140 grams of cheese per month. Babies received less bread but were allotted milk, whereas the elderly were not. Other than the babies, the total caloric allocations were inadequate. To stave off hunger, the French substituted new foods, such as turnips and rutabagas, which had formerly been fed only to cattle. Ersatz coffee was brewed from roasted grain and chicory, and drunk without milk or sugar. Odette had to spend much of her day waiting in lines and strategizing where to find food for the twins.

But that would soon be the least of her worries.

A New Order

Even before most refugees could return to their homes, the Vichy government was moving rapidly to assert its authority and to take the country on a different path. Within days of being established, it started to enact restrictive laws concerning various "undesirables" that emulated those of Germany. It promptly adopted foreign policies that aligned France more closely with Germany and distanced her from her former close ally Great Britain. And in a quest to rally national unity around the regime, Vichy launched propaganda campaigns against de Gaulle, the Free French, and the British who were continuing the war.

Beginning just days after Vichy received its sweeping powers, a succession of laws were enacted that were explicitly nationalist and anti-Jewish. On July 12, the regime ordered that only those individuals with a French father could be a member of a minister's cabinet. A few days later, this policy was extended to public servants and teachers, and subsequently to dentists, doctors, pharmacists, and lawyers. On July 22, the government announced that it was reviewing the status of citizenship of those who had been naturalized since 1927—a measure that would disproportionately affect Jews. On August 13, secret societies were banned, a measure aimed at Freemasons, who were viewed as part of a wider "Jewish-Bolshevik conspiracy." Public servants had to swear that they had never belonged to any such group.

Driven by both Nazi policy and homegrown anti-Semitism, a series of measures was enacted that progressively stripped Jews of their rights. On August 17, Jews who had fled to the southern zone were barred from crossing the demarcation line into the occupied zone. On August 27, the government repealed a decree that barred publication of material inciting racial hatred. Collaborationist newspapers were thus free to express anti-Semitic vitriol. *Au Pilori* (literally, "to the pillory") ran a series of articles in August on the theme of "The Jews must pay for the war or die."

On September 27, the Germans ordered that a census of the occupied zone be taken to obtain a full accounting of its Jewish population. Jews were obliged to register with the authorities, to carry specially marked identity cards that read *Juif* for men or *Juive* for women, and Jewish-owned businesses had to display a placard identifying themselves as such. The explanation offered for the ordinance on the front page of *Le Matin* was: "For some time, the Jews who have returned to Paris and other large centers are showing themselves to be particularly arrogant. They seem to have no conscience whatsoever of their heavy responsibility in the events that have led France to catastrophe . . . The ordinance . . . is to control their activity."

These actions were preludes to a sweeping French law, the Jewish Statute (Statut des Juifs) of October 3, that both defined Jewishness (any person who had three Jewish grandparents, or who was married to a Jew and had two Jewish grandparents) and barred French Jews from a host of professions, including holding public office; teaching; being officers in the military; editing or managing newspapers, magazines, or other periodicals (except scientific ones); directing, producing, or distributing films; or working in theater production or radio.

The new regulations upended the Monod household. Although an atheist, Odette was fully Jewish under the law. Indeed, her mother, Berthe Buna Zadoc-Kahn, was the daughter of Zadoc Kahn, who had been chief rabbi of France. The twins were not considered Jewish on account of their father's Protestant pedigree, but the census ordered by the Germans applied to Odette and the members of her family who remained in France. They were to register by October 20 with the deputy prefect of the arrondissement in which they lived, and to disclose their home address, nationality, place and date of birth, marital status, profession, and how long they had continuously stayed in France.

Almost 150,000 people in the greater Paris area complied, about 90 percent of the estimated Jewish population, of which slightly more than half were native French citizens. Odette's mother and her sister Lise declared themselves to the authorities. Jacques even had to register with the authorities as the spouse of a Jewish person.

The anti-Jewish measures multiplied. In early October, the government authorized prefects to arrest and intern "any foreigner of the Jewish race," and Algerian Jews were stripped of the French citizenship that had been granted them since 1870. In order to exclude further Jewish (as well as British) influence on French culture, the German

Propaganda Ministry compiled a list of more than two thousand works by 842 authors for French publishing houses to purge. The publishers cooperated by promptly withdrawing Freud, Einstein, Brecht, Mann, Malraux, Aragon, and even Shakespeare.

All of these acts were dramatic reversals from France's prewar objections to the persecution of Jews within Germany. France had welcomed refugees from the Nazi purges and from other countries where Jews were oppressed, and had allowed them to live and work freely. But now there was no outcry. Of course, with the press and public gatherings controlled by the Germans and Vichy, there were no permitted means of public demonstration. It was also the case that Jews made up less than 1 percent of the French population. Given the general populace's burdens in the aftermath of the country's collapse, the anti-Jewish campaign was of lesser concern to most French, and heartily welcomed by some.

Camus, however, was outraged. He wrote Francine, "All the Jews are being thrown out of our office, even those who have returned from fighting the war." He felt that journalism was becoming dishonorable, because of both the new regulations and the propaganda that the newspaper was publishing. He declared, "So I am going to choose another profession," and asked Francine to look into obtaining some land to farm in Algeria.

The poisonous atmosphere had thwarted his own writing. He had resumed work on his essay *The Myth of Sisyphus,* but could not find the fire that he needed. "I have no joy, not even in writing," he wrote a friend. The prospects for publication of his essay, or of his novel and play, were very dim. "I don't plan to publish anything in France for many years," he said, then added sarcastically, "apart from praise of the family."

Limited to expressing privately his support for his Jewish colleagues, he wrote to a college friend from Algiers who had introduced him to Francine and who was stripped of her teaching post. Camus also wrote to Irène Djian, another Jewish friend: "All of this is particularly unfair and particularly abject, but I want you to know that the thing is not looked upon with indifference by those not directly involved. In other words, this is the time to show solidarity with you . . . This is what I am saying and what I will say each time it is necessary: Let the wind pass, it cannot last, it will not last if each one of us in his place just calmly affirms that this wind smells bad."

Traitors?

As it was promulgating one discriminatory law after another, Vichy was also waging a propaganda war against de Gaulle, his Free French movement, and their British backers. Vichy acted swiftly to discredit de Gaulle. On August 3, a military tribunal in Clermont-Ferrand convicted de Gaulle in absentia of treason and desertion in wartime, stripped him of his rank, and sentenced him to death. Those who joined de Gaulle were thus following not merely a renegade general but a condemned traitor.

With respect to Britain, Vichy's purpose was to rally national unity by laying blame elsewhere for France's present misfortune. It was also a calculated move based on the expectation that Britain would soon fall to Germany, and the belief that distancing France from her former ally would curry favor with the victor. Indeed, some such as Laval openly hoped "ardently that the British would be defeated" and asserted that "France had suffered too often as a result of British dishonesty and hypocrisy."

Laval, Pétain, and other Vichy ministers capitalized on recent painful events to rouse anti-British sentiment. The evacuation at Dunkirk, for example, was portrayed as the British abandonment of France and a catalyst to her defeat. Fresher still was the outrage over the British attack on the French fleet at Mers-el-Kébir in Oran, Algeria. While undertaken to preempt the use of the fleet against them, it was condemned as an act of utmost treachery. Laval declared, "France has never had and never will have a more relentless enemy than Great Britain. All of our history attests to it."

Britain's role in carrying out a joint operation to Dakar with de Gaulle and the Free French was reported as yet further evidence of British "perfidy." To have both enabled de Gaulle—who was described in the press as the "ex-Frenchman" and the "ex-general traitor"—to lead the operation against a French territory, and for the British to have participated directly, was an act of "horrifying impudence."

Churchill backed the Dakar mission, dubbed Operation Menace, because the British wanted to capture or neutralize what remained of the French naval forces in order to reduce the threat to its navy. Dakar was a major port with a number of Vichy vessels based there. De Gaulle thought that he could persuade Vichy forces in French West Africa to join his Free French forces, particularly in the face of a large Allied task force. If, however, de Gaulle could not convince Dakar to come over to the Free French, then the plan was to take the port by force.

The operation did not unfold as de Gaulle had hoped. The Allies dropped leaflets on the city and sent a landing party under a white flag, but the High Commissioner and Vichy forces did not yield to his appeals. The landing party was fired upon. Shells and torpedoes were then exchanged between the Allied ships and Vichy cruisers, submarines, and coastal batteries. Not wanting to pit Frenchman against Frenchman (many of his troops knew or even had relatives among the Vichy sailors they faced), de Gaulle called off a landing assault and left the rest of the operation up to the British. Over the course of three days, the British ships engaged the Vichy fleet and shore batteries. Each side inflicted substantial damage on the other, with two Vichy submarines sunk. The Allied task force then withdrew without taking the port.

The operation was an embarrassing setback for de Gaulle and for Churchill, but it was a bonanza for Vichy propagandists. They had shown that de Gaulle was unwelcome and impotent, and that Vichy forces could repulse an attack by France's former ally. Moreover, as a reprisal, Vichy launched two bombing missions on British-owned Gibraltar.

Pétain publicly congratulated the High Commissioner in Dakar:

> France follows your resistance to the partisan betrayal and British aggression with emotion and confidence. Under your authority, Dakar offers an example of courage and fidelity.

All of metropolitan France is proud of your attitude and the resolve of the forces that you command.

As for de Gaulle and Britain, Radio Française snarled: "By the will of the traitor de Gaulle, the blood of our sailors and soldiers is shed in Dakar. Let us remember!"

With the battle at Dakar, Vichy was estranged further from France's former ally, and acted in a manner that was more like that of an ally of the Reich. These movements were confirmed in a radio address on October 10, in which Pétain outlined the elements of the "New Order" and the "National Revolution" he wished to bring about in France. Pétain blamed France's defeat not only on the military but also on the weaknesses and defects of the former political regime. The new regime, Pétain explained, "must free itself of traditional friendships and enmities." The marshal declared that France was prepared to pursue "international collaboration" in all fields "with all of its neighbors." Turning to Germany specifically, he elaborated: "She knows besides that, whatever the political map of Europe and the world may be, the problem of Franco-German relations, milked so criminally in the past, will continue to determine her future." Pétain said that after its victory, Germany can choose between "a traditional peace of oppression" and "a new peace of collaboration."

If there was any confusion or doubt about the direction in which Pétain's line of reasoning was taking France, it was shortly made plain. In late October, Hitler went on a rare diplomatic trip on which he was to meet with Francisco Franco of Spain and Benito Mussolini of Italy. One aim of the trip was to secure agreements that would further isolate Great Britain in the war. Hitler had some bargaining chips in the form of the respective territorial concerns of Spain, Italy, and France in Africa, and possible interests in Britain's holdings once it was defeated. As the recent battle at Dakar attested, France had a potential role to play in Hitler's plans.

So, on October 22, Laval was summoned to meet Hitler at Montoire, near Tours, as the Führer was on his way to the Spanish border. Laval labeled France's declaration of war on Germany a crime, and explicitly proposed collaboration between the two countries. In order to further explore the substance of that collaboration, it was suggested that Pétain could meet with Hitler on his return trip on October 24. Pétain had sought such a face-to-face meeting for some time and went to Montoire. The meeting was not announced beforehand in the press.

The two men met in front of Hitler's train in the Montoire station and then talked for an hour and a half in Hitler's car. Hitler was reported to have told the marshal, "I know that personally you did not want war and I regret meeting you in these circumstances." Recounting the British attacks at Oran and Dakar, the latter led by a general who had disowned his country, Pétain indicated that the French territories in Africa were one potential domain for collaboration. After a review of the overall military situation, Hitler expressed his desire for a speedy end to the war and the establishment of a new European community opposed to the British. He explained that he had invited Pétain and Laval to a meeting to ascertain France's inclinations with regard to this community and collaboration with Germany.

Six days after the meeting at Montoire, Pétain explained his intentions in meeting the Führer to the public in a short radio address:

> Frenchmen,
>
> Last Thursday I met the Chancellor of the Reich . . . This first meeting between the victor and the vanquished marks the first righting of our country.
>
> I went freely at the invitation of the Führer. I did not submit to any "diktat," any pressure on his part. A collaboration was envisaged between our two countries. I accepted it in principle. The means will be discussed later . . .
>
> It is in a spirit of honor and in order to preserve the unity of France—a unity that has lasted for ten centuries—within the new European Order, which is being built, that I today embark on the path of collaboration . . .
>
> This collaboration must be sincere. It must be at the exclusion of all thoughts of aggression, it must entail a patient and confident effort.

The marshal then added prophetically, "This policy is mine. The ministers are responsible only in front of me. It is me alone that history will judge."

A key exhibit in that judgment would be the published photographs of Pétain at Montoire, shaking Hitler's hand.

Pétain meets Hitler at Montoire, France, October 24, 1940. Six days later Pétain announced he was taking "the path of collaboration." (AP Images)

CHAPTER 8

AN HOUR OF HOPE

The miserable have no other medicine, but only hope.
—WILLIAM SHAKESPEARE, *Measure for Measure*

THE VICHY PROPAGANDA CAMPAIGN SUCCEEDED TO A DEGREE AS reverence for Pétain stifled skepticism. But Vichy's larger aims of rallying the country and its colonies to its leadership, of stoking anti-British sentiment, and of destroying de Gaulle's Free French movement were undermined by three forces in the fall of 1940.

The first was the Royal Air Force. From July to October, the RAF successfully (albeit narrowly) repulsed the Luftwaffe's attacks on Britain's air defenses and cities during the Battle of Britain. Contrary to Pétain's expectations, four months after France's fall, Britain was still free and Hitler had suffered a major setback.

The second was de Gaulle's successful appeals to a number of territories that rallied to the Free French. In late August, Chad, Cameroon, the Congo, Ubangi-Shari, and Equatorial French Africa rallied to the side of the Free French, followed in September by French Polynesia and New Caledonia.

The third and perhaps most powerful force working against the Vichy and German propaganda efforts was the BBC. Ever since de Gaulle's first radio broadcast on June 18, the BBC had been waging a battle of the airwaves, or *la guerre des ondes,* against Vichy-controlled Radio Nationale and German-controlled Radio Paris, as well as countering the heavily censored, collaborationist newspapers. Every evening since July, French citizens who tuned their sets to the BBC's "Radio Londres" at 8:15 p.m. heard news about developments of the war in Britain and in Africa—both good and bad—that they weren't getting from Vichy- and German-controlled sources.

The BBC devoted a half hour to broadcasts in French, as well as providing several brief news bulletins during the day. Maurice Schumann, another Frenchman who escaped France on the SS *Batory*, began his regular evening broadcast *Honneur et Patrie* on July 18, 1940. And from September 6 on, a team of journalists produced the program *Les Français parlent aux Français* (The French speak to the French) during which de Gaulle made frequent appearances to repeat his call to arms, to assail the legitimacy and policies of the Pétain government, and to announce the growing ranks of the Free French.

Listeners learned not only that de Gaulle had succeeded in recruiting colonies to his cause, but that the British were not losing the war and were instead making a gallant, defiant stand. The collaborationist Paris newspapers enthusiastically reported the tonnage of the bombs dropped on London each day and ran pictures of various targets in flames. They gave no sense that the tide of the battle had turned in mid-August. But the BBC did. At the height of the aerial combat, the German communiqués reported 143 British planes downed with just 32 German planes "that did not return to base," whereas the actual totals were 59 British and 120 German planes lost. In fact, German losses exceeded those of the British on almost every day of the campaign through October.

Well aware of the contradictory reports emanating from Radio London, Vichy and German sources tried to counter them. On August 17, two days after the costliest day for the Luftwaffe over Britain, *Le Matin* was compelled to run an editorial urging "that the French regain their common sense." The article stated:

> The war between Germany and England has entered a decisive phase and the result is not in doubt. However, some French seriously forecast England's victory over Germany! Those French are certainly not asking themselves where are France's interests . . .
>
> It is England that led us to catastrophe. And one asks oneself how is it that there are still people in France who believe or who make believe that an English victory would be profitable for our country?
>
> One despairs of French common sense.

"Common sense" was a common refrain, and also the suggested antidote to any sympathies expressed for de Gaulle. In the middle of the battle of Dakar, *Le Matin* decided that it must address those readers who had sent in letters declaring that the true French were those who had confidence in England, and who had written "Vive de Gaulle!":

> Thus, de Gaulle, traitor to his country, in the service of England, and condemned to death . . . still finds here defenders who have the audacity to write "Vive de Gaulle!"
>
> One despairs over the common sense of certain of our compatriots.

Those French who understood English could also tune in to the BBC Home Service and sense Britain's defiant, confident mood. A few days after witnessing the pivotal air battles over Britain from an RAF operations room, Churchill captured that mood when he addressed the House of Commons:

> We have not only fortified our hearts but our Island . . . The people have a right to know that there are solid grounds for our confidence which we feel, and that we have good reason to believe ourselves capable, as I said in a very dark hour two months ago, of continuing the war "if necessary alone, if necessary for years". . .
>
> It is quite plain that Herr Hitler could not admit defeat in his air attack on Great Britain without sustaining most serious injury. If after all of his boastings and bloodcurdling threats and lurid accounts trumpeted around the world of the damage he has inflicted, of the vast number of our Air Force he has shot down, so he says, with so little loss to himself . . . if after all this his whole air onslaught were forced after a while to peter out, the Führer's reputation for veracity of statement might be seriously impugned . . .
>
> We believe that we shall be able to continue the air struggle indefinitely and as long as the enemy pleases . . .
>
> The gratitude of every home in our Island, in our Empire, and indeed throughout the world, except in the abodes of the guilty, goes out to the British airmen who, undaunted by odds, unwearied

in their constant challenge and mortal danger, are turning the tide of the World War by their prowess and by their devotion. Never in the field of human conflict was so much owed by so many to so few.

The German losses were, in fact, unsustainable. Their attacks petered out by October 31, even if the exaggerated claims did not. Hitler quietly postponed, then abandoned, his plans for an invasion of Britain.

Churchill's eloquent and passionate resolve was a far cry from what the French heard during the critical phase of the Battle of France, and directly opposed what they were hearing and reading from domestic sources. Despite Dunkirk, despite Mers-el-Kébir, and despite Dakar, the British were earning the admiration of the growing number of French who tuned to the BBC. And if there was any doubt about Churchill's inclination toward France, he addressed the French directly, in good French, on Radio London on October 21. Delivered during an air raid on London, his opening lines set the tone of his fourteen-minute-long broadcast:

> Français!
>
> C'est moi, Churchill, qui vous parle. Pendant plus de trente ans, dans la paix comme dans la guerre, j'ai marché avec vous et je marche avec vous encore aujourd'hui, sur la vieille route. Cette nuit, je m'adresse à vous dans tous vos foyers, partout où le sort vous a conduit. Et je répète la prière qui entourait vos Louis d'or: «Dieu protège la France».
>
> [Frenchmen! It is me, Churchill, who speaks to you. For more than thirty years, in peace and in war, I have marched with you and I am marching still along the same road. Tonight I speak to you at your firesides, wherever you may be, or whatever your fortunes are. I repeat the prayer around the *louis d'or* [a French gold coin]: *"God protect France."*

Churchill then went on to warn the French of Hitler's ultimate aims: "I tell you what you must truly believe when I say that this evil man, this monstrous abortion of hatred and defeat, is resolved on nothing less than the complete wiping out of the French nation . . . All Europe, if he has his way, will be reduced to one uniform Bocheland."

He urged the French to "rearm your spirits before it is too late" and to "have hope and faith, for all will come right." He then made a request: "What we ask of you at this moment in our struggle to win the victory which we will share with you, is that if you cannot help us, at least you will not hinder us. Presently you will be able to weight the arm that strikes for you, and you ought to do so." He vowed to "never stop, never weary, and never give in" until "the task of cleansing Europe from the Nazi pestilence and saving the world from the new Dark Ages" was complete.

Of course, not every French home owned a radio set. And neither could every one that did own a set always receive BBC broadcasts. And so a ritual had emerged across the country, in villages and cities, where people gathered together—in homes, in the streets, around cafés, to listen to the evening broadcast at 8:15.

The authorities took note and enacted countermeasures. The first step was for the Germans to attempt to jam the broadcasts, but this was of limited effect because the BBC broadcasted on several different bands. The next step was for the Germans to order that in the occupied zone, citizens could listen only to broadcasts from its areas of control—Poland, France, Holland, Belgium, Norway, and the Reich. Vichy followed suit by enacting a law on October 28 that outlawed listening to the BBC and all "antinational broadcasts" in public places. One effect of these measures was to arouse greater interest in Radio London.

For those who were opposed to the armistice, who were suffering under the Occupation, who were terrified by the new exclusionary laws, and who were appalled by the regime's expanding rapport with the occupiers, the news coming from London was the only source of moral support and hope. And to a small number, it was an inspiration to act.

Armistice Day

Léon-Maurice Nordmann was one of those who was very eager to do something to strike back at the Nazis. A successful lawyer at a young age, he had been politically active in the left-wing Socialist Party. Nordmann had opposed the Munich agreements and the policy of appeasement. When war was declared, he requested a place in a combat unit. He was very disappointed when, on account of his poor eyesight, he was assigned to the meteorological service. With his office in the plush Hôtel Georges V in Paris, Nordmann was frustrated and felt very guilty about serving in such comfort, just a short distance from his home, while others were off fighting the war.

Since his demobilization, Nordmann had been seeking out like-minded friends to see what might be done to rally against the occupiers. Very likable, a good conversationalist with an easy smile, Nordmann had a lot of very bright friends, including lawyer André Weil-Curiel and fellow music lover Jacques Monod.

Weil-Curiel served during the war as a liaison officer with the British Army and wound up being evacuated at Dunkirk. Stranded in England after the armistice, he immediately joined de Gaulle's Free French. In late summer, he was sent back to France to gather intelligence on the situation in Paris and to promote the Free French. Once back in Paris, Weil-Curiel sought out Nordmann. They and a third lawyer, Albert Jubineau, formed a group they dubbed Avocats Socialistes (Socialist Lawyers). Their primary aim, and the aim of other nascent groups involving Paris's intelligentsia, was to gather and spread accurate information about what was happening inside and outside France. They hoped to build a clandestine organization that would recruit friends and colleagues and form links with other groups.

Monod joined Nordmann's group in October 1940.

By early November, after the news of the Pétain-Hitler meeting, Weil-Curiel and Nordmann were wondering what it would take to provoke Parisians to begin to express their discontent. For their part, they were looking to make some public act of defiance against the authorities. Armistice Day, November 11, the day commemorating those who sacrificed and served in World War I, provided an ideal opportunity to arouse patriotic feelings.

There were rumblings about some form of demonstrations for the eleventh. The Young Communists organization in Paris disseminated a leaflet urging students to celebrate the memory of their fathers and brothers killed in the war. Another leaflet was spread among Paris high schools urging students to rally at the Place de l'Étoile at 5:30 p.m. The head of the police caught wind of the plans. On November 10, the newspapers carried an official notice: "Public organizations and private enterprises in Paris and the Department of the Seine will work normally November 11. Commemorative ceremonies will not take place. No public demonstrations will be tolerated."

Radio London, however, thought otherwise. Maurice Schumann closed his program by urging all French people: "On the graves of your martyrs, renew the oath to live and die for France." Another commentator added his appeal to veterans to maintain their hope, and to go to the Tomb of the Unknown Soldier and other monuments to pay their respects: "On November 11, to oppression and to cowards, you all say: NO!"

Meanwhile, Nordmann, Weil-Curiel, and friends ordered a huge wreath and made a three-foot-long calling card, wrapped in blue, white, and red ribbons. They addressed the card in large capital letters: LE GÉNÉRAL DE GAULLE. Before dawn on November 11, they put their tribute into a Citroën truck and drove through the empty streets to the Place de la Concorde, then up the Champs-Élysées to the statue of Georges Clémenceau, the prime minister who led France to victory in 1918. At 5:30 a.m., they quickly placed the wreath and the card at the foot of the statue and hurried away.

A good number of Parisians were able to see the card before a police patrol took it away. Word spread quickly about the gesture all over Paris. Throughout the day, citizens came to place more bouquets at the statue, as well as at the Tomb of the Unknown Soldier, at the statue of Joan of Arc near the Tuileries, and at a plaque near Notre-Dame that commemorated British sacrifices in World War I. The French police were generally accommodating and allowed the bouquets to be placed.

But late in the afternoon, as the school day ended, the mood changed. Students and teachers formed ranks and marched up the Champs-Élysées. Some carried a wreath shaped into the Cross of Lorraine, some wore red-white-and-blue ribbons. Some started to sing the verboten "Marseillaise." Some shouted "Vive la France!" and "Down with Pétain!" The procession grew to several thousand people who started to pack in around the Arc de Triomphe.

The Germans were surprised for once—their orders had been defied. Around six p.m., armed soldiers on motorcycles, in trucks, and on foot hurried to the scene and began to disperse the crowd with clubs. Some students were beaten. Shots were fired into the air. The Germans closed off the streets and chased the marchers away from all of the monuments. More than a hundred people were arrested; some were taken to the Cherche-Midi and Santé prisons. Most would be released in the following few weeks.

The Germans were furious. The demonstrations were deemed "incompatible with the dignity of the German Army." They closed the universities, required that every student register with the police, and dismissed the rector of the Sorbonne.

Resistance!

Nordmann and Weil-Curiel were elated by the first show of defiance to the authorities since the Occupation began. They were immediately searching for other things they could do to rally opposition. The Avocats Socialistes were in contact with other groups. René Creston, a childhood friend of their comrade Jubineau, was part of another clandestine group that was organized around the Musée de l'Homme (Museum of Mankind). Organized by linguist Boris Vildé, anthropologist Anatole Lewitsky, and librarian Yvonne Oddon, the Musée group was focused on helping French and British soldiers evade the Germans and get to London, on gathering information, and on countering German and Vichy propaganda. Creston put the lawyers in contact with the Musée group, and the two quickly decided to work together.

The Musée network had in turn grown to include a group that called itself the Friends of Alain Fournier, a supposed literary club that wanted to spread information by establishing a clandestine press and distributing tracts in the occupied zone. The group included Agnès Humbert, art historian at the neighboring Musée des Arts et Traditions Populaires; Jean Cassou of the Museum of Modern Art; educator Marcel Abraham; writer Claude Aveline; publishers Albert and Robert Emile-Paul; Jean and Colette Duval, a teacher and writer, respectively; and Monod's neighbors at 30 rue Monsieur-le-Prince.

Cassou was to be the editor in chief. However, they needed some way of printing and distributing many copies of whatever they put together. The Germans controlled the supply of paper and ink as well as all of the presses. Cassou made contact with Paul Rivet, the director of the Musée de l'Homme, who offered the use of an old Roneograph duplicating machine that had been stored in the basement. Rivet introduced Cassou to Oddon and Vildé, and Cassou in turn introduced Humbert to Rivet. It was decided that Humbert would be the liaison between the two groups.

Vildé warned Humbert of the consequences if the network was exposed. "Many of us will be shot," he said, "and all of us will go to prison." Humbert laughed, but he knew that Vildé was probably right. Life had already changed for several in the group: Cassou and Humbert were dismissed from their posts without any explanation in October, and Rivet was sacked from the Sorbonne and the Museum not long after the November 11 demonstrations.

The two groups decided to collaborate on putting out a newspaper that would relay the news coming from the BBC about the real state of affairs in France and overseas, and support de Gaulle. They called themselves the National Committee of Public Safety, and after some discussion they settled on naming the newspaper *Résistance*. The first page was to be written by Vildé, Lewitsky, Oddon, and others, while the next three were composed by Aveline, Cassou, and Abraham. Humbert was the typist; Lewitsky, Nordmann, Monod, and others volunteered to distribute copies.

The first issue appeared on December 15, 1940. The editorial read:

> Resist! This is the cry that comes from the hearts of all of you who suffer from our country's disaster. This is the wish of all of you who want to do your duty. But you feel isolated and disarmed . . .

> *Resistance* is here to speak to your hearts and minds, to show you what to do.
>
> Resistance means above all to act, to be positive, to perform reasonable and useful things . . . The method? Group yourselves in your homes with those you know. Choose your leaders. They will find other groups with which to work in common . . . Find resolute men and enroll them with care. Bring comfort and decision to those who doubt or who no longer dare hope. Seek out and watch those who have renounced our country and betray it. Meet together every day and transmit information useful to your leaders . . . Beware of inconsequential people, of talkers, and of traitors. Never boast, never give yourselves away. Face up to the moment. Later we shall tell you how to act.
>
> In accepting our responsibilities as your leaders, we have promised to sacrifice everything, staunchly and pitilessly, for this mission . . . We have only one ambition, one passion, one wish, to bring about the rebirth of a pure and free France.

An Hour of Hope

De Gaulle wanted to seize on the momentum of the November demonstrations and the simmering undercurrent of discontent. In order to further rally solidarity with his cause and to encourage more disobedience to the occupiers, he came up with an idea for New Year's Day—a countrywide silent demonstration against the authorities. His spokesman Maurice Schumann broadcast the appeal on Radio London on December 23:

> January first will offer to all French the opportunity to demonstrate that they are united in their grief and in their hope.
>
> For the majority of French people, directly or indirectly repressed by the enemy, such a demonstration can only be silent. But, in accomplishing it in silence, it will only be more effective.
>
> On January first, from two o'clock to three o'clock in the non-occupied zone, and from three o'clock to four o'clock in the occupied zone, all French people will remain in their homes and local

> shelters. All French people will remain indoors, either alone, or with family and friends. During this hour of contemplation, everyone will think together of liberation. This will be the hour of hope.
>
> Everything should be done, everywhere, discreetly and firmly, so that this silent protest of our crushed country takes on an immense scope.

The second issue of *Résistance* appeared on December 30. The six-page paper summarized several months of developments: the rallying of colonies to the Free French, battles in North Africa, British bombings of Germany, Roosevelt's reelection in the United States, Germany's pillage of France's economy, anti-Semitism, and more. A notice repeated de Gaulle's appeal to remain indoors on January 1.

De Gaulle himself repeated the call on New Year's Eve, with even greater force and optimism:

> The hour of hope of January 1, during which no good French person will appear out of doors, wishes to say:
>
> Our provinces belong to us! Our land belongs to us! Our men belong to us! Those who take our provinces, who eat the wheat of our lands, who hold our men prisoner, they are the enemy!
>
> France expects nothing of the enemy, except this: that he leaves! That he leaves defeated! The enemy entered into our country by force of arms. One day, the force of arms will chase him from our lands. He who laughs last, laughs best!
>
> This is what all French people wish to show the enemy in observing the hour of hope.

The call was repeated on New Year's Day up until 2:45 p.m. on the BBC. The authorities attempted to counter it by luring shoppers outside with sugar and potatoes—made specially available without a ration ticket. They were ignored. That afternoon, in Paris, Lyon, Marseille—across all of France—the streets were largely empty.

A Dangerous Game

For Nordmann, however, the hour of hope morphed into days of panic. Mistakes had been made; he was a wanted man.

In the preceding days, Nordmann had met a number of young aviators associated with an aero club in Aubervilliers, a northeastern suburb of Paris. The men were hoping to find a way to England to join the Free French. One of their group had previously met Nordmann and introduced him to the others. While Nordmann explained that passage to England was not possible at the time, he and the fliers discussed how de Gaulle's cause could be advanced by the dissemination of favorable propaganda.

Three of the fliers went to Nordmann's home on the boulevard Arago one night, where he gave them several copies of the first issue of *Résistance* to distribute. The fliers then proposed that they could print and distribute additional copies. Since the number of copies of the first issues had been limited to a few hundred, Nordmann found their proposal attractive. He visited the aero club to verify that the group had the equipment and supplies to do so. Nordmann gave one of the members, nineteen-year-old Albert Comba, a list of more than twenty names and addresses to which to deliver the copies.

On Sunday evening, December 29, the group printed another four hundred copies. Unfortunately, they were not careful enough about disguising their activity. The noise of their Roneograph and the coming and going of the fliers at the club attracted the curiosity of a neighbor whose home overlooked the club's courtyard. He approached the group to find out what they were doing. The man asked for copies of the newspaper. Thinking he was sympathetic, the fliers gave him two.

However, the neighbor promptly tipped off the police, receiving in return some gasoline. The police decided to conduct an immediate search of the aero club that Monday. They quickly found the Roneograph hidden under sheet metal, with traces of black ink still on it. They also found the names and addresses of the club members and went straight to Comba's home to search it. There they found a suitcase with thirteen envelopes containing about twenty issues each of

Résistance. They tracked down Comba at work and arrested him. The other fliers were also promptly picked up and arrested. Under intense questioning, the men divulged that they had become involved in printing the newspaper through contact with Nordmann and Weil-Curiel.

Comba was desperate to minimize the damage. The police had not found the list of names in his wallet, which was at home. Between interrogations, he convinced a guard to go to his home, find the list, and destroy it. One of the other fliers also persuaded the guard to telephone Nordmann to tip him off that the Aubervilliers group had been discovered. The guard did phone Nordmann, but when he went to Comba's and took the wallet, he did not destroy the list. When the interrogations resumed, Comba admitted under pressure that the guard had the wallet, which the guard then turned over with the list of names.

sie d'une liste au
é COMBA

Continuant l'information,

Rapportons qu'à la suite des déclarations du nommé COMBA, nous procédons à de nouvel les recherches en vue de découvrir la liste sur laquelle selon lui se trouveraient les noms de personnes à qui les tracts doivent être diffusés.

Cette liste n'a pu être découverte dans le logement qu'il occupe mais une fouille effectuée sur sa personne l'a fait découvrir à même la peau.

Il s'agit de trois feuillets manuscrits, portant plusieurs noms avec un chiffre inscrit en regard.

Nous relevons ces noms et adresses comme suit:

Pierre LAMBERT I7 rue Henri Turot, Paris
I9 eme arrt (on peut laisser à l'appartement)
Le chiffre 50 est inscrit en regard.
Albert NAUD I98 Bd Voltaire-I00-(on peut laisser à l'appartement)
Pierre ARON 39 rue du Sentier- I0.
René Georges ETIENNE I46 Bd Magenta -20
Me JUBINEAU 8 place d'Iéna I6 eme -50- (on peut laisser à la concierge)
Jacques I00
G? DOYEN II rue Anatole de la Forge XVII - 5
Jean Victor MEUNIER 86 rue de Miromesnil VIII- I5.
François LEMARET 36 Ave du Roule-Neuilly- 5
Lucien Vidal NAQUET 65 bis rue de Varenne VII-30.
Andrée Marty CAPGRAS 69 rue de Dunkerque-I0.
Henri GOLDSCHUL 71 rue de Rennes - I5.
Dr Léon BOUTBRIER 71 rue de la Glacière -Montgeron (Seine et Oise)......I0
Dr LONGUET 48 rue des Accacias Alfortville...I0
Jacques MONOD Laboratoire de Zoologie -Sorbonne Faculté des Sciences, de préférence entre I4 heures et I5 heures 20
Maurice LACROIX 6 rue Quatrefages V.......I0
Robert WERNER 38 rue du Bois de Boulogne (Neuilly) (I0)
Dr Jean HAMBURGER 97 rue de Prony XVII....I5
Mlle Geo FERRY 3 rue Abbé Rousselot.......I0
Guy BAISSIN I rue Jean Louis Forain-Le Chesnay (SetO)..................................I0
Mlle JULIE ESPIN 4 bis rue Descombes......I0

List of names obtained by interrogation of Albert Comba by Paris police, December 30, 1940. Jacques Monod is noted to receive twenty copies of the illegal newspaper *Résistance*. (Dossier BA 2443, Archives de la Préfecture de Police in Paris)

On the list were twenty-two names, most with addresses, followed by the number of copies they were to receive (anywhere from five to one hundred). The names on the list included lawyers and physicians, and the fifteenth entry read:

> Jacques MONOD Laboratoire de Zoologie-Sorbonne Faculté des Sciences, de préférence entre 14 heures et 15 heures . . . . 20 [underscore in original]

The next day, New Year's Eve, the police organized searches of each address on the list. They contacted the Sorbonne and learned that Monod resided at 26 rue Monsieur-le-Prince. Three inspectors went first to the apartment building, where the concierge directed them to the fifth-floor apartment. Neither Jacques nor Odette was home; their housekeeper answered the door. The three men searched the apartment for copies of the illegal newspaper or any other incriminating materials. They found no copies of *Résistance* or, they reported to their superiors, any other evidence that "Mr. Monod is involved or was involved in any political activity whatsoever."

The inspectors then proceeded to Monod's third-floor laboratory at the Sorbonne in order to inform him of the search that they had just conducted and to ask him to make some declarations concerning the matter in which he had been implicated. They were informed that Monod had left and that it was not known when he would return.

The searches of the other people named on the list also generally produced nothing, except a couple of instances in which the newspapers had been received but had either been returned or remained unopened. Two addressees had also received urgent messages advising them to destroy the newspapers.

The first stage of the investigation complete, the fliers were moved to prison, warrants were issued for Nordmann and Weil-Curiel, and the subsequent investigation was turned over to the Germans on January 6.

Nordmann was terrified. He asked Vildé to find him a place to hide and an escape route out of Paris. Vildé asked another member of the network, the wealthy and beautiful Countess Elisabeth de la Bourdonnaye, known as Dexia to her friends, to take Nordmann in while preparations were being made. She readily agreed, and without asking his name, gave him a bedroom in her home in Paris and brought him his meals. Nordmann realized the danger in which he was placing the countess, and he thanked her profusely. Dexia was well aware of the risks of harboring a fugitive, so she was vigilant. When she saw German police in the courtyard the next morning, she alerted Nordmann, who snuck out by the front door and did not return until after nightfall.

Concerned that Nordmann was endangering Dexia, Vildé arranged for Nordmann to travel to the unoccupied zone with another member of the group, Albert Gaveau. On January 13, Gaveau and Nordmann went to the Montparnasse station and boarded a train for Douarnenez, the location of another safe house. At the Versailles-Chantiers station, the plainclothes members of the German secret police (Geheime Feldpolizei) boarded the train, just after Gaveau had gone off to the bathroom. They grabbed Nordmann, but Gaveau got away.

The prisoner was transferred to a car and driven back toward Paris. As the car slowed near an intersection, Nordmann jumped out and tried to run away. One of the policemen opened fire, hitting Nordmann in the leg and knocking him to the ground. The lawyer was recaptured, bandaged, and taken to Santé Prison.

Gaveau, it would turn out, was working for the Germans. Nordmann was carrying a list of sympathizers that Weil-Curiel had given to him, The Gestapo had most of the names of the Musée network.

EARLY ONE MORNING, Monod was in his laboratory at the Sorbonne, setting up an experiment on bacteria, when the Gestapo burst into the building, blocked the stairwells, and hustled upstairs to find him.

As the agents began to question him, Monod noticed their discomfort at being in the lab. He recalled later, "The police networks—including the Gestapo—detested laboratories . . . They were afraid of radioactivity, they were afraid of microbes, of viruses, and they hated working there." Their fears worked to Monod's benefit, for they did not search his lab carefully enough. Behind his laboratory notebooks and pipettes, there were several incriminating papers. After further questioning, Monod was released.

OTHERS IN THE Musée group were not nearly so fortunate. On February 8, Albert Jubineau was arrested by the Germans. On February 10, Lewitsky and Oddon were arrested by the Gestapo at their apartment, while Alice Simmonet, another member of the group, was arrested with her husband at a restaurant. On February 11, René Creston surrendered when German agents came looking for him and others at the Musée de l'Homme. Vildé and Dexia eluded capture for another month; Agnès Humbert was arrested two months later. Cassou and Aveline escaped to the South of France, and Rivet to South America, but altogether seventeen members of the Musée network were arrested. The first significant resistance network in the occupied zone had been crushed in a matter of just a few months.

The demise of the Musée network, the arrest of his friend Nordmann, and the visit he received from the authorities underscored crucial lessons for Monod: resistance was a very dangerous business, and much greater care would be essential to preserving the secrecy and security of any group.

CHAPTER 9

WAITING AND WORKING

The man of genius is he and he alone who finds such joy in his art that he will work at it come hell or high water.

—STENDHAL, *The Life of Haydn*

THE LIBERATION THAT DE GAULLE PROMISED AND THE VICTORY that Churchill had pledged to share with France were at best faraway dreams in early 1941. The Free French had rallied or taken some territories, and the British had gallantly defended their homeland, but the Allies (which included Canada, Australia, and New Zealand) did not have the men or materiel to contemplate the retaking of metropolitan France. The United States continued to maintain its official nonintervention policy, enforced by the Neutrality Acts of 1939 and earlier years, and the Soviet Union remained bound to its nonaggression pact with Germany.

Those contemplating resistance and all of those hoping for the liberation of France faced, therefore, a long, indefinite wait until some distant time when the Allies might gather enough strength to attempt to dislodge the Germans. For Camus, that meant returning to his writing and making a living. For Monod, that meant finally finishing his doctorate and hopefully finding some worthwhile problem to pursue after that.

ADAPTATION

Ever since his demobilization, and despite the Nordmann debacle and the interruption from the Gestapo, Monod had been trying to get his experiments and his degree back on track. Throughout the fall of 1940, Monod filled the pages of his notebooks with measurements of how bacteria grew when fed various sugars. Day after day in his lab at the Sorbonne, he made up media with different kinds and amounts of sugars, then inoculated the flasks with bacteria. He took measurements of the density of the bacteria at different points in time and plotted their growth on little squares of blue graph paper. The experiments were tedious and repetitive, and the bacteria did not always cooperate. On one page of his notebook, after observing that his cultures contained giant clumps rather than a fine slurry of bacteria, Monod scrawled, *"Nécessité absolue Trouver origine de cet emmerdement."* [Absolute necessity to find the origin of this pain in the ass.]

The bacteria typically showed an exponential growth rate when fed a single sugar, such as glucose, as its sole energy source. That is, after some time lag, the bacteria divided continuously until the sugar was exhausted. Monod found that, in general, the density the bacteria achieved depended upon the amount of sugar provided. That was not surprising, as the relationship had been observed before with various crude nutrient broths. Monod's incremental contribution was to perform the experiments with precisely defined nutrients.

Monod then had the very simple idea to test how bacteria behaved when fed two sugars at once. He wondered, for example, whether they would grow better or faster on two different sources of energy than on just one.

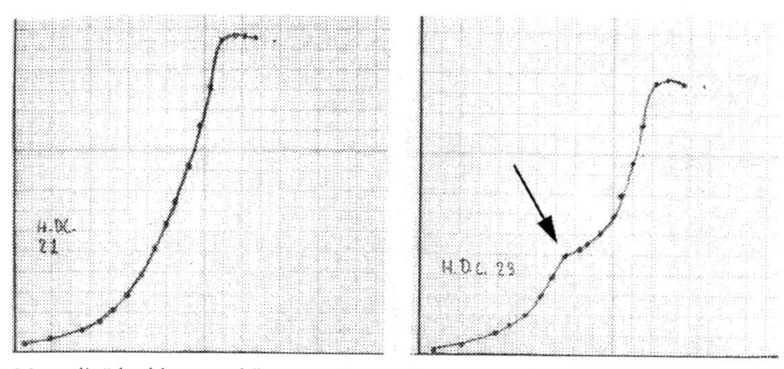

Monod's "double-growth" curve. Graphs from Monod's research notebook (winter 1940) on which he plotted growth over time. On the left, bacteria grew exponentially in the presence of two particular sugars. On the right, bacteria grown in the presence of two different sugars grew exponentially for a time, then paused (arrow) before resuming growth (double-growth). Monod's pursuit of this puzzling phenomenon would lead to a Nobel Prize twenty-five years later. (Archives of the Pasteur Institute)

The answer was not so simple. While the bacteria grew similarly when fed glucose and some sugars, Monod measured a puzzling response to certain other combinations. Specifically, instead of one growth phase, the bacteria exhibited two phases: they first grew exponentially, then stopped growing for a period of time before again resuming exponential growth. Monod was stumped. Why did the bacteria pause if they were able to utilize both sugars?

Baffled by this peculiar behavior, Monod went to consult André Lwoff at the Pasteur Institute. Monod had first encountered Lwoff a decade earlier at the Station Biologique at Roscoff, where Lwoff had introduced the young zoology student to the powers of microbiology. Lwoff was now head of the Microbiology Department, and Monod respected his vast knowledge. He showed Lwoff his biphasic growth curves. "What could that mean?" Monod asked.

Lwoff hesitated, then said, "That could have something to do with enzyme adaptation."

"Enzyme adaptation? Never heard of it!" Monod replied.

Lwoff gave Monod a bunch of papers on the phenomenon, a book on microbiology by Émile Duclaux—a close colleague of Louis Pasteur's and his successor as director of the Institute—and a copy of a PhD thesis by a Finn, Henning Karström, who had coined the name for the observation in 1930. What Duclaux, Karström, and others had noticed was that in various microbes the production of certain enzymes—proteins that carry out chemical reactions in cells—depended upon the nutrients present in the medium. It appeared as if the bacteria or yeast cells "adapted" to the presence of a nutrient by making an enzyme that would then break the nutrient down so that it could be used.

Lwoff suggested to Monod that his biphasic growth curves were due to the bacteria, after a lull, adapting to the second, less preferred, food—much like Parisians had to do during that hard, cold winter. How the bacteria adapted was a complete mystery, however, as essentially nothing was revealed by the earlier researchers or known in 1941 about how the activities or production of enzymes were controlled. Monod decided on the spot that this would be his quest: he would get to the bottom of enzyme adaptation.

Excited by his discovery, or rather his rediscovery, of the phenomenon, he dove into the literature Lwoff had given him. Enlightened by previous work, he designed new experiments to determine whether in fact the two growth cycles that he observed and called "double growth" or "diauxy" were in fact a manifestation of enzyme adaptation. He devoted the second half of his thesis to his investigations and possible explanations.

For the first time Monod was fully engaged, both intellectually and experimentally, in a bona fide scientific mystery. In trying to decipher whether the bacteria were using each sugar successively, he showed a knack for devising simple but revealing experiments. For example, just by changing the ratio of two sugars in the medium, from 1:3 to 1:1 to 3:1, Monod shifted the length of each of the two growth curves proportionally, as would be expected if one sugar were being used at a time.

Monod's experiments convinced him that the lag before the start of each growth curve was, in fact, a case of enzyme adaptation—of the necessary enzyme not being active or produced until some time had elapsed. But he perceived that there also must have been another mechanism at work, because he needed to explain why one sugar was not used until the other was depleted. He proposed that one sugar must somehow inhibit the use of the other. He then devised a series of tests that confirmed that one group of sugars inhibited adaptation to another set of sugars.

Monod did not know the mechanisms of enzyme adaptation or inhibition—that would require much more general knowledge about the making of proteins than was available in 1941—but his thesis displayed a striking command of the hypothetico-deductive scientific method: Monod laid out alternative explanations, deduced what should be observed under each explanation, and devised experiments to corroborate or negate each possibility.

Lwoff was impressed. The Sorbonne was not. After Monod's PhD ceremony, a member of the jury told Lwoff, "What Monod is doing does not interest the Sorbonne."

Happiness in Hell

Camus, too, was focusing on finishing and publishing his work—his cycle on the absurd that was now five years in the making—after finally finding his way out of France.

After a couple of months in Clermont-Ferrand, Camus had moved with *Paris-Soir* to Lyon. Francine Faure joined him there in November 1940, where she became both his secretary and his second wife. Camus's divorce from Simone Hie had been finalized, which freed him to make good on his promise to marry Francine. He did so promptly in December. Strained for money, they by necessity had a simple ceremony. The two exchanged brass wedding rings at the town hall, with Camus's comrade Pascal Pia, who worked as editorial secretary at the paper, as a witness.

The newspaper was struggling, however, and Camus was let go the same month. With no reason to stay in Lyon, and funds running very low, the newlyweds headed for Oran, Algeria, where they could live rent-free in an apartment owned by Francine's parents. Francine took up substitute teaching while Camus worked on finishing *The Myth of Sisyphus,* the third part of his trilogy. After his isolation and all that had transpired in France, Camus was energized by being back in Algeria among friends and familiar places, even though it, too, was governed by Vichy.

Camus intended for *The Myth of Sisyphus* to complement and illuminate *The Stranger* because he could more explicitly address the realm of the absurd in essay form than in fictional narrative. Its jolting opening lines announced his chief concern: "There is but one truly serious philosophical problem, and that is suicide. Judging whether life is or is not worth living amounts to answering the fundamental question of philosophy. All the rest—whether the world has three dimensions, whether the mind has nine or twelve categories—comes afterwards. These are games; one must first answer."

Writing as war engulfed western and northern Europe and North Africa, and as the fear of another mass slaughter was becoming realized, Camus underscored the immediacy of the question: "I see many people die because they judge that life is not worth living. I see others paradoxically getting killed for the ideas or illusions that give them a reason for living (what is called a reason for living is also an excellent reason for dying). I therefore conclude that the meaning of life is the most urgent of questions."

Camus asserted that meaning had to be approached, first and above all, in the light of the absurd condition of human existence—of the conflict posed by the human desire for meaning and the total indifference of the universe to that desire. And, second, one had to consider meaning in the face of the obvious fact of a finite lifetime and a certain death. Integrating these two elements, the central question for Camus thus became: If everyone is destined to die and the universe could not care less, how can life have any meaning?

"Man is mortal," wrote Camus. "One can nevertheless count the minds that have deduced the extreme conclusions from it." In his essay, Camus sought to push those conclusions as far as his reasoning could take him. He wanted to explore "a universe suddenly divested of illusions and lights," in which "man feels an alien, a stranger." He identified and analyzed three possible responses to the absurdity of existence: suicide, taking a leap of faith, or embracing the absurd and living life fully in the face of it.

For Camus, suicide was a confession that life "is not worth the trouble" and denial of any reason for living. He rejected it on the grounds that it was merely a way out of the question. Likewise, an appeal to faith, to God, to something outside of known experience was also a complete elusion of the question, an act of "philosophical suicide" (but one committed nevertheless by some illustrious thinkers who had previously confronted the same question). Camus sought to "live without appeal" to such either religious or philosophical inventions.

Rather, in the lucid recognition of the absurd condition and the abandonment of "divine fables," Camus saw freedom—to think, to create, and to live in a world "of which man is the sole master. What bound him was the illusion of another world." The hero of Camus's absurd world was Sisyphus. In the last chapter of his essay, Camus used the myth to make the connection between embracing the absurd and finding happiness. Sisyphus had been condemned by the gods to rolling his rock repeatedly up the mountain, only to have it roll back down each time. But for Camus, it was Sisyphus's scorn of the gods, his hatred of death, and his passion for life in the face of such a futile struggle that illustrated his essential point: man knows himself to be the master of his days. He concluded: "The struggle toward the heights is enough to fill a man's heart. One must imagine Sisyphus happy."

Having begun with the question of suicide, this tubercular Algerian without a franc to his name and living under a regime that valued neither freedom nor human life, crafted what he believed was "a manual of happiness."

On February 21, 1941, Camus recorded in his notebook, "Finished *Sisyphus*. The three absurds are now complete. Beginnings of liberty."

Liberty perhaps, but not prosperity. Camus's two previous works—*The Stranger* and *Caligula*—were also unpublished; he was out of work and almost unknown as an author. Camus had no publisher for his trilogy, and even if he had had one, publication was difficult for anyone at the time. There was a severe paper shortage, as well as the censors to get past.

And before Camus could be published, there was the question of whether what he had written had any merit. While Francine copied *The Myth of Sisyphus* for distribution—by hand, as they had no typewriter, Camus sent copies of *The Stranger* and *Caligula* to former teacher Jean Grenier and to Pascal Pia in Lyon. Grenier gave the first a positive review: "*L'Étranger* is very successful, especially the second part, despite the troubling influence of Kafka. The pages about prison are unforgettable." But he did not like the play: "Perhaps in the theater it will seem different."

Pia loved the novel: "Very sincerely, it has been a long time since I have read something of this quality. I am convinced that sooner or later *L'Étranger* will find its place, which is among the best. The second part—the pretrial investigations, the trial, the prison—is a demonstration of the absurd put together like a perfect mechanism."

Pia had several connections in the literary and publishing world, and assigned himself the task of getting his friend published. It was important to secure endorsements from established literary figures, so he gave the manuscript to André Malraux's brother-in-law to pass on to the Goncourt Prize–winning novelist, and to writer Jean Paulhan, former editor of the *Nouvelle Revue Française* (*NRF*), the most influential French literary magazine prior to the war. (Paulhan was also a member of the Musée de l'Homme network; he was arrested in early May 1941 and released after a week.) The *NRF* was published by Gaston Gallimard, and Pia thought that Gallimard would be interested in publishing Camus if Paulhan and Malraux offered favorable recommendations.

The two luminaries were both impressed with *The Stranger.* Malraux wrote to Pia: "*L'Étranger* is obviously an important thing. The power and simplicity of the means which finally force the reader to accept his character's point of view are all the more remarkable in that the book's destiny depends on whether this character is convincing or not. And what Camus has to say while convincing us is not negligible."

Paulhan reported: "I read *L'Étranger* in one sitting . . . It's very fine, frankly very good. Germaine [his wife] and I were gripped by it."

Pia thought that Camus's three works should be published together, so when copies of *The Myth of Sisyphus* had been made, he also sent those along to Malraux and Paulhan. Malraux wrote directly to Camus after reading the essay: "The link between *Sisyphe* and *L'Étranger* has many more consequences than I supposed. The essay gives the other book its full meaning." Malraux told Camus that he was going to recommend to Gallimard that he publish the books at the same time: "What matters is that with these two books together, you take your place among the writers who exist, who have a voice, and who will soon have an audience and a presence. There are not that many."

Reaching that audience, however, would pose many challenges. Publishing any book, let alone two, in France's crippled economy was difficult. Malraux reminded Camus, "The problem of paper being scarce remains."

There was also the matter of the German censors. *The Myth of Sisyphus* included a chapter on Franz Kafka. Because Kafka was a Czech Jew, his works were considered harmful and undesirable. They had been banned in Germany and withdrawn from publication in France.

Gallimard told Camus that he indeed wanted to publish *The Stranger* and *The Myth of Sisyphus*, but he had to request reluctantly that the Kafka chapter be excised. Camus had no option but to comply.

And there was also uncertainty about how Camus's stark analysis yet positive message in *The Myth of Sisyphus* might be received by readers living under such oppressive conditions—conditions that would worsen considerably by the time the book appeared in print in October 1942. Camus condensed his message for the paper band that bookstore patrons would see wrapped around his book: "Sisyphus, or Happiness in Hell."

CHAPTER 10

THE TERROR BEGINS

The best political weapon is the weapon of terror. Cruelty commands respect.

—HEINRICH HIMMLER, Reichsführer of the SS

IN JUNE 1941, HITLER FLOUTED HIS NONAGGRESSION PACT WITH Stalin and launched a massive surprise invasion of the USSR. Dubbed Operation Barbarossa, it was an enormous gamble. Hitler was betting that German forces could subdue the Soviets in a blitzkrieg, as they had done to western Europe, before the Russian winter set in. The consequences of success or failure would be great. Either the USSR would fall and its people and lands would be subjugated by the Reich, or it would hold and Hitler would face the strategic and logistical challenges of conducting a war on two fronts.

Either outcome had important implications for the prospect of France's liberation. On the one hand, if Hitler prevailed, the Reich could become even stronger as it exploited Soviet resources and manpower. And in that case, their Vichy collaborators also stood to gain. Alternatively, if the campaign bogged down, the eastern front would engage troops and drain resources that might otherwise be used to strengthen and defend the western front against a potential Allied invasion.

One of the immediate consequences of Hitler's invasion of the USSR was to free French Communists to act against the Occupation forces. The French Communist Party (Parti Communiste Français, or PCF) was very disciplined and faithfully adhered to policies dictated by Moscow and disseminated by Comintern, the international Communist organization. For several years prior to the war, the PCF had been a part of the "popular front" of left-wing parties resolutely opposed to Fascism. The nonaggression treaty between Germany and the Soviet Union, signed just a week before the invasion of Poland, had thus stunned French Communists, and the PCF was promptly banned after the start of the war. The invasion and subsequent occupation of France then put the Party and its members into the irresolvable dilemma of balancing their patriotic duty to defend France with their loyalty to Moscow.

As a result, Communist loyalties were suspect, and their voices and actions during the first year of occupation were somewhat restrained, particularly in the occupied zone.

In the non-occupied zone, Communists were more openly hostile to Vichy, as Vichy was to the Communists. Vichy dissolved the PCF in late September 1940, and thousands of Communists were arrested in Paris in the first week of October.

After the invasion of the USSR, Comintern reversed course and instructed the PCF to launch an armed struggle against the Germans. Weakened by the mass arrests, the absence of its leaders, and disillusionment over the temporary détente with Hitler, very few members responded at first. One who did was twenty-one-year-old Pierre Georges, a veteran of the Spanish Civil War, and later known as Colonel Fabien. Georges had been arrested in the mass roundups of Communists the previous fall but had escaped. On August 19, two of his comrades had been executed after having been arrested in a demonstration. On the morning of August 21, 1941, Georges entered the Barbès-Rochechouart Métro station in Montmartre and fired two shots at a young German naval cadet standing on the platform, killing him. Georges then slipped away into the crowd.

The shooting of a German officer in broad daylight surprised and alarmed the authorities. Suspecting that the attack was intended to arouse anti-German resistance, some commanders urged swift reprisals. Vichy government representatives told the German administration that it blamed the attack on Communists and promised to try, convict, and sentence several Communists to death. Three men were executed just one week after the Métro killing. In the meantime, Gen. Otto von Stülpnagel, military commander in France (Militärbefehlshaber in Frankreich, or MBF), decided to issue a warning that was published in newspapers and broadcast on radio: "Beginning August 23, all Frenchmen taken into custody, either by the German authorities in France or on orders originating with them, will be regarded as hostages. Should any further criminal action occur, hostages will be shot in a number corresponding to the seriousness of that action."

But Communists in particular were not deterred. Their calculation was that their violent acts showed their solidarity with the Soviet Union and that German reprisals would alienate the French population. Another attack followed soon after the Métro killing when, on September 3, two Communists wounded an officer at the entrance to the Hôtel Terminus. Von Stülpnagel ordered the execution of three more Communist hostages held in custody. In response to three more minor attacks over the following week, von Stülpnagel ordered the execution of ten additional hostages. On September 16, a captain was killed on the boulevard de Strasbourg. Von Stülpnagel ordered that twelve hostages be shot.

Hitler was not satisfied with the magnitude of the reprisals. After the first two attacks, he thought that the killing of three hostages was "much too mild." Through an intermediary, he told von Stülpnagel that "a German soldier is worth more than three communists." Hitler thought that a ratio of one hundred hostages to one German was more appropriate. Therefore, he urged that fifty more hostages be executed, that another fifty be executed if the perpetrator was not caught promptly, and that von Stülpnagel arrest another three hundred hostages and execute one hundred of them if another German was assassinated.

Von Stülpnagel was deeply concerned that such mass reprisals would backfire, alienating the population and stoking greater anti-German feelings. He even asked to be recalled if such orders were to be enforced. But Berlin was deaf to such considerations and concerned only that retribution be swift and overwhelming.

On October 20, two of Fabien's comrades gunned down the Feldkommandant of Nantes outside the city's cathedral. He was the highest-ranking German killed up to that time. Hitler learned of the attack within a few hours. Von Stülpnagel asked to delay any executions while the crime was investigated. Hitler ordered that fifty hostages be executed immediately, fifty more twenty-four hours later, and another fifty twenty-fours after that if the assassins had not yet been caught. Forty-eight hostages, the majority of whom were Communists, were selected and shot on October 22. The consequences for attacking Germans were underscored in the next day's newspaper by the long list of the names of those executed.

But the executions were not over. The previous day, there had been another killing when a young Communist shot a civilian military adviser in Bordeaux. In reprisal for that murder, another fifty hostages were selected and shot on October 24.

The groups of hostages who were shot included a wide assortment of Frenchmen. Some had been convicted of serious crimes against the German occupiers, such as sabotage, but others were guilty of no more than being a Communist, some as young as seventeen-year-old Guy Môquet. Vichy officials shared von Stülpnagel's worry that the killing of innocent Frenchmen would provoke and not subdue resistance. On the evening of October 22, Pétain addressed the nation by radio: "Frenchmen, two shots have been fired at officers of the army of occupation: two are dead . . . This morning, fifty Frenchmen have paid for these unspeakable crimes with their lives. Fifty more will be shot tomorrow if the guilty are not caught . . . Frenchmen, your duty is clear: the murders must stop. According to the Armistice, we set down our arms, we do not have the right to strike Germany in the back. The foreigner who ordered these crimes knows well that this is clearly murder."

Still, the attacks continued, some aimed at Germans, some at French policemen. A hand grenade was thrown into a Wehrmacht canteen; another grenade attack was on a military traffic post, yet another on boulevard Montparnasse; and there were shootings on boulevard Magenta, rue de la Seine, boulevard Malesherbes, and rue des Maronites. Altogether, there were sixty-eight incidents in Paris or elsewhere by the end of 1941.

Seeking additional deterrents against such attacks, Hitler issued the Nacht-und-Nebel Erlass (Night-and-Fog Decree) on December 7, 1941, for dealing with resisters and other "offenders" in occupied territories. The name was an allusion to Richard Wagner's opera *Das Rheingold,* in which the character Alberich makes himself invisible so as to torment his subjects. The directive acknowledged: "Within the occupied territories, communistic elements and other circles hostile to Germany have increased their efforts against the German State and the occupying powers since the Russian campaign started. The amount and the danger of these machinations oblige us to take severe measures as a deterrent." The order reiterated that "the adequate punishment for offences committed against the German State or the occupying power which endanger their security or a state of readiness is on principle the death penalty." But Hitler and the High Command thought that, in addition to summary executions, there would be some intimidation value to making resisters simply disappear into the "night and fog" of German concentration camps, without a trial, and without their families knowing their destination or fate. Making opponents disappear would also circumvent the potential backlash of publicly announced executions.

THE VAST MAJORITY of French people condemned both the assassinations and the German reprisals, and very few were in favor of resistance. Nonetheless, the surge in Communist-led resistance gave rise to the formation and expansion of affiliated resistance groups. Although he was not a Communist, and despite the potential consequences of being affiliated with them, Monod joined a Communist-inspired university organization toward the end of 1941. The total number of casualties or damage inflicted by the first wave of armed resistance was, in fact, militarily and strategically insignificant. The attacks were isolated and uncoordinated. If resistance was to become a more substantive factor in France, many more citizens would need to participate. Recruitment was a key priority. Therefore, one focus of Monod's group and of his activity was to recruit students to the cause.

In addition, Monod helped put together tracts that were published from time to time. Working in the recesses of his old laboratory at the Sorbonne, Monod and several colleagues wrote and duplicated the text and disseminated it among students and academics.

As if Monod needed any reminder of the risks he was taking, in January 1942, his former comrade Léon-Maurice Nordmann and other members of the Musée network went on trial. While Nordmann had at first been charged only with distributing *Résistance* and received a modest sentence, he was subsequently charged with espionage, as were several other members of the network. Despite a lack of evidence, the virulently anti-Semitic prosecutor zealously pursued Nordmann.

The timing of the trial was unfortunate. After the cycles of killings and reprisals that had taken place in recent months, the Germans had no inclination to show any mercy. The prosecutor got the convictions he sought, and ten members of the group were sentenced to death. On February 23, 1942, Nordmann, Boris Vildé, and five other members of the network were lined up in front of a firing squad at Fort Mont-Valérien outside Paris and shot.

In spite of the increasing violence in the capital, the worsening shortages, the difficulty in travel, Parisians had to find ways to continue some kind of normal family life. The Monod twins, Philippe and Olivier, born just a month before the invasion of Poland, were already two and a half years old but had still not met their paternal grandparents in Cannes. Odette could not cross the demarcation line legally, and Jacques would not risk having her cross on foot or by night. If the twins were going to make a visit south, their father would have to take them.

As Easter 1942 approached, Odette made all of the preparations for the boys' journey and stay while Jacques secured a pass (*laissez-passer*). It was an exhausting twenty-six-hour trip, with a change of trains in Marseille. Jacques got no sleep while he kept an eye on the two boys, who behaved perfectly throughout the long night and morning. Jacques's mother, Sharlie, was thrilled to see her son coming up the terrace steps of their home, a rucksack on his back, dirty with dust and grime, and "half-dead with fatigue" but so "happy and proud" to be carrying one clean, fresh rosy-cheeked boy in each arm, dressed in little American overalls.

Charlotte kissed the boys, who, not at all shy, kissed their grandmother's hand and said, *"Bonjour, grandmaman"* as though they knew her. Jacques set the boys down, and they quickly discovered and befriended a pet cat.

During the twelve-day stay, Charlotte was relieved to find that Jacques was "the same vivid, energetic creature he always was." She felt that his determination and hard work must be an inspiration to Odette and to his friends. Jacques said that he was resolved to bringing the boys back during the summer, along with their mother. But that plan, and daily life for Odette, was about to get more problematic.

The SS Takes Charge

As military commandant for all of France, von Stülpnagel was primarily concerned with the security of the troops stationed there, and with maximizing French economic cooperation in order to support the larger German war effort. Up until December 1941, he had not done anything to advance Nazi racial policies. He did not support the confiscation of Jewish property, for example, as he believed such activities dishonored the military, alienated the population, and diverted the military's attention from more important industrial objectives. The persecution of the Jews was primarily a matter for (1) the agents of the security service—the Sicherheitsdienst (SD) of the Schutzstaffel (SS)—who were under the overall command of Heinrich Himmler, and (2) the Vichy government, which had acted on its own accord in imposing increasingly repressive measures. Through 1941, the SD had been limited by a shortage of manpower and a lack of executive authority of the kind vested in von Stülpnagel.

But following an attack in early December, and the issuance of the Night-and-Fog Decree, a new twist was added to the reprisal order: the mass execution and deportation of Jews, a job that would be handed over to the SD. While he did not object to deportations, von Stülpnagel was convinced that mass shootings were unwise and wrote as much to his superiors in Berlin:

> *I intend to order only a* limited *number of executions and will adjust the number to suit the circumstance.*
>
> *At least under the present circumstances, I can no longer arrange* mass shootings *and answer to history with a clear conscience*

because of my knowledge of the entire situation, the consequences that such hard measures could have on the entire population, and on our relationship with France.

Berlin replied by insisting upon adherence to Hitler's policy of mass reprisals. Von Stülpnagel then tendered his resignation, explaining that without the trust of his superiors to use his own judgment "the position of the MBF in the occupied area becomes more difficult, leads to weighty conflicts of conscience, and undermines my self-confidence, energy, and determination . . . I can withdraw to private life with clear conscience, confident in the knowledge that I served my people, country, and opponents with complete unselfishness and fulfilled my duties to the best of my ability."

The timing of von Stülpnagel's departure coincided with larger developments unfolding on both of Germany's extended war fronts. Two months earlier—after the Japanese attack on Pearl Harbor on December 7, 1941—the United States had entered the war. While the Americans were preoccupied with Japan in the Pacific, it was just a matter of time before U.S. involvement became a factor in Europe and North Africa. More significantly, a Soviet counteroffensive drove the Germans back from Moscow. The prospect of a much more protracted war demanded greater war production in Germany and the occupied territories, and more manpower. It also heightened the sense of urgency surrounding the Nazi campaign against the Jews. The latter required men who would not question policies emanating from Berlin, those who were more in step with Nazi ideology than professional soldiers like von Stülpnagel.

Command in France was therefore reorganized. While a new MBF replaced Otto von Stülpnagel (his cousin Carl-Heinrich von Stülpnagel), responsibility for the security of German forces was transferred from the Army to the SS, which then had authority over all German and French police and was thus given free rein to pursue Nazi racial policies. On June 1, 1942, Hitler appointed SS Brigadeführer Carl Oberg, a member of the Nazi Party since 1931 and one of Himmler's men, as senior SS and police leader. Oberg, who spoke little or no French, had earned a reputation for savage repression while serving in

Poland. He was given the direct authority to carry out the Nazi campaign against Jews, Communists, and resisters. Under Oberg's command, genocide came to France.

The Yellow Star

Oberg's assumption of command coincided with the introduction of the eighth ordinance governing Jews in the occupied zone. The law, to take effect on June 7, 1942, stated, in part:

> 1. A DISTINCTIVE SIGN FOR THE JEWS
>
> 1. Jews over the age of six are prohibited from appearing in public without wearing the yellow star.
> 2. The Jewish star is a star with six points having the dimensions of the palm of a hand with a black border. It is made of yellow fabric bearing the inscription "Juif" in black letters. It must be worn visibly on the left side of the chest, attached firmly to clothing.
>
> 2. PENALTIES
>
> Infractions of the ordinance will be punished by imprisonment and/or a fine. Police measures, such as the internment in a Jewish camp, can be added or substituted for these penalties.

The law greatly satisfied the most ardently anti-Semitic collaborationist newspapers. On the day after the law took effect, *Le Matin* declared that the sight of the yellow star provoked Parisians to think, "Never would one have thought that there are so many Jews in Paris." The newspaper remarked: "Another surprise was awaiting Parisians in the afternoon, in finding an important number of Jews who were walking about, talking in the cafés, meddling in the lines in front of theaters and cinemas, or else simply taking the Métro. And that was only the part of the Jewish population that was outside! One must not forget that there were, in 1941, 1,200,000 Jews in France, of which more than 350,000 were in Paris and its suburbs."

The number of Jews in all of France was actually only about 310,000, with about half that number in the greater Paris region. Nonetheless, the anti-Semites would not suffer the sight of Jews in public places for very long. Jews were already subject to an eight p.m. to six a.m. curfew imposed in February 1942, and yet another ordinance in early July barred them altogether from parks, cinemas, cafés, restaurants, libraries, or theaters. They could shop for food only from eleven a.m. to noon, and for other goods from three to four p.m., and they were to ride only in the last train car on the Métro.

The yellow star made the spotting of violators easy for the police. Those who dared to try to circumvent the law and not display the insignia risked being arrested or being denounced by Nazi sympathizers. Nevertheless, Odette and Jacques decided that she would not wear the yellow star.

With all of the restrictions, it was difficult for Jews to live in Paris and not to be in violation of some ordinance. Just to get a glass of water from any establishment on a hot summer day or to walk through the Luxembourg Gardens was grounds for arrest.

But even such extreme prohibitions paled in comparison to what unfolded the week after Jews had been banished from public venues.

Roundups

At four in the morning on Thursday, July 16, 1942, about 4,500 French policemen fanned out across Paris and its suburbs and began knocking on doors. Working in pairs, each carried a handful of index cards bearing the names of 27,388 Jews who were to be arrested that day. All of those targeted were foreign-born men and women between the ages of sixteen and sixty-five. They were told to bring their identification and ration cards, food for at least two days, and a number of personal items including one pair of shoes, two pairs of socks, a sweater, two shirts, sheets and blankets, eating utensils, a drinking glass, and toiletry items. Children were to accompany parents unless an older grandparent remained at the home. Otherwise, the police were to make sure that the household utilities were shut off when they left.

Over the course of two days, 12,884 people were arrested, almost one-third of whom were children. Accompanied by policemen, entire families walked in broad daylight through the streets carrying suitcases to assembly points in each arrondissement or suburb. Those who did not have children under sixteen were then bused to the Drancy internment camp northeast of Paris; those with children were taken to the Vélodrome d'hiver (Vel d'hiv), a sports stadium near the Eiffel Tower in the fifteenth arrondissement. Onlookers who saw the green-and-beige city buses with their windows closed and guarded by policemen understood that something dramatic was happening.

But neither those arrested nor their fellow Parisians knew what fate awaited them.

For the nearly five thousand taken to Drancy, they found a filthy, overcrowded nightmare with insufficient water or toilets. The conditions at the Vel d'hiv were even worse, for the glass-covered stadium was poorly ventilated and unbearably hot in July. There was no place to wash and little to drink, there were no mattresses to sleep upon, and the bathrooms had been closed to prevent escape. The authorities had made no preparations, as the prisoners were not expected to stay at either facility for long.

Indeed, just three days after the beginning of the roundup, about one thousand prisoners were bused from Drancy to Le Bourget train station and put on a train bound for the Auschwitz concentration camp in Poland. Upon their arrival two days later, 375 men were separated from the other prisoners and gassed. Four more trains left Le Bourget with four thousand more deportees from Drancy on July 22, 24, 25, and 29.

The prisoners at the Vel d'hiv were taken to the Pithiviers and Beaune-la-Rolande camps in the Loiret. Then, on July 31 and August 3, 5, and 7, more than four thousand parents and older children were loaded onto four trains, also bound for Auschwitz. Almost half were gassed upon arrival. Fewer than forty would survive the next three years.

THERE HAD BEEN roundups before, but not on this scale, not in this manner, and not with the direct purpose of extermination. Previous roundups were largely a matter of Vichy demonstrating its commitment to Nazi racial policies. But the July 16–17 roundup had been ordered by the Germans, who were seeking 100,000 French Jews in 1942 as a first stage of the "Final Solution" in France. Laval and Secretary-General of Police René Bousquet negotiated with Oberg and representatives of Himmler and Adolf Eichmann to pare the figure down to 32,000. The raids were carried out by the French police, both as a matter of needed manpower and so as to avoid inciting anti-German feelings. But unlike previous roundups, these included women and children—entire families. And within just three weeks, more than nine thousand had been sent to Auschwitz.

Terror reverberated throughout both the immigrant and native French Jewish communities. Lise Bruhl, Odette Monod's sister and also fully Jewish under the law, was very fearful that she would be arrested. Lise was married to Georges Teissier, a biologist and colleague of Jacques who, like Jacques, was both a Protestant and active in the Resistance. The couple and their three daughters also lived in the fifth arrondissement, near the Val-de-Grâce Hospital. During the first two years of the Occupation, the families saw a lot of each other. Françoise, their middle daughter who turned fifteen in 1942, often took the twins to the park while Odette and Lise shared addresses where they could get hard-to-find vegetables and other foods.

At the time of the July roundups, Lise had stopped sleeping at home. While the girls were technically not Jewish, and even had fake baptism certificates, Lise had decided that if the children were arrested, she would turn herself in to the authorities in exchange for them. Françoise was, in turn, terrified that her mother was going to be arrested. Every knock or noise in their apartment building put her on edge.

After the roundups, the Bruhl sisters decided to get out of Paris. Lise left with her youngest daughter in order to live in the countryside with a sister-in-law, using the cover story that the girl was so thin that she could not bear the restrictions in the capital. Odette and the twins went to Cannes (in the nonoccupied zone) to live with Jacques's parents. To do so, she had to cross the demarcation line into the nonoccupied zone, which was forbidden for Jews to do. In order to pass as non-Jewish, Odette obtained a false identity card, using the last name Brulle instead of Bruhl.

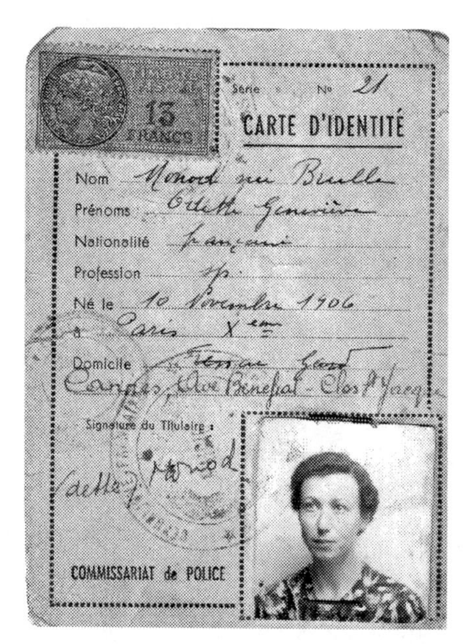

13 FRANCS

Série N° 21

CARTE D'IDENTITÉ

Nom Monod née Brulle

Prénoms Odette Geneviève

Nationalité française

Profession s/p

Né le 10 Novembre 1906

à Paris Xème

Domicile [illegible]

Cannes, Ave [illegible] - Clos [illegible]

Signature du Titulaire : Odette Monod

COMMISSARIAT de POLICE

Odette Monod's false identification card. Odette changed the spelling of her maiden name, Bruhl, to "Brulle" to conceal her Jewish identity. (Courtesy of Olivier Monod)

GIVEN THE OPPORTUNITY, many French sought to live in the non-occupied zone. Restrictions were fewer. For example, Jews were not required to wear the yellow star. Food was also more available. And German soldiers were rarely encountered.

That, however, was soon to change.

CHAPTER 11

THE PLAGUE

All were separated from the rest of the world, from those they loved or from their routine. And in that withdrawal they were obliged, those who could, to meditate, and the others to live the life of hunted animals.

—Albert Camus, *Notebook IV*

Camus also sought refuge in the nonoccupied zone in 1942, but for different reasons from most. In January, while in Oran, he suffered a relapse of his tuberculosis. He thought he had been cured, but he was suddenly coughing up blood, sweating, and feeling weak. Francine raced out of the apartment to find Albert's doctor. After a rough night, Camus told Francine's sister, "I thought it was all over for me this time."

The disease was in his left lung for the first time. The standard care for TB included extended bed rest and pneumothorax therapy in which the lung was deliberately collapsed by the injection of air. The painful procedure enabled lesions to heal, but the lung would reinflate, so the treatment had to be repeated every two to three weeks. Camus's doctor started the periodic treatments, but then his office was closed by the Vichy authorities' enforcement of a quota on Jewish physicians. Camus's injections continued from a colleague's office.

Another recommendation to aid Camus's recovery was that he seek a change in climate, specifically France's mountain air. Fortunately for the impoverished couple, Francine's aunt's mother ran a boardinghouse in the tiny village of Le Panelier, just outside the town of Chambon-sur-Lignon in the Massif Central. Francine and her sister spent summers there as children. The Camuses applied for a travel pass that would allow them to go after Francine finished up her teaching in Oran in July. They arrived at Le Panelier in August.

As the fresh air and somewhat more plentiful food were helping him regain his strength, Camus saved what energy he had for writing. He had been working, on and off, for more than a year on a novel that was to be part of a new cycle on the theme of revolt. Just as *The Stranger* was to illustrate the theme of the absurd, Camus's intent with *The Plague* was to portray the choices humans faced when confronted with evil. Under a header entitled "Beginning," Camus wrote in his notebook on October 23: "*The Plague* has a social meaning *and* a metaphysical meaning. It's exactly the same. Such ambiguity is in *The Stranger* too."

The novel was to be an allegory for the Nazi occupation of France and the response of the French to it. Camus reminded himself: "The first thing for a writer to learn is the art of transposing what he feels into what he wants to make others feel." To convey his feelings, Camus chose to set his story in Oran during an outbreak of the bubonic plague. The Nazis had long been referred to in France as *la peste brune* (the brown plague) on account of their brown shirts, and the Occupation itself had acquired the same name.

Camus summarized his overall purpose in his notebook: "I want to express by means of the plague the stifling air from which we all suffered and the atmosphere of threat and exile in which we lived. I want at the same time to extend that interpretation to the notion of existence in general. The plague will give the image of those who in this war were limited to reflection, to silence—and to moral anguish."

His form was to be that of a "chronicle" of the plague, as reported by a narrator-witness. Camus had been researching all he could learn about past plagues, and about how the disease was manifested. He learned of at least one striking similarity between major plagues in history and the current blight in France. He jotted in his notebook:

> 1342—The Black Death in Europe. The Jews are murdered.
> 1481—The plague ravages the South of Spain. The Inquisition says: The Jews.

Oran itself had, in fact, been stricken several times with outbreaks of the plague, and had also experienced a typhus epidemic in 1941 while Camus was living there. Camus drew upon his knowledge of the town in developing the story's setting, a town that Camus wrote, "let us admit, is ugly . . . How to conjure up a picture, for instance, of a town without pigeons, without any trees or gardens, where you never hear the beat of wings or the rustle of leaves—a thoroughly negative place, in short?" It was a town where "everyone is bored, and devotes himself to cultivating habits." A town so ordinary, Camus explained, that "it will be easily understood that our fellow citizens had not the faintest reason to apprehend the incidents that took place in the spring of the year in question [Camus stated that it was 194–] and were (as we subsequently realized) premonitory signs of the grave events we are to chronicle."

The story begins with a Dr. Rieux coming across a dead rat on the second-floor landing of his office, the first of thousands that would be seen over the coming days, but whose significance goes unheeded by the population.

FRANCINE LEFT FOR Algeria in October in order to resume her teaching. Wanting to maximize his time in the mountains before the winter weather arrived, Camus planned to stay at Le Panelier until the end of November, telling a friend, "After that I will return to Algiers and settle there no doubt, unless something unforeseen happens." He asked Pascal Pia to book him passage on a steamer that was to leave from Marseille on November 21.

On the night of November 7, the unforeseen happened. Dubbed Operation Torch, three Allied task forces comprising more than 600 ships and carrying more than 60,000 troops made a surprise landing on the coast of French North Africa, in Algeria and Morocco. Despite some Vichy opposition, the Allies quickly took Casablanca, Algiers, and Oran, and controlled some 1,300 miles of coastline.

It was a major Allied victory. Radio London spoke with a new confidence. On November 10, the evening broadcast of *Les Français parlent aux Français* declared, "In the space of five days, a world accustomed for three years and two months to seeing Germany send its armies over the continents and oceans has seen a sudden and abrupt change in the face of things."

However, with the Allies now facing southern France across the Mediterranean, Hitler gave the order for the German Army to cross the demarcation line on November 11 and to occupy all of France. Formally, it was a breach of the armistice. Hitler offered his explanation in a published letter to Pétain: "The circumstances being such, I have the honor and the regret of informing you that, in order to remove the danger that threatens us, I am forced, in concert with the Italian government, to give the order to my troops to go through France by the most direct route to occupy the Mediterranean coast."

The plague had arrived in southern France. The day of the German invasion, Camus wrote in his notebook: "Caught like rats!"

Francine was now in Allied-held territory, but Camus was in German-occupied France. They were cut off from each other. There would be no passage, or even any direct mail. In his novel, when the state of plague is declared, the prefect orders the town gates closed. In a draft version of what would become part of *The Plague*, Camus wrote:

> In short, the time of the epidemic was mainly a time of exile. One of the most striking consequences of the closing of the gates was in fact the sudden separation that it imposed upon people who were not prepared for it. Mothers, children, spouses, lovers who a few days beforehand believed they were undertaking only a temporary separation, who had kissed one another on the platform of a station with two or three remarks, certain of seeing one another again in several days or weeks . . . all at last found themselves cut off without recourse, prevented from seeing or communicating with one another.

Stranded, an "exile," and still weak from TB, Camus had no alternative but to focus on writing. He noted: "Make separation the big theme of the novel." That general sense of separation was being compounded across France by another, growing symptom of the brown plague: slavery.

Workers for Germany

With ongoing offensives in North Africa and the Soviet Union, and defensive installations to build on the western front, the German war machine had enormous and increasing demands for materiel and the manpower to produce it. As the Soviet campaign claimed heavy losses—almost 1.3 million German casualties by June 1942—workers in Germany were drafted to replace them. The support and expansion of the Reich therefore relied more and more on labor from the occupied territories. French prisoners of war were employed in German factories and on German farms, and much of France's domestic production had been coopted for German needs. Some 275,000 French laborers were building airfields and fortifications on the Atlantic Coast, and another 400,000 worked in French armament factories that supplied the Germans.

But that was not enough to satisfy demand, or Hitler. He appointed longtime Nazi Party member Fritz Sauckel as plenipotentiary for the mobilization of labor, and gave him the power to secure all available labor throughout German-held lands. In mid-1942, Sauckel and Hitler decided that France could afford to supply an additional 350,000 workers to Germany.

Sauckel discussed the situation with Vichy prime minister Laval, and proposed to furlough 50,000 French POWs if France sent 150,000 workers to Germany. Laval agreed to the deal and announced the Relève program in late June 1942. It was portrayed as a patriotic volunteer program through which workers could secure the release of fellow countrymen who had been held for two years.

The response was tepid. Only 53,000 workers volunteered in the first three months of the program, too few to satisfy Germany's demands. So the Germans asked Vichy to address the shortfall by passing a compulsory labor law that would require able-bodied men to work. Vichy complied by instituting laws in September 1942 that required men between the ages of eighteen and fifty, and unmarried women between the ages of twenty-one and thirty-five, to work at least thirty hours per week. The number of workers sent to Germany tripled in

October and quintupled in November relative to that in September, bringing the total to 239,000 in the second half of 1942.

That was still not enough. German manpower was stretched by reversals on two fronts: the Allied landing on the coast of French North Africa that prompted the occupation of all of France, and therefore tied up yet more soldiers; and more losses and surrenders on the eastern front. In late November, the Soviets' Operation Uranus encircled 300,000 Axis (mostly German) troops around Stalingrad.

Germany demanded yet another 250,000 French workers. Laval agreed under the same three-workers-for-one-POW formula. But drumming up sufficient workers was difficult. Laval ordered prefects to undertake a census of all men aged twenty-one to thirty-one. And on February 16, 1943, he established the Service du Travail Obligatoire (STO; Obligatory Labor Service), which required Frenchmen aged twenty to twenty-two to go to work in Germany for two years in lieu of their customary military service.

The eligible young men who received summonses to report for duty faced a terrible dilemma. Should they obey, they would have to leave their families and France, and work for their occupier. Moreover, with the Allies now bombing German factories, and the tide of the war potentially changing, they might be risking their lives in Germany. On the other hand, if they did not obey, they would be outlaws subject to arrest and perhaps convicted to forced labor anyway. The compulsory-labor law, the census, the STO, and the departures of workers for Germany aroused widespread bitter resentment toward Laval and the Vichy regime as so many families, most of whom had suffered some separations or losses already, faced yet more anguish.

The regime's deep complicity in the sending of French labor to Germany also provoked the relentless fury of Radio London, whose broadcasts encouraged workers on a nearly daily basis to defy Vichy's schemes. "Workers and bosses, peasants and bureaucrats, in order for France to live, not a man for Germany!" was the message on January 23. "One's sacred duty is to do everything to remain on French soil," said another appeal. "Frenchmen, do not go there!" was a refrain in March. Resistance newspapers echoed the cry. *Le Franc-Tireur,* published by the movement of the same name, declared: FRENCHMEN! STAND AGAINST SLAVERY.

—~—

MANY MEN DID go to Germany—more than 60,000 in January 1943, almost 80,000 in February, and about 110,000 in March. But many who were summoned did not report for work and went on the run. Those who evaded the STO and the authorities became known as *réfractaires*. While more than 600,000 French workers were sent to Germany in 1942–43, more than 200,000 became *réfractaires*. The Resistance took pride in their defiance of the STO; the headline of the March 15, 1943, issue of the Resistance newspaper *Libération* stated in large bold type:

LA JEUNESSE FRANÇAISE RÉPOND: MERDE.
[The French youth answer: Shit.]

EXILES

Réfractaires could not return to their homes without risking being caught by the French police. While some tried to cross the Spanish or Swiss borders, and generally failed, most focused on the challenges of finding shelter and food somewhere in France. Unable to obtain a legitimate ration card, many wound up wandering the countryside, looking for work on farms. Successful evasion of the authorities, then, depended upon the assistance given by others.

By coincidence, one of the most hospitable hideouts for *réfractaires* and for others trying to stay out of the reach of the authorities was the Vivarais-Lignon Plateau region, where Camus was living. A good distance from any major city, the plateau contained many small villages, such as Le Panelier, in which residences and farms were scattered. It was difficult country to traverse, and villagers could easily spot the approach of any officials. Most important, the local Protestant clergy and townspeople were resolutely committed to sheltering anyone sought by the Vichy authorities or, after the occupation of the southern zone in November, by the Germans.

The longer Camus stayed in the region, the more he learned of the scope of clandestine activities in the area. In Le Chambon-sur-Lignon,

Protestant pastors André Trocmé and Edouard Theis inspired their flock of just a few thousand to shelter several thousand Jewish children and adults, as well as escaped prisoners, *réfractaires,* and members of the Resistance, over the course of the war. One of Camus's Jewish friends from Algeria, André Chouraqui, lived in a nearby village and was involved with a group called the Oeuvres de Secours d'Enfants (OSE), which ran orphanages and found homes for the children of refugees who had been killed or deported. The two men met regularly and shared Algerian food, and Chouraqui translated parts of the Bible dealing with plagues for Camus.

At his boardinghouse, Camus met and befriended Pierre Lévy (alias: "Fayol"), a Marseille Jew and member of a resistance group called Combat, to which Pascal Pia also belonged. Fayol and Camus listened to the BBC together, and Fayol and his wife, Marianne, showed Camus how to get letters through to Francine in Algeria via Portugal. Camus received his first reply from Francine in late March 1943, almost six months after the beginning of their separation.

As the months passed, Camus painted the picture of everyday life in occupied France in his fictional tale of the plague in Oran. The city's trapped inhabitants were forced to endure shortages and long lines waiting for food, the rationing of gasoline, and temporary power blackouts. Camus also described "isolation camps" that quarantined people suspected of having the disease. Camus knew about the internment camps that held Jews who had been rounded up, and about the deportations. In his notebook, he jotted: "In the chapter on the isolation camps: the relatives are already separated from the dead—then for sanitary reasons children are separated from their parents and the men from the women. So that *separation becomes general.* All are forced into solitude."

Camus did manage to make forays into Paris. On one trip, he shared a section of his novel with Jean Paulhan, who in turn shared it with Jean Lescure, who was interested in putting together an anthology of writings by French authors "concerned with man and freedom." The subversive volume could not be published in France, so Lescure had the manuscripts smuggled out to Switzerland, including a draft chapter from Camus entitled "Exiles in the Plague" (*Les Exilés dans*

La Peste), and other pieces by Jean-Paul Sartre, André Maurois, Paul Valéry, and Jean Paulhan. The preface to the collection informed readers that these writers were breaking their silence to affirm "the dignity of a conception of man" that, however crushing the military and political defeat of France had been, could not be suppressed. Three thousand copies of *Domaine Français* were printed and smuggled back into France, wrapped in plain paper, and given away. It was Camus's first contribution to the literature of resistance, and would be far from his last.

One passage read:

> They felt the profound sorrow of all prisoners and exiles which is to live with a memory that serves no purpose. Even the past, of which they thought incessantly, had only the taste of regret. For they would have wished to be able to add to it all that they regretted having left undone while they might have done so, with the man or woman whose return they now awaited—just as in all their activities, even the relatively happy ones, in their lives as prisoners, they tried vainly to involve the one who was absent. Impatient of the present, hostile to their past, and cheated of their future, they were like those whom man's justice, or hatred, forces to live behind prison bars.

It would be four years before most French readers could encounter such a passage (in Camus's novel). But for all those who had suffered because of the plague—the millions separated from their families or from their homeland by the German invasion, by the chaos after the collapse of the country, by the demarcation line, and by the invasion of the southern zone, or having been taken prisoner, interned in camps, sent to Germany to work, or deported, or whose family members were so exiled—Camus's words would resonate with the memories of their heartache.

In the third year of occupation, however, great unknowns remained: How could towns and the country in general get rid of the plague? And when would the infection end?

CHAPTER 12

BROTHERS IN ARMS

Plague. All fight—and each in his way. The only cowardice is falling on one's knees.

—ALBERT CAMUS, *Notebook IV*

THE ALLIED LANDINGS IN ALGERIA AND MOROCCO CREATED THE opportunity for the Free French forces in North Africa to join with them to try to rout the Germans and Italians from the continent altogether. De Gaulle ordered Col. Philippe Leclerc, a bold, charismatic commander who had proven his talents in the African desert, to advance from Chad through Libya to link up with the British in Tripoli. Taking numerous Italian-held forts and towns on their 1,300-mile trek, the French then joined up with General Bernard Montgomery's British 8th Army to force the Germans out of Tunisia. In late March 1943, Gabès, Tunisia, became the first French city in North Africa to be liberated and reoccupied by Free French troops.

The propaganda value of the French conquests was priceless. De Gaulle went on the BBC to boast: "With the victory of our Chad troops, the enemy has seen rise, once again, the flame of the French war that he had believed extinct in the disaster and the betrayal, but which has not ceased for a single day to burn and grow larger under the breath of those who did not despair. For them, this victory is not only a brilliant feat of arms, it is also one of the harbingers of this new France—a hard and proud France that is being built in this trial."

While North Africa was securely in Allied hands, France proper remained under the Nazi grip, which only tightened as the tide of the war turned against the Germans elsewhere. De Gaulle believed that the time to rally metropolitan France toward resistance had arrived. He sought to rouse the French youth in particular:

Certainly, it is on the youth of France that the suffering of the country weighs most heavily. Physically, it is they especially who lack all of which our country has been stripped. Of ten boys and girls living at home, nine do not eat enough to satisfy their hunger. Morally, they feel, more cruelly than their elders do, the humiliation of their families and their country. What inspires anger and disgust in their twenty-year-old souls is the presence of the enemy, forced labor, repression . . . But you, you are the sons and daughters of a great nation . . .

The enemy is there, with its strength, its police, its propaganda. He is there, defiling our soil, poisoning our air, dishonoring our homes, outraging our flag. He is there, half-defeated, trying to compensate for the victories he is lacking by the oppression of unarmed people. Youth of France, it is now or never to do anything that can be done to hurt the invader, while waiting for the power to destroy him. It is up to you, especially, who bears the hard and great duty of the war. It is you that the enemy aims for first, he who, right now, wants to mobilize you to work to his profit. Do everything to escape him, and if that is impossible, to deceive him, to ruin him, to disappoint him. Group yourselves with discipline into the resistance organizations that are the Fighting France from within. Follow instructions . . .

Young men, young women of France, courage! This is the hour of greatest effort. It is at this cost that the chains will fall, that the prison will open, that the sun will reappear. It is at this cost that you will find the joy of being in the world, the passion to live and to give life, the right to sing and laugh, the pride of being free in a glorious country. Listen to your heart. It contains the future of France!

"Marchal," the Communist Franc-Tireur

Monod also thought that the time had come for more direct action. Grand speeches and articles in newspapers were not going to make the Germans leave. If the Resistance was going to be of any military significance in expelling the occupiers, it needed not only to expand in scale but also to have more of its members take up arms. Thanks in large part to the STO, the ranks were indeed growing. But if Monod himself was going to have a more significant influence, he had to join a group that was committed to armed resistance. He would be taking a dangerous step, as the consequences for resistants had been expanded under Kommandant Oberg's command in Paris to include their families. In a published notice, the senior SS officer and police leader had declared:

> I have noticed that it is especially the close relatives of the assassins, saboteurs, and agitators who have helped them before or after their action. I have decided then to deal the most severe punishments not only to the assassins, saboteurs, or agitators themselves once arrested, but also in case they are on the run, to the families of the criminals, if they do not appear in ten days at a German or French police station. As a consequence, I announce the following penalties:
>
> 1. All of the male relatives up and down the family line as well as the brothers-in-law and cousins over the age of eighteen will be shot.
> 2. All female relatives to the same degree will be sentenced to forced labor.
> 3. All of the children, up to the age of seventeen, of the men and women affected by these measures will be placed in a reformatory.

It was common practice for the Germans to question, threaten, or arrest the relatives of those they were seeking. Since Odette and the children were living away from Paris with Jacques's parents in Cannes, none of them were likely to be in immediate danger should Monod get caught. With his family safe, or at least as safe as he might hope them to be under the circumstances, Monod mulled over how to escalate his efforts in the Resistance. Several of his Sorbonne colleagues were members of the Francs-Tireurs et Partisans (FTP), a militant organization that grew out of the first Communist efforts to organize armed attacks against the Germans. One spring evening, Francis Cohen, a former laboratory colleague and also an FTP member, came to Monod's apartment to try to recruit him into the group. After talking long into the night, Monod agreed to join the FTP.

Monod soon learned that there were many factors that undermined the effectiveness of the Resistance—a lack of weapons, a lack of funding (which was needed to obtain arms and to support resistants who were living on the run and had no income), and a lack of coordination among the many different Resistance groups. Communications were also difficult and fraught with risks. Because each Resistance organization grew out of the efforts of a small number of people in different regions of the country, and because it was imperative to maintain secrecy so that most people knew only a few others in the same group, there was little communication across the groups. There were also the political divisions, largely carried over from before the war. De Gaulle was an unknown when he left France and was looked upon suspiciously by the Communists, for example. While he purported to speak for all of France that yearned to be free, what happened if and when liberation came was of deep concern to those who were risking their lives inside France to bring that about. The Communists were not about to simply hand over the reins of the country to one who did not share their worldview.

But the first objective that had to be achieved before any such concerns could be relevant was to get the Germans out of France. So most parties and organizations recognized a need for coordination in anticipation of an Allied invasion to retake the country. The first attempt at coordination was the formation of the Conseil National de la Résistance (National Council of the Resistance). De Gaulle's delegate, Jean Moulin, succeeded in gathering representatives of eight Resistance groups, six political parties, and the major trade unions to a meeting in Paris in late May. At the meeting, the representatives agreed to unite behind de Gaulle as France's authorized representative among the Allies.

Such gatherings were not only difficult to coordinate but also dangerous, as they presented great targets for the Gestapo. Indeed, just four weeks after the first council meeting, Moulin and a number of other Resistance leaders were arrested together at a meeting in a Lyon suburb. As had become standard practice, Moulin was tortured—by the notorious SS officer Klaus Barbie—in an attempt to extract information about Resistance operations and leadership. Moulin died without divulging any secrets.

In the FTP, Monod soon learned that Communist suspicions extended to anyone who, like himself, was not a member of the Party. They would not allow non-Communists any say in the planning or decisions within the FTP. Monod had long held serious reservations about Communism, and he had told Francis Cohen that he had no illusions about the French Communist Party. But he wanted to get more deeply involved in the FTP, so after a sleepless night he put his reservations aside and joined the Communist Party.

Monod kept up the appearance of a normal routine. He continued to conduct his experiments at the Sorbonne, and he also continued to play and conduct music. Monod met the great blind organist André Marchal, who played at the historic Saint-Germain-des-Prés church and had a close following of many young musicians, especially students at the National Conservatory of Music (Le Conservatoire). Marchal had an organ in his home, at which he held musical gatherings every Tuesday evening. Marchal introduced Monod in turn to Norbert Dufourcq, a professor of music history at the Conservatoire who had founded a youth chorus called Le Mouvement Musical des Jeunes (The Young People's Musical Movement). Dufourcq, knowing of Monod's experience with his La Cantate chorus before the war, asked him to direct the chorus. Monod accepted, and the students were soon rehearsing Bach, in German, which was generally not being performed anywhere during the Occupation. Monod reminded the students: "Do not forget that before becoming the language of Hitler, German was the language of Johann Sebastian Bach." That was bold talk at the time, which, along with his conducting, earned Monod the admiration of many students.

Monod was to conduct the chorus on May 21, 1943, in a concert at the Conservatoire, accompanied by a student orchestra. Odette made the long trip all the way to Paris from Cannes to see the performance, leaving the children behind with their grandparents. She arrived in time to see the dress rehearsal two days before, and saw that her husband was clearly enjoying himself. She reported to her in-laws, "He is in great form and I admire the sang-froid and the ease with which he commands together the orchestra, the singers, and the organ." The all-Bach program included three cantatas directed by Monod, and a concerto for violins performed by the student chamber orchestra.

Monod's new conductorship would turn out to be short-lived, as Marcel Prenant, chief of staff of the FTP and chair of comparative anatomy and histology at the Sorbonne, had a different job for the maestro. An infantry officer when he was wounded in the previous war, and a committed anti-Fascist between the wars, Prenant was stationed near Sedan before being taken prisoner in May 1940. He was released in 1941 because of his veteran status, and joined the Resistance. In 1943, he was put in charge of cooperation with the other Resistance organizations on military matters. Toward the end of September, he recruited Monod's brother-in-law, Georges Teissier, also a biologist, to become his deputy. Teissier and another Sorbonne FTP member, Charles Pérez, in whose laboratory Monod had conducted his thesis research, recommended Monod to Prenant. Prenant charged Monod with recruiting members with military training who could in turn train and lead action units.

In the fall of 1943, the question on everyone's mind was: When are the Allies coming? While no one knew the timetable, the leaders of the Resistance anticipated that it would launch into widespread action at the time of the landing and mount an insurrection against both Vichy and the Germans.

Another uncertainty concerning the anticipated invasion was where it would take place. The Allies might come from the west (from England) or the south (from North Africa or Italy). Cannes, on the Mediterranean coast and fewer than forty miles from the Italian border, could become a battle zone and need to be evacuated. In early October 1943, Odette and the children left Cannes and moved back north, not to Paris, but to Saint-Leu-la-Forêt, a small suburb about twelve miles north of the city. It was thought to be relatively safer than Paris itself for Jews or, in Odette's case, for a Jew living under a false identity. Her sister Lise had moved there beforehand with her daughter Françoise and found Odette a house to rent on the same street. Monod and Georges Teissier (Lise's husband) took the train to Saint-Leu to be with their families. In order to manage his priorities, Monod established a new routine. He spent every Wednesday night until Thursday morning, and Saturday evening until Monday morning, in Saint-Leu being a husband and father. The rest of the time he divided his efforts between being Monod the scientist at the Sorbonne and being "Marchal" the FTP organizer. The latter alias sometimes took him far from Paris.

Dangerous Liaisons: "Martel" in Switzerland

Whatever success Monod might have in enlisting military-trained recruits into the FTP, their effect would be limited by the shortage of weapons. It was a problem that plagued all Resistance groups across the country, and the FTP in particular. Marcel Prenant had made repeated appeals to de Gaulle's representatives for weapons to be delivered by parachute drops, and while he received many promises, he had not secured weapons. Monod, however, had one special contact in another group whom he thought might be able to help the FTP—his brother Philo, who was working for the Resistance across the border in Geneva, in neutral Switzerland.

Ten years older than Jacques, Philo was an experienced, well-traveled attorney. After a few years working for the famous Sullivan & Cromwell law firm in New York, he had returned to Paris to work in the firm's Paris office. He was vehemently opposed to the Munich Pact in 1938. When war was declared in 1939 he was assigned to the Blockade Section of the Foreign Ministry, watched the government unravel in the course of the German invasion, and followed it to its makeshift quarters in Bordeaux. Philo heard de Gaulle's broadcast on June 18, 1940, and decided immediately that he would follow the general's lead.

When he was demobilized in July 1940, Philo headed to his parents' home in Cannes. He wanted to join the Resistance right away, but he could not at first find a group to join. This was both a matter of the nascent groups being very small and secretive, and of the fact that Philo knew very few people in the south. It was not until François Morin, a friend as well as a member of Jacques's doctoral examination committee, visited Philo that he found a way in. Morin, aka "Forestier," was a member of Combat who was responsible for military matters. Philo told Morin of his desire to become an active resistant, and Morin introduced him to Claude Bourdet, one of the founders of the movement. Bourdet invited Philo to join Combat and, using the alias "Martel," Philo took charge of the Cannes-Antibes region. Philo subsequently replaced Bourdet as head of the Alpes-Maritimes region.

Philo's connections with Sullivan & Cromwell proved to be especially valuable. In November 1942, his former supervisor in the Paris office, the American Max Shoop, came to visit him in Cannes on his way to neutral Switzerland. Philo completely trusted Shoop and disclosed his activities in the Resistance. Shoop appeared to take a special interest in what Philo had to say, and for good reasons. Philo did not know it at the time, but Shoop had joined the American Office of Strategic Services (OSS), the forerunner of the CIA, and was working with another Sullivan & Cromwell alumnus, Allen Dulles, who was stationed in Bern, Switzerland, gathering intelligence for the Americans.

Three months later, Shoop sent a message to Philo. As he had suspected, his friend Shoop's visit had not been simply that of a tourist passing through. Shoop promised American aid and requested a report on the military strength of the Secret Army, an organization comprised of Resistance fighters from three groups—Combat, Liberation-Sud, and Franc-Tireur (a separate group, not the FTP)—and headed by François Morin. Philo brought the news of Shoop's interest to a meeting with Henri Frenay, Morin, Bourdet, Pierre Guillain de Bénouville, and other principals of Combat in Saint-Clair, on the edge of Lyon. "Martel" told the gathering that Shoop "wanted to know what kind of military assistance we could offer the Allied cause."

Philo was the hero of the day. Starved as they were for funding, arms, and communications gear, Combat's leaders rejoiced at the Americans' interest. Frenay suggested that, since the Americans were his contacts, Philo go to Switzerland and work out further arrangements. De Bénouville, who was in charge of "external relations," was to establish the courier network for communications between Lyon and Switzerland, and a reliable means of slipping back and forth across the border.

Philo and de Bénouville soon met with Shoop and Dulles, who was posing as an adviser to the American embassy, in Bern. The Frenchmen handed over an organizational chart of the Resistance and outlined their needs, which included (1) funds for maintaining an office in Geneva; (2) 25 million francs per month to develop the organized resistance; (3) additional funds to cover guerrilla actions, including sabotage; (4) reserve funds to take care of *réfractaires*; (5) arms, explosives, and food rations; and (6) assistance with establishing radio and air links.

The Americans listened attentively and took voluminous notes. They were astonished by the lack of material aid coming from London. With Washington's approval, they offered their radio facilities in Geneva so that the groups inside France could communicate with de Gaulle. They immediately agreed to fund a bank account so that a delegation could be set up in Geneva, which would be organized by Philo. Technically, he would be representing not Combat, but Les Mouvements Unis de la Résistance (MUR; the United Movements of the Resistance), which included the same three movements behind the Secret Army. Philo soon secured the commitment of 37 million francs for their operations and was asked to identify one hundred parachute-drop sites for arms. De Bénouville set up a system of transferring funds from Switzerland to banks inside France.

The MUR provided intelligence reports to the Allies via the Swiss delegation, which quickly established the utility of the Resistance as an already "landed army." That is, the MUR wanted to be viewed as a military organization that was already in place within France ahead of the anticipated Allied landing. The intelligence reports also demonstrated the importance of the Geneva office. Two pouches per week were sent to Geneva from Lyon that contained reports on sabotage missions and enemy troop movements, detailed maps, and results of Allied bombing missions to help with the design of subsequent missions. The pouches also compiled information on life in France, which was shared with the press and reported to the world. Couriers from Switzerland in turn sent requests to the Resistance for specific intelligence, as well as consignments of arms, munitions, explosives, and sheets of counterfeit ration cards.

The FTP, however, was not part of the MUR. The formation of the latter organization was aided by the fact that its three constituent organizations all started in the south, in the unoccupied zone, and thus each had much greater freedom of movement than their counterparts in the north. Indeed, while the MUR had the Secret Army, the FTP was by necessity a guerrilla organization using very small teams to carry out sabotage and other actions using hit-and-run tactics. The FTP had not competed successfully with the other organizations for London's attentions, and its members were suffering for lack of arms.

In October 1943, de Bénouville called for a meeting of the major Resistance groups in Geneva in order to discuss the problems of the lack of weapons and money, and to present the situation to the Americans to see what they could do. Jacques Monod was chosen as the FTP liaison to the MUR and made contact with de Bénouville. He was already well apprised of the Geneva delegation's activities through his brother. Philo and de Bénouville raised the possibility of helping the FTP, so Jacques proposed to his superiors that he go to Switzerland to try to secure arms shipments. They agreed.

It was very risky crossing the border. The Swiss tightly controlled their border to avoid being swamped with refugees and to preserve their neutrality. On the French side, there were both customs agents and German patrols on the lookout for *réfractaires,* refugees, black marketeers, and resistants either trying to get out of or back into France. Jacques knew of the dangers and did not want to chance being caught with any compromising papers. After chorus rehearsal one night, he walked toward the Métro along with Geneviève Noufflard, a twenty-three-year-old student at the Conservatoire and member of the chorale, as the two had done on several previous occasions. Noufflard had noticed that Jacques carried a very fine, soft Moroccan leather briefcase. He turned to Noufflard and said, "I have something to ask you . . . I am leaving tonight to [do] something dangerous; and I would like you to keep this for me." He handed Noufflard the briefcase, adding that, if he did not come back in a number of days, "Please see that my wife is told."

Noufflard understood and agreed.

De Bénouville had set up a system for sneaking into Switzerland through the border town of Annemasse. After arriving at the train station, Jacques and other delegates were led across the border by French customs agents who had been recruited by de Bénouville. They were received on the other side by members of Swiss border security, Services de Renseignements Suisse (SR), who were complicit in the scheme, and then were taken the few miles to Geneva, where they were welcomed by Philo and the Geneva delegation.

Along with Jacques and de Bénouville, the meeting brought together Gen. Pierre Dejussieu ("Pontcarral"), chief of staff of the Secret Army; Louis-Eugène Mangin ("Grognard"), military delegate of the French National Committee; Pierre Arrighi ("Charpentier"), representing military groups of the northern zone; Marcel Degliame ("Fouché" or "Dormoy"), chief of L'Action Immediate, comprised of several militant groups; and Emmanuel d'Astier de la Vigerie, of the MUR. The delegates brought a number of reports with them to meetings with the Allies, including a table summarizing the forces that were available presently for direct action and the projected effectiveness of the Secret Army and the Resistance by the time of an Allied landing. Of the roughly 200,000 men who could fight in the Secret Army and the FTP, fewer than 15,000 had any kind of arms. To attain their potential effectiveness, they would need altogether more than 80,000 submachine guns or automatic pistols, a quarter of a million grenades, and 40 tons of explosives per month. Jacques was promised arms for the FTP and given the instructions on how to inform his contacts of where the parachute drops were to take place.

After the meeting, the men were to go back over the border in pairs. Jacques made it safely, but Degliame and Dejussieu did not. Two Swiss officers, who had accompanied the two Frenchmen to the border and shown them through a passage in the barbed wire, reported that they had been picked up by a German patrol on the other side. De Bénouville sent out the alert through his network, as the two men carried vital knowledge of the whole picture of the Resistance, and no one could predict what they might divulge under torture. Action teams began plotting how to free them from the Germans—risky missions that were certain to incur casualties and retributions. Fortunately, because the prisoners were carrying large amounts of money, their captors took them at first as merely smugglers and did not realize they had two major leaders of the Resistance in their hands. Quick-thinking French customs agents persuaded the Germans to hand the criminals over to French police for prosecution. After paying a heavy fine, Degliame and Dejussieu were released and took the next train to Paris.

"Bauchard" the Journalist

With a little inspiration from his friends, Camus was also moving into action. During his stay at Le Panelier, he visited nearby Lyon to see Pascal Pia, who was deputy to Marcel Peck, chief of the Lyon region for Combat. Pia introduced Camus to his friend Francis Ponge, a Resistance poet who stayed at an apartment belonging to the sister of René Leynaud, another poet, who was one of Combat's leaders within Lyon. Camus struck up friendships with both men, exchanging books and writings. When Camus went into Saint-Étienne for his pneumothorax treatments, he arranged to meet Leynaud there and to wander the grim little city with him for hours. Leynaud told Camus that he had stopped writing, that he would do that "afterward"—meaning after the war. In the little apartment in Lyon, they often talked about literature, right up until curfew time, when Leynaud would excuse himself to go sleep in a safe house.

Through Ponge and Leynaud, Camus met René Tavernier, who ran a literary review out of his home in a suburb of Lyon, which also served as a clandestine meeting spot for a group of Resistance writers called the Comité National des Écrivains (CNE; the National Writers' Committee). Camus joined such notable fellow members as Jean Paulhan, François Mauriac, and the poet Louis Aragon; the latter even lived at the house for more than a year during the Occupation. While Camus was too sick to participate in the field, his contact with so many resistants and fellow writers inspired him to compose an article in July 1943 for *La Revue Libre,* a periodical published by the Franc-Tireur resistance movement. In his anonymously published "Letter to a German Friend," Camus addressed a fictional former comrade from whom he had been separated for the previous five years. He spoke of their respective love of country and foretold the defeat of Germany:

> We shall meet soon again—if possible. But our friendship will be over. You will be full of your defeat. You will not be ashamed of your former victory. Rather, you will longingly remember it with all your crushed might. Today I am still close to you in spirit—your enemy, to be sure, but still a little your friend because I am withholding nothing from you here. Tomorrow all will be over . . . I want to leave you a clear idea of what neither peace nor war has taught you to see in the destiny of your country.

After recounting all that France had to suffer up till that time—"humiliations and silences, with bitter experiences, with prison sentences, with executions at dawn, with desertions and separations, with daily pangs of hunger, with emaciated children, and, above all, the humiliation of our human dignity"—Camus asserted that France, not Germany, would prevail. He wrote:

> I belong to an admirable and persevering nation which, admitting her errors and weaknesses, has not lost the idea that constitutes her whole greatness . . . I belong to a nation which for the past four years has begun to relive the course of her entire history and to take her chance in a game where she holds no trumps. This country is worthy of the difficult and demanding love that is mine. And I believe that she is decidedly worth fighting for since she is worthy of a higher love. And I say that your nation, on the other hand, has received from its sons only the love it deserved, which was blind. A nation is not justified by such love. That will be your undoing. And you who were already conquered in your greatest victories, what will you be in the approaching defeat?

Camus's opportunity to fight for the country he loved arrived later that fall, after he moved to Paris. While he had hoped to find a way to Algeria, it was a job as a reader at his publisher Gallimard that brought him back to the capital. And it was his old friend Pia who made the key introductions that brought Camus officially into the Resistance. Hunted by both the Gestapo and by Vichy, Pia (aka "Renoir") had moved to Paris in August. At the time, Combat was looking for an editor to ramp up publication of its namesake newspaper. With his experience, Pia was a natural choice, but he had been given new duties as secretary-general of the MUR. Pia thought right away of Camus and decided to introduce him to Jacqueline Bernard, who was the executive secretary of the section of Combat involved in publishing and circulating the newspaper. Bernard was an early adherent of Combat, having been recruited in 1941 to type some of its early bulletins. She was the daughter of well-off Jewish parents who lived in Lyon (in the former unoccupied zone); her whole family was committed to the cause. Within a few days of her joining, her father, Colonel Bernard, had given 50,000 francs to the movement. Her brother Jean-Guy joined Combat as well.

Camus's first meeting with Bernard was a typical clandestine rendezvous. Bernard's parents' former maid was the concierge of a building on the rue de Lisbonne. She allowed the group to use the back room of her small ground-floor apartment. In addition to Bernard ("Auger" or "Oger"), Combat's printer André Bollier ("Vélin") was present. Pia brought Camus, who introduced himself as "Bauchard." Bernard noticed how undernourished the pale man in the worn-out suit looked, but that was not exceptional in food-rationed Paris. Sitting at the end of a narrow table, "Bauchard" explained that he "had already done a little journalism" but that he would be happy to do page layouts, write articles, or whatever would make him useful. He listened carefully to Bernard's and Bollier's explanations of how the paper was produced. Neither Bernard nor Bollier had any idea that they were speaking to a writer whose recent books had been widely discussed in literary circles, nor did Camus know whom he was addressing. And just as well, for the Gestapo was crawling all over Paris, constantly arresting resistants, often by following people to or from meetings just like the one in the concierge's apartment and then extracting the names and addresses of their companions under torture. Camus and his new associates left the meeting without incident, and "Bauchard" would indeed prove to be a useful addition to *Combat*'s staff.

CHAPTER 13

DOUBLE LIVES

Whatever there be of progress in life comes not through adaptation but through daring.

—HENRY MILLER, "Reflections on Writing"

THE ATMOSPHERE IN AND AROUND PARIS HAD GROWN INCREASingly bleak and tense toward the end of 1943. For Parisians, the winter would be the most miserable of the Occupation, as supplies of food and coal dwindled to their lowest of the war. For the Germans, the reversal of their momentum had erased their once smiling and courteous demeanor. The Allies' advances in North Africa, Italy, and the USSR meant that the second battle of France was just a matter of time—but when? "When are they landing?" was the question on everyone's mind, and even lips.

The waiting was agony. For those trying to stay out of sight of the Germans, like Odette Monod and Lise Teissier, living in a strange town among strangers, there was the constant fear of being denounced. For members of the Resistance, it was a matter of holding on against the unrelenting efforts of the Gestapo as their comrades were arrested, deported, or executed. Just a month after the summit in Geneva, Pierre Arrighi ("Charpentier") was arrested at the Café Le Triadou on the boulevard Haussmann and sent to Buchenwald concentration camp. Marcel Peck, Pia's Combat chief, who had escaped the Nazis three times previously, was also snatched in November and disappeared without a trace. In December, the pressure was further intensified when the Germans insisted that Pétain revamp his government with stronger men who were willing to do whatever was necessary to crush the Resistance. He appointed Joseph Darnand, the head of Vichy's paramilitary force (the Milice), as secretary-general of the forces for the maintenance of order. A ruthless, dedicated anti-Semite, Darnand had even joined the Waffen SS a few months earlier. Special courts-martial were set up under Darnand's authority to try resistants. The penalty for being caught with arms was immediate execution. In an interview with *Paris-Soir,* Darnand declared an all-out war on the rural guerrilla groups of the Resistance known as the *maquis:* "The bands of *maquis* will be attacked everywhere, with sufficient numbers of forces and the means necessary." Since his agents spoke French, they were feared perhaps even more than the Gestapo.

In the face of such increasing pressure, Jacques Monod tried to juggle his double life. He continued his laboratory work, but not at the Sorbonne. A colleague who knew of his Resistance activities and who was a member of the Réseau Vélites, a Resistance group operating out of the nearby École Normale Supérieure, was arrested in November. Fearing that his identity might be disclosed, and knowing that the authorities were aware because of the Nordmann affair in late 1940 that Monod worked at the Sorbonne and lived nearby, it was too risky for him to frequent the university. André Lwoff, who knew of Monod's clandestine activities (because Monod had connected him with the Resistance), offered Monod space in his laboratory in the attic of the Pasteur Institute. Monod shifted his experiments to the more esteemed institution and even managed to give a seminar there in December. The Institute also provided some presents with which Monod surprised the twins. Always happy to see their father on his overnight visits to Saint-Leu-la-Forêt, they were delighted when he arrived home, put his arms on the table, and shook out a white laboratory mouse from each sleeve of his overcoat.

Monod did decide, however, that his conducting job was too much, and stepped down, much to his and Dufourcq's disappointment. He reported to his parents, "I am living a terribly austere and absorbed life, with, thank God, the joy of being able to be very often with Odette and the little ones."

Entrer Dans le Bain

Monod did not lose contact with his entire chorus, however. Geneviève Noufflard tracked him down in January with a bold request—she wanted to join the FTP and to work with him. The young flutist and singer was unwilling merely to wait for liberation; she thought that it was very important to get ready for the battle ahead. She wanted to join a group that was committed to "immediate action" against the Germans—including sabotage and combat. Ever since the episode with Monod's briefcase, she knew that he was involved in serious work.

Monod tried to dissuade her, telling her of the terrible risks involved—of arrest, torture, deportation, and death. But Noufflard was very well aware of the dangers, for what Monod did not know was that the music student had long been supporting the Resistance and had many friends and acquaintances who were involved. Right after the fall of France, she and her parents retreated to Toulouse, in the nonoccupied zone, where they hid escaped prisoners and men wanting to go fight with de Gaulle. Noufflard snuck back and forth across the demarcation line several times, sometimes while smuggling documents. René Parodi, a founder of the movement Libération-Nord, was a family friend who was subsequently arrested and died after being tortured at Fresnes Prison. After returning to Paris to resume her studies, Noufflard visited regularly with Franz Stock, the German chaplain of the notorious prison, posing as the fiancée of one prisoner in order to try to learn the fate of a friend's brother. Her parents' home in Paris—which she shared in late 1943 with her sister Henriette, a medical intern, as well as her father, a painter (her Jewish grandmother and mother were living in Normandy under false names)—had served as a safe house for all sorts of characters on the run from the authorities. Despite being just half a block away from the Hôtel Matignon, the headquarters of the collaborationist government and Laval's residence, the grand eighteenth-century *maison* on the elegant rue de Varennes was a haven for Jewish refugees, *réfractaires*, Resistance members, and downed Allied airmen.

The arrival of the first aviator, a member of a B-17 crew named John Spence, caused a great deal of excitement in the household. Among the first American fliers to be hidden by the Resistance and escorted out of France, Spence's experience was typical of the more than 1,500 Allied airmen who evaded capture by the Germans and escaped from France to freedom. His journey required the assistance of a large number of

French people: ordinary citizens whom he just happened to encounter; committed members of Resistance networks; and unaffiliated sympathizers, such as the Noufflard sisters. And because of the large number of people involved, many of whom knew one another barely if at all, evasion work was one of the most dangerous kinds of Resistance activity. Harboring or aiding the escape of an enemy soldier was cause for deportation or execution.

On the morning of January 23, 1943, Second Lt. Spence's plane, the *Green Hornet,* was one of twenty-one B-17s of the 303rd Bomber Group that took off from Molesworth, England, to bomb the Lorient port area and the Brest U-boat pens in Brittany. After dropping their bombs, they were hit by flak, lost two engines, and fell behind the formation. They were quickly pounced on by about a dozen enemy Focke-Wulf Fw 190 fighters. Spence, who was the navigator, fired away until his gun went out. His pilot tried to take evasive action, but the plane lost another engine, so he rang the bailout alarm. Spence jumped out at about 4,500 feet and parachuted into a soft mud field. He injured his ankle but was able to walk, so he rolled up his chute and started running east. He soon encountered a handful of peasants, as well as his engineer, Sidney Devers.

The two airmen gave away their chutes and kept walking. They were given a drink of cider at a farmhouse before reaching the town of Paule. Another Frenchman gave them food and clothing before they continued farther, and they were fed again by another farmer in Glomel, eventually sleeping in a haystack on the outskirts of Bonen. They walked for three more days, bypassing a German garrison of 1,000 men, and getting fed in turn by several families. Finally, they met an English-speaking Catholic nun who gave them a note to present at the Château de Quellenec near the village of Saint-Gilles-Vieux-Marché. They were received at the eighteenth-century castle by a sixty-seven-year-old woman, Simone de Boisbossel, the countess of Keranflec'h, who gave them a hot bath, food, and a warm bed before taking them the next day by car to a train station, where they used money in their escape kits to buy train tickets to Paris. The ticket agent ignored Spence's Tennessee accent and gave them their tickets without saying a word. Then, as the men were boarding, he whispered to Spence: *"Bon voyage."* The countess guided the men to her daughter's home just outside Paris.

Spence and Devers were handed over to the care of Robert Ayle, who was a "helper" with the "Comet Line," one of the Resistance networks that escorted soldiers, aviators, and others sought by the Germans out of France to neutral Spain by various routes. Once in Spain, British officials would get them to Gibraltar and then to England. Ayle was to make the arrangements for Spence's and Devers's passage. It was, however, a dangerous journey on foot through the Pyrenees. Only two weeks earlier, Andrée de Jongh, the Belgian woman who cofounded the Comet Line in 1941, was arrested near the Spanish border with three evaders in the course of attempting her thirty-third crossing. A new route had to be scouted before Devers and Spence could make their attempt. In the meantime, for security reasons, the fliers were given false identity papers and moved around to different houses in Paris. In order to strengthen his ankle for the rigorous midwinter hike, Spence was taken on walks around the city and looked after by reliable friends of the organization.

A colleague of Henriette Noufflard brought Spence to spend the day on the rue de Varenne. The sisters were very excited to actually see an American, in the flesh, in their house! Americans had joined the Allied bombing effort only a few months earlier. In the anguish of the Occupation, Spence represented the status of the airmen who were risking their lives to strike back at Germany. The sisters made sandwiches with everything they had in their bare pantry, and managed to find some cream for their last box of tea. Both Henriette and Geneviève spoke excellent English, so Spence was able to entertain them with stories of his prior bombing missions, his bailout, and his evasion. After a long afternoon of conversation, it was Geneviève's responsibility to walk Spence back that evening to the home of Dr. Jules Tinel on the boulevard Saint-Germain. Tinel had sheltered a succession of evaders, and his son Jacques was also one of Comet's agents. Noufflard led Spence through mostly empty back streets while they spoke quietly in English. She shifted to French when they approached other pedestrians, to which Spence replied by smiling and chuckling.

Two weeks after arriving in Paris, Spence, Devers, and three other evaders were escorted south by train by Jean-François Nothomb, who had scouted the new route through the Pyrenees. While such a large group would appear to increase the risk, the strategy was to fill a six-person compartment on the train so that no other passengers were present. After some snags in Spain, Spence made it safely back to England on March 15, 1943.

Within less than a year, however—by the time of Geneviève and Monod's conversation about her joining the FTP, Robert Ayle had been arrested by the Gestapo, along with de Jongh's father (both would face the firing squad in March 1944); Jules Tinel had been arrested and imprisoned; his son Jacques had died at Mittelbau-Dora prison camp; and Spence's escort Nothomb was arrested that January, tortured, and imprisoned—just a few of the hundreds of Comet members who were arrested or the more than 150 who perished for aiding complete strangers in the Allied cause.

THE NOUFFLARDS WOULD continue to shelter whoever was sent their way, including another American bomber's navigator, but Geneviève was determined to join an organization and *entrer dans le bain* (get into the bath), in the slang of the Resistance. She told Monod that she was aware of the risks and had thought her decision through. She was, in

Geneviève Noufflard, from her wartime identity card. (Courtesy of Geneviève Noufflard)

fact, very nervous, as the prospect of being tortured terrified her. She explained to Monod that no one could know how one would react in that circumstance. She knew people who had been tortured and could not have predicted who would have buckled and who'd remain silent.

Monod told her, "Okay, all right, come tomorrow."

GENEVIÈVE WAS PUT to work right away as a "liaison agent." The Resistance could not use the telephone or the mail for fear of being intercepted, so people were used to transmit information. In order to exchange orders, letters, or documents, liaison agents met in the streets, or at cafés, or in churches, or in offices—anywhere where two people could have a casual encounter without attracting notice. Noufflard walked or rode a bicycle to each rendezvous. It was too dangerous to take the subway when carrying any document because the trains and stations were the frequent targets of police controls and searches.

To conceal whatever she was carrying, Noufflard hid items in her clothes, or in shopping bags under a load of vegetables, or under feminine articles that a potential searcher might be too embarrassed to handle. It was difficult to pass documents hand-to-hand in the street without being noticed, so papers were often disguised as mail, enclosed in envelopes with imaginary addresses and even stamps on them. These were best passed inside a post office, or by innocently asking the contact for the favor of mailing a letter. Monod showed Noufflard one of his favorite places for the safekeeping of sensitive papers: just outside the door of his lab at the Sorbonne there was a mounted giraffe whose hollow leg bones accommodated documents that the Gestapo would have loved to find.

The job required a great deal of discipline. Normally, everyone Noufflard was to meet had been personally introduced to her by an FTP member she had also met. If she was meeting someone she did not know, it was critical that she recognized the right person. That was done by giving some prearranged sign or code words. So, for example, when Geneviève was to meet someone whom she was told would be carrying a certain type of wine, she was to say to him: "Where is the Parc des Princes?" And the reply was to be: "You are there, mademoiselle." For extra safety, they each also carried bus tickets with numbers on them.

The work entailed constant danger. It was most critical to make sure that she was not being followed, for if the Gestapo or the Milice tracked her, she would lead them to other agents and to their bosses, and all would be arrested. The Germans had agents in the streets, and knowing that everyone was compelled to work, they watched for revealing behavior. It was dangerous even to wait for someone on a sidewalk. She might be asked for papers and searched, or those watching might wait to see whom she was waiting for and follow them. The rule was not to wait much more than five minutes at a given rendezvous. There was usually a backup meeting, called a *repêchage* (literally, "re-fishing"), if one missed an appointment.

Next to the constant anxiety, the hardest strain was on her memory. Noufflard could not write down any details, and neither could she meet people at the same place twice, just in case either was observed the first time. So she had to memorize all of the times, places, people, and aliases to meet at her appointments, sometimes up to fifteen different meetings in the course of a day. Then she had to memorize a whole new set of instructions for the next day.

Arrests

Despite the emphasis on security in the FTP, there were lapses. Marcel Prenant, alias "Auguste," was living a double life similar to Monod's. Along with his role in the FTP, he had a laboratory at the Sorbonne, his Jewish wife and mother-in-law were living outside of Paris, and he was staying away from the family apartment on rue Toullier near the Sorbonne. He was living instead in Port Royal Square, with his friends and fellow resistants Vladimir and Joulia Kostitzine, the latter of whom worked in his lab. On the morning of January 28, 1944, he had a rendezvous on a quiet street in the Paris suburb of Gentilly with a delegate who went by the name of "Robin." The FTP man was responsible for one of the most important regions of the country—the sector that included Normandy, one of the potential sites of the hoped-for Allied landing. Prenant saw Robin each week, and during the interval Robin would visit his sector. In this particular encounter, Prenant was planning to reproach Robin, as he had learned that the delegate was marking his meetings down in a notebook.

Prenant entered the empty street but saw no sign of Robin. Unconcerned, he kept walking until he saw four men come out from behind some doors, pistols drawn. One of them said, "German police." Prenant knew it was hopeless. They handcuffed him and put him in a waiting car. Inside, he saw Robin, collapsed in the corner, having clearly been beaten. Prenant knew instantly that he was a victim of Robin's notebook. They were driven along the Left Bank of the Seine until they reached the rue des Saussaies, headquarters of the Gestapo in Paris.

Prenant was pretty sure what would come next. He was seated in front of an officer, behind whom he could see a collection of truncheons of various types. Leaning back against the display, standing mute, arms crossed, were three scar-faced brutes with their eyes riveted upon him. Prenant had thought many times about the attitude he would take if he fell into the hands of the Gestapo. He knew that he was done for; his single concern was not to give up any information that could be used to grab any of his comrades or used against the FTP in general. Prenant had no compromising documents on him. But his lab was a different matter, and he was sure that it was already being searched. One consolation of that fact was that the search would alert his friends that he had been arrested.

The interrogation began. The officer asked what each of the keys on his keychain were for, what the abbreviations meant on the papers he had in his pockets, and what his role was in the Resistance. The trick was to give away small things, either facts that the Gestapo knew already or that would at least give the appearance of being truthful. Prenant answered each question, including admitting that he was part of the FTP. After a brief lunch break, the officer informed Prenant that they had searched his lab and his home, where they arrested his son. "But you don't live at your home?" the officer asked.

"No," Prenant replied.

"Then where do you live?" his interrogator asked.

"I will not tell you," Prenant said. He did not want to say where he was living, or at least he wanted to hold out until his roommates had sufficient time to get away.

"We shall see about that," the Gestapo man said.

The officer was determined to find out who and where Prenant's associates were. After trying to coax the address out of Prenant, he said, "That's enough! Either you give me this address immediately, or I will put you in the tub. Do you know what that is?"

Prenant had heard of *la baignoire* as a method of torture in which one's head was held underwater until just before drowning. The officer gave him another chance; Prenant refused. He and his henchmen then took Prenant upstairs to the fourth floor of the building and into a small bathroom. After asking for the address one last time, which Prenant refused again, the men seated him on the rim of the tub and tossed him in. One of the men then grabbed Prenant's ankle chain and yanked it up high, plunging his head under the very cold water. They let him up after a short time, then submerged him again for much longer. Prenant deliberately thrashed about, splashing his torturers. They let him up again, then plunged him down again, and continued to repeat the process. After the eighth cycle, Prenant decided that he was risking an escalation of the torture and that he had bought enough time for his friends to escape. When the officer asked if he was ready to answer, he said "Yes," and gave the Port Royal Square address. While Prenant was allowed to dry off, the other men ran off to his hideout. They found it empty and did not find any incriminating papers in Prenant's room. The Kostitzines had taken documents out of his briefcase and torn them into little pieces before flushing them down the toilet and making their getaway. After a brief interrogation the next day, Prenant was transferred to Fresnes Prison; he was eventually sent to Neuengamme concentration camp.

Malivert

The loss of Prenant was a blow to the FTP, and while the leadership had confidence in Prenant's ability to defy his interrogators, no one could be certain whether the damage would be limited to Prenant and

Robin. The arrests ratcheted up the anxiety and tension throughout the organization, but just like other Resistance groups that had suffered arrests, the movement had to continue and the lost leaders had to be replaced. The FTP promoted Georges Teissier, Monod's brother-in-law, to take Prenant's place as chief of staff. At the same time Monod was given a new role as FTP's delegate to a new organization, the French Forces of the Interior (Forces Françaises de l'Intérieur; FFI), which united the various Resistance military groups: the Secret Army (itself a combined force of Combat, Libération-Sud, and Franc-Tireur); the ORA (L'Organisation de Résistance de l'Armée); and the FTP. Created in large part in anticipation of the battle for liberation, the FFI was designed to coordinate Resistance activities both before and after the Allied landing. Gen. Marie-Pierre Koenig of de Gaulle's staff in London was its overall head, and Pierre Dejussieu ("Pontcarral") was appointed its chief within France.

There was a commandant in charge of each of roughly twenty regions of France. And in each region, there were officers in charge of different military departments such as intelligence (Deuxième Bureau; G2) and operations (Troisième Bureau; G3). Monod was made head of operations for the Paris region. The Deuxième Bureau's agents were to collect information on potential targets. These would be passed to Geneviève Noufflard, who would then bring them to Monod, who was in turn responsible for selecting which actions would be taken and for giving orders to action teams.

Monod discovered that he could not maintain his double life for long. Along with his new responsibilities, and on the heels of Prenant's arrest, came news of the arrest of another colleague from the Réseau Vélites, Raymond Croland, who knew Monod's identity and activities. The cofounder of the network and a member of the same department as Georges Teissier at the École Normale Supérieure, Croland was carrying out experiments throughout the war on the induction of mutations in bacteria and fruit flies by X-rays. Late in the afternoon of February 14, while Croland was waiting for a contact from London in the office of his fourth-floor biology laboratory, five French and German policemen in plainclothes arrived at his building, blocked off the entrances and exits, and arrested him.

The establishment of the FFI also increased Monod's exposure because, as it was another new Resistance superorganization, the FFI and those within it instantly became priority targets for the Gestapo. Monod decided that he had to plunge fully into clandestine life (*l'illégalité*)—sleeping at different addresses and staying away from the apartment on rue Monsieur-le-Prince. He also changed his alias. The creation of the FFI would not instantly solve long-standing problems such as the shortage of weapons. Monod was skeptical that he or his bureau would have the means to make an impact on the enemy. So he chose an ironic pseudonym, named after a character in Stendhal's novel *Armance* who was impotent: "Malivert."

He kept working on the weapons problem. He made another trip to Switzerland, but this time it was in midwinter and, worse, the area around Annemasse (the usual crossing point) was swarming with Darnand's militiamen. Monod, de Bénouville, and "Miranda," the defense chief of the *maquis,* took an alternate route on foot through the snowy Jura Mountains. De Bénouville was very impressed with and appreciative of Monod's calm demeanor; with the collar of his Canadian jacket turned up against the wind and cold, Monod looked "as calm as though he were on his way to the subway." On their way back, the men found bicycles in the town of Morvillars and pedaled to Belfort, which was full of German soldiers. They reached their train, blended in with the other passengers, and made it back to Paris again safely.

Such adventures and the requirements of a clandestine lifestyle were not compatible with continuing work in the lab, even if Monod was hiding out in the attic of the Pasteur. Just as he had in the winter of 1940, Monod had to step away again from his research. The timing was unfortunate, for, despite all of the distractions, he had been making important progress. During the winter of 1943–44, he had begun a new set of experiments with Alice Audureau, a graduate student in Lwoff's laboratory at the Pasteur. Probing further into the phenomenon of double growth that he had rediscovered in his thesis, Monod and Audureau followed up on reports that strains of the *E. coli* bacterium that did not metabolize lactose could be coaxed into doing so by being grown on media containing lactose. There was a conflict in the literature as to whether this was due to an effect on an enzyme that was promoted by the presence of the sugar, as most thought, or due to a rare mutation. Audureau and Monod found that such strains were indeed mutants, which showed that the ability to metabolize lactose was a genetically determined characteristic of the bacteria, not a property promoted by the sugar. Moreover, such mutants were not rare. In fact, Audureau isolated several from a stool sample taken from her boss, Lwoff, which she and Monod mischievously dubbed *E. coli* ML, which stood for *mutabile in Lwoffi* or for *merdae Lwoffi,* depending on the audience. Genetics, and especially bacterial genetics, was in its infancy at the time, so the use of genetic investigations to clarify a physiological puzzle like enzyme adaptation was novel and powerful.

Monod managed to squeeze in a few more days in the lab before his Resistance work overtook him completely.

As his clandestine activities accelerated, he only dropped into the lab for odd supplies, like rubber stoppers. Monod needed them as silencers—not for weapons, but for a duplicating machine. He and Geneviève Noufflard sometimes had to make thousands of copies of bulletins or other documents. It was very difficult to do so without attracting attention.

Noufflard's building also housed offices of the Department of Agriculture, and she learned that they had a splendid duplicating machine. To use it, they had to sneak in at night during the curfew hours and carry out the job without being seen or heard. Her concierge was in on the scheme and agreed to be the lookout.

She and Monod found that they had to move the machine into the corridor so that they could not be seen from the street. But there were people living on the floors below, and both the moving and operation of the machine made a lot of noise. If Monod and Noufflard were caught red-handed printing communiqués on the latest FTP activities, they would each be in dire straits. So Monod plucked two large rubber stoppers from flasks in the lab, cut them into four pieces, and made "shoes" for each of the four feet of the machine. At midnight, he and Noufflard took their own shoes off and, carrying ink, stencils, and two suitcases full of paper, crept into the offices, moved the machine, and made their copies. Finishing at dawn, they moved the machine back, packed up their copies and spent supplies, and lugged them out. The policeman on the street only saw a very affectionate couple who appeared to be heading off on a trip.

Combat

Camus was able to maintain a double life as a staff member of *Combat* and as a writer on the Paris scene. With so many friends in the Resistance, he found a variety of ways to contribute. Soon after he settled into a room at the Hôtel Mercure, just two hundred yards or so away from the Hôtel Lutétia, home of the German Wehrmacht, he sent a note alerting the Fayols that he was sending to them one of his friends from Oran who had also been trapped in France. This particular friend was Jewish, so Camus indicated that she "has delicate health from a hereditary infection." His new job on the reading committee at Gallimard gave him an office in which to work on his own articles and to hide documents for others. It also put him into contact with a number of other Gallimard writers, some of whom, like André Malraux, were wanted men. One day, Camus asked quietly around the office for help in lodging a "very important person." Waiting outside on the street was Malraux, who was escorting the VIP—Capt. George Hiller, a British officer with the Special Operations Executive (SOE) who was arranging parachute drops to the Resistance. An editor took in Hiller.

In the course of his first few months back in the capital, Camus actually became *more* visible as he wrote for the stage and socialized with other writers and artists. He had met Jean-Paul Sartre when he attended the latter's play *Les Mouches* (*The Flies*) on one of his journeys into Paris while staying at Le Panelier. As he settled into his new job, he began meeting regularly with Sartre and his companion Simone de Beauvoir in the cafés near Saint-Germain-des-Prés. Both the short, walleyed existentialist and the pioneering feminist were charmed by Camus's good humor and enthusiasm, and impressed by his talents and intelligence. In talking about the war, politics, and the theater, Camus discovered that he had much in common with the couple, despite his much more humble upbringing.

The Café de Flore was a favorite haunt for the trio. Besides its clientele and its convenient location in the heart of the Left Bank, just a few blocks from the Gallimard Building, the Flore had the added appeal in coal-limited Paris of a wood-burning stove. It was there at their first meeting together that Sartre proposed that Camus direct his new play *Huis Clos* (*No Exit*) and perform the male lead. Camus accepted, and they began to rehearse in de Beauvoir's room at the Hôtel la Louisiane. However, the venture was abandoned when the female lead, for whom the play had been commissioned by her husband, was arrested because of her contacts with the Resistance.

Often dining and drinking together, the three unconventional writers developed a sense of solidarity, and Camus was introduced to a wider circle of interesting artists. Camus was put in charge of a cast that included de Beauvoir and Sartre and that was invited to stage a public reading of a surrealist play that Picasso had written in the 1920s. Performed in a mutual friend's living room, the standing-room-

Camus, Sartre, and de Beauvoir in Picasso's apartment. This photo was taken at a reunion of the cast and audience of the reading of Picasso's play, staged in March 1944. Beauvoir is standing at the far right, Picasso is standing third from the right, Sartre is seated at the far left, Camus is to his right. (Bibliothèque Nationale de France)

only audience for *Le Désir attrapé par la queue* (Desire caught by the tail) included Picasso himself, as well as the painter Georges Braque, and a variety of poets, directors, and actors. In appreciation for their efforts, Picasso invited the illustrious cast back to his apartment.

CAMUS WAS FOLLOWING his own prescription and living fully—working at Gallimard by day, seeing his friends in the evenings, working for *Combat,* and carving out a few hours for his own writing. He was working on the rewrite of his play *Le Malentendu* (*The Misunderstanding*), part of his cycle of the absurd, about a man who returns home from living overseas to discover his mother and sister have been taking in lodgers and killing them. He hoped to have it staged that summer.

Camus even found room for a new love. Among the audience the night of Picasso's play reading was twenty-two-year-old Maria Casarès, who, despite her youth, was already a noted and experienced actress on the Paris stage. At the time, she thought Camus was a fine actor, but she could not have imagined that she would soon be tapped for the female lead in his new play. She met Camus at a work meeting in the director's apartment and was immediately drawn to his "extraordinary presence" and his air of vulnerability. Camus was struck by Casarès's dark beauty and her expressive eyes, and was deeply impressed by her talent. Her background was equally admirable. Born in Spain, she was the daughter of Santiago Casares y Quiroga, who served briefly as prime minister and minister of war of the Republic in 1936 until the Civil War broke out. Though just fourteen at the time, Maria volunteered as a nurse in Madrid hospitals, tending to the wounded. She and her mother and father then escaped to France just before the border was closed. After the Germans invaded France, Maria entered Le Conservatoire and began acting. Her acclaimed debut in the title role of *Deirdre of the Sorrows* in 1942 put an end to her formal studies as she became instantly in demand.

An intense and public affair started soon after she and Camus met. Francine was in Algeria, so Camus took Maria around Paris to his favorite cafés and restaurants. One reason why Camus was able to live such a public life while a member of *Combat* was that, unlike Monod's scientific colleagues, none of his colleagues on the newspaper staff knew his true identity. He had no past associations with anyone (except Pia). "Bauchard" simply appeared at staff meetings and did his job. Camus also had protection beyond his pseudonym. The network had provided him with false papers in the name of "Albert Mathé," a journalist who was born near Paris. Jacqueline Bernard was stunned when he eventually told her who he really was.

It was not the case that Camus and his comrades were in any less danger than other resistants. If anyone was followed to the concierge's back room or caught with printing materials, everyone was exposed. Indeed, the Combat movement suffered many blows from the Gestapo during the winter and early spring of 1944, and some landed very close to Camus.

On January 28, the same day that Marcel Prenant was taken, Jean-Guy Bernard and his wife, Yvette Bauman, were expecting Jean-Guy's sister, Jacqueline, for dinner at their apartment on the rue Boissy-d'Anglas. Jean-Guy was head of the N.A.P.-Railways, a branch of the Resistance in charge of infiltrating and interfering with rail transport. Yvette was in charge of social services for Combat and the MUR. Jean-Guy went to answer the door, expecting Jacqueline, but she happened to be running late. The door was pushed in by two Gestapo agents, who then took the couple away. Jacqueline had escaped arrest merely by chance. Yvette Bauman was subsequently deported to Ravensbrück; Jean-Guy died in transport to Auschwitz.

On March 8, the Gestapo caught up with *Combat*'s printer, André Bollier (Vélin), in Lyon. It was actually the second time he had been captured; he had promptly escaped from the French police two months earlier and gone completely underground. Although just twenty-three years old, Bollier was a vital part of the movement. He had devised the scheme in which the newspaper was printed not only at a plant in Lyon but at more than a dozen other presses around the country, which ensured its wide distribution. He also set up a phony company in order to import newsprint stock . . . *from Germany.* He did not shrink away from action, either. He had led a small armed band that broke *Combat* cofounder Berty Albrecht out of a psychiatric hospital in December 1942. After being tortured repeatedly and threatened with execution, Bollier escaped from a military hospital and resumed printing *Combat*.

In late March, the Gestapo struck again at the heart of the movement. Pierre de Bénouville had a network of people and a handful of apartments in Paris that handled daily courier messages, composed coded replies, typed them out, and shipped packets regularly to Switzerland. The Germans caught one of de Bénouville's men in Annemasse with documents that compromised his typist. She managed to alert the network while the Gestapo searched her apartment, but not everyone was warned in time. Claude Bourdet, who had become head of Combat when Henri Frenay went overseas, and whom Camus had also met, rang at the typist's apartment and was greeted by a revolver pointed at his face. The Germans found more addresses and began to set traps. Miranda was grabbed off the street on his way to a meeting. Alain de Camaret ("Nizan"), de Bénouville's longtime friend and lieutenant who knew everyone in the organization, was snatched at another apartment. The Gestapo almost nabbed de Bénouville as well. He and his wife packed their bags in a hurry and left their apartment just in time. They took refuge with a friend in Paris who was not in the Resistance and would not be in any address book the Germans might find.

THE GERMANS' AND Joseph Darnand's aims were to crush the Resistance. Yet, despite the loss of so many commanders as well as foot soldiers, movements carried on. The Comet Line was rebuilt several times, de Bénouville rebuilt his network, and *Combat* continued to publish. Indeed, in March 1944, Camus authored his first (anonymous) editorial for the newspaper, in which he urged readers, particularly those who had been bystanders up to that point, to join the Resistance and to battle the Germans. Under the headline "Against Total War, Total Resistance," Camus wrote: "You cannot say, 'This doesn't concern me.' Because it does concern you. The truth is that Germany has today

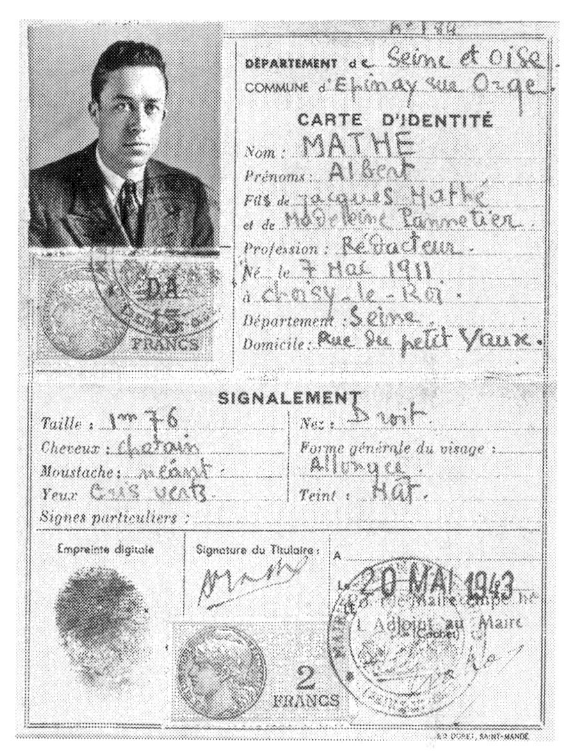
DÉPARTEMENT de Seine et Oise.
COMMUNE d'Epinay sur Orge.

CARTE D'IDENTITÉ

Nom : MATHE
Prénoms : Albert
Fils de Jacques Mathé
et de Madeleine Pannetier.
Profession : Rédacteur.
Né le 7 Mai 1911
à Choisy-le-Roi.
Département : Seine.
Domicile : Rue du petit Vaux.

15 FRANCS

SIGNALEMENT

Taille : 1m76
Nez : Droit
Cheveux : Chatain
Forme générale du visage : Allongé.
Moustache : néant.
Yeux : Gris verts.
Teint : Mat.
Signes particuliers :

Empreinte digitale
Signature du Titulaire :
A
Le 20 MAI 1943
Pour le Maire empêché
L'Adjoint au Maire
2 FRANCS

Camus's false identity card, in the name of Albert Mathé, writer. All of the information on the card—birth date, place, parents—is false. (Courtesy of Collection Catherine et Jean Camus, Fonds Camus, Bibliothèque Méjanes, Aix-en-Provence, France. All rights reserved.)

not only unleashed an offensive against the best and proudest of our compatriots, but it is also continuing its total war against all of France, which is exposed in its totality to Germany's blows."

Recounting recent episodes of German reprisals, Camus continued:

> These dead Frenchmen were people who might have said, "This doesn't concern me." . . .
>
> Don't say, "I sympathize, that's quite enough, and the rest is of no concern of mine." Because you will be killed, deported, or

tortured as a sympathizer just as easily as if you were a militant. Act: your risk will be no greater, and you will at least share in the peace at heart that the best of us take with them to the prisons . . .

Total war has been unleashed, and it calls for total resistance. You must resist because it does concern you, and there is only one France, not two. And the incidents of sabotage, the strikes, the demonstrations that have been organized throughout France are the only ways of responding to this war. That is what we expect from you.

CHAPTER 14

PREPARATIONS

If they attack in the west, that attack will decide the war. If this attack is parried, then the whole story is over.

—ADOLF HITLER, December 20, 1943

NO MATTER HOW MANY MIGHT JOIN THE EFFORT, THE RESISTANCE could not hope to dislodge the Germans on its own. The French were vastly outgunned and outnumbered. In the spring of 1944, the Germans had fifty-eight divisions in France comprising some 750,000 soldiers, whereas of the roughly 40,000 resistants in the greater Paris region, fewer than 1,000 were armed. The ultimate military value of the sorts of actions that Camus urged, as well as the intelligence being fed to Switzerland and London, was to help prepare the way for an Allied landing by damaging German capabilities. The sabotage of railroad, telephone, and electricity lines was intended to hamper German war production, the movement of supplies, and the mobility of troops, while intelligence gathered on the locations of fighting units, supply and munitions depots, and key factories within France was used to guide the Allied bombing effort.

The impact of sabotage carried out by the Resistance was substantial, especially on the railroads. One argument that Resistance leaders had been repeating to Allied commanders was that their small action teams were, in certain circumstances, much more precise and effective at destruction than high-altitude bombing. For example, it was much easier for a few men or women who knew the local landscape and railroad timetables to sever a rail line, derail a train, or blow up a locomotive than for a bomber to hit such small targets from 20,000 feet. Moreover, such acts of sabotage avoided the collateral damage of bombing raids on civilian lives or property, as well as avoided risking aviators' lives. Statistics supported those claims. From January to

March 1944, the Resistance destroyed more than 800 locomotives, compared to 387 hit by the Allied air campaign. Altogether, from June 1943 to May 1944, the Resistance destroyed 1,822 locomotives and 2,500 freight cars and damaged at least 8,000 more.

The cutting of rail lines was also widespread, as small amounts of well-placed explosives were sufficient to destroy short sections of track. In October and November 1943, Vichy police reported more than 3,000 attempted attacks on the rail system, with more than 400 successful acts in November causing major damage, including more than 130 derailments. Individual lines were hit repeatedly, as soon as possible after repairs to prior damage had been made. For example, the Paris–Brest rail line, which carried supplies for the fortification of the German defenses along the so-called Atlantic Wall, was cut twenty-four separate times by saboteurs from January to May 9, 1944.

No one in the Resistance knew how long such efforts would need to be sustained, however, because no one in France knew the time or the place of the planned Allied landings. Not even de Gaulle in Algiers or General Koenig in London was told of the plans or the date. With arrests of Resistance leaders so frequent, Allied planners deemed it too dangerous to disclose any specific information. Instead, the FFI and other, unaffiliated groups were told to listen to the BBC on the first, second, fifteenth, and sixteenth of each month for prearranged coded messages that would alert them that invasion was imminent. They were then to listen for subsequent confirmatory messages that would indicate that the landings would be taking place within forty-eight hours. The latter would also be the signals to each region or group to begin previously planned operations against rail, road, and communications targets, dubbed Plan Vert (green), Plan Tortue (tortoise), and Plan Violet (purple), respectively.

Monod and other operations men thus had the dual tasks of planning and carrying out actions for some indefinite period prior to the landings while also developing plans for the crucial hours, days, and weeks after the invasion. As months passed without the Allies landing and without any signal from the BBC that the event was imminent, and as arrests of leaders and comrades mounted, the daily tension and anxiety of Resistance work took a toll. It was difficult not to feel discouraged at times. One evening in April, Monod and Noufflard returned to her home to finish some work, exhausted again from an already long day. Suddenly, sirens went off and they heard explosions in the distance. They hurried up to the roof of the building and looked east over the city. They were transfixed by the bright flashes of one explosion after another lighting up the night sky. A continuous roll of thunder shook the building, amplified by the roar of plane engines and the crackling of antiaircraft fire. It was a massive raid on marshaling yards just outside the city. "Here they come!" Noufflard thought, hoping that such large-scale and systematic destruction, so close to the city, meant that the invasion might not be too far off.

Marshaling yards that fed as many as ten or twelve rail lines and covered many acres were key targets that were best attacked by heavy bombers. The combined effect of Resistance saboteurs and Allied bombers was to reduce rail traffic dramatically across France in the first half of 1944. The primary military effects of that reduction were to curtail the flow of supplies such as concrete and steel for the construction of German defenses along the west coast, and to force the diversion of thousands of defense construction personnel to repairing the railroads.

For Three Hours They Shot Frenchmen

Tracks were repaired, and destroyed equipment was replaced. The Resistance was thus engaged in a continuous race with the Germans, forcing them to make repairs and to supply replacements, while hoping to escape reprisals. But the reprisals grew increasingly brutal. On April 1 and 2, 1944, a horrific episode unfolded in Ascq, a small village in the Pas de Calais region of northern France near Lille. A train carrying soldiers of the 12th SS Panzer Division Hitlerjugend ("Hitler Youth") from Belgium toward Normandy was blown partly off the tracks as it approached the station at Ascq just before eleven p.m. Two cars were derailed, but no one on the train was injured. Nonetheless, the troops were now vulnerable to Allied air attack, and enraged.

All German units in the west had been issued new orders in February concerning how to respond to such "terrorist" acts: they were to seize civilians in the immediate area and to burn down any houses from which they took incoming fire. The commanding officer of the Hitlerjugend convoy took matters further, as Camus recounted in his second anonymous editorial in *Combat*, entitled "For Three Hours They Shot Frenchmen":

> At around 11 that night, M. Carré, the station chief at Ascq, having been awakened at his home by night shift personnel, was on the telephone dealing with the situation when a German transportation officer entered his office screaming, followed by a number of soldiers who used their rifle butts to beat M. Carré along with M. Peloquin, a senior clerk, and M. Denache, a telegrapher, who also happened to be on the premises at the time. The soldiers then withdrew to the office doorway and from there fired on the three prostrate employees with submachine guns . . . Then the officer led a large contingent of troops into the town, broke down the doors of the houses, searched them, and rounded up some sixty men, who were marched to a pasture opposite the station. There they were shot. Twenty-six other men were also shot in their homes or thereabouts . . .
>
> The executions did not stop until officers of the general staff arrived on the scene. The killing went on for more than three hours.
>
> Whether it is possible to conjure up vividly enough an image of a scene described in such blunt language I do not know.

Camus declared that the massacre of eighty-six men at Ascq showed how the enemy "is increasing his efforts, outdoing himself, each time descending a little deeper into infamy and a little further into crisis . . . beyond what anyone could have imagined." He closed his editorial by expressing the hope that "the image of this little village soaked in blood and from this day forth populated solely by widows and orphans should suffice to assure us that someone will pay for his crime, because the decision is now in the hands of all the French, and in the face of this new massacre we are discovering the solidarity of martyrdom and the power that grows out of vengeance."

While Camus intended for his descriptions of German reprisals to arouse further support for the Resistance, such incidents often had the opposite effect. The Resistance was resented by many in the general population and blamed for the increased repression by the Germans and the Milice, and the grief that brought. The German policy of reprisals was in fact intended to provoke anti-Resistance sentiment, and it worked to a significant degree.

Nor, despite four years of occupation, was the general population necessarily looking forward to an Allied invasion that could bring widespread destruction. The Allied bombing campaign was a potential preview of what was to come. The air campaign against urban targets such as marshaling yards was escalated in April at the direction of Gen. Dwight Eisenhower, Supreme Commander of the Allied Forces in Europe. Eisenhower and his planners saw the bombing campaign as critical to the success of the invasion, but they knew that there would be considerable civilian casualties. Attacks on installations at Le Havre and Lille in April left thousands homeless, heavy night raids on April 21 over the marshaling yard at Saint-Denis and the Gare de La Chapelle killed 640 Parisians, and 400 civilians were killed three days later in a daylight raid on the marshaling yard at Rouen.

Both the Germans and Vichy sought to exploit such incidents and to raise doubts about the success of an invasion in their propaganda. After bombs fell on Paris, the collaborationist press labeled the Allied airmen *les gangsters de l'air* and declared: "Montmartre and the northern suburb of Paris have suffered from the most violent Anglo-American terrorist bombardment since 1940." On April 26, 1944, Marshal Pétain made his first and only wartime visit to Paris, ostensibly to attend a service at Notre-Dame for the victims of the recent bombardments. Despite all that had transpired in the preceding four years, he was cheered by large, enthusiastic crowds along his route, and spoke to more than ten thousand people gathered before the Hôtel de Ville. Two days later, he addressed the entire nation by radio. Under some pressure from the Germans, he condemned the Resistance and identified Bolshevism as France's principal enemy:

> Our country is experiencing days that will count to be among the saddest that it has known. Excited by foreign propaganda, too many

of her children are placing themselves at the hands of unscrupulous masters who want to create a climate that is the forerunner of the worst disorder. Odious crimes that spare neither women nor children grieve the countryside, the cities, and even the peaceful and hardworking provinces. The government has the responsibility to bring an end to this situation and is working toward that. But it is my duty to put you on guard personally against this threat of civil war that is destroying what the foreign war has spared to this point. Those who are pushing France onto this path are invoking the claim of liberating her. This alleged liberation is the most deceptive of the mirages that you could be tempted to believe . . . True patriotism is only expressed by total loyalty. Those who, from afar, are hurling you instructions for disorder want to lead France into a new adventure the outcome of which there can be no doubt. Frenchmen, whoever among you—public servant, serviceman, or simple citizen—participates in the Resistance compromises the future of the country. When the actual tragedy has ended and when, thanks to the defense of the Continent by Germany and to the united efforts of Europe, our civilization will be definitively sheltered from the danger of Bolshevism that weighs on her, the hour will come when France will again find her place.

Building and Breaching the Atlantic Wall

The Germans, however, had been preparing furiously to prevent that mirage of liberation. On November 3, 1943, Hitler had issued Directive Number 51 to his generals, which stated that while Germany had struggled for the previous two and a half years against the threat of Bolshevism, and "the threat from the East remains . . . an even greater danger looms in the West: the Anglo-American landing! . . . If the enemy here succeeds in penetrating our defenses on a wide front, consequences of staggering proportions will follow within a short time." The Führer had concluded: "Only an all-out effort in the construction of fortifications, an unsurpassed effort that will enlist all available manpower and physical resources of Germany and the occupied area, will be able to strengthen our defenses along the coast in the short time that still appears to be left to us." Hitler foresaw the need to counterattack quickly with mobile reserves in order to "throw the enemy back into the sea." He ordered his generals to formulate and deliver their plans to him within twelve days.

Hitler immediately asked Field Marshal Erwin Rommel, the "Desert Fox" who led the North Africa campaign, to make an inspection tour of the Atlantic defenses. Rommel surveyed the entire coast from the North Sea to the Pyrenees Mountains, and was shocked by what he saw. With the exception of the Pas-de-Calais area, where many expected the invasion to come, the fortifications were largely incomplete and entirely inadequate. Rommel was given command of the Atlantic defenses on January 15, 1944, and hurled himself into the task. Understanding that Allied air superiority would hinder movement of units inland, Rommel concluded that the decisive battle would be at the beaches, where the Allies would be weakest, and that he should concentrate his efforts on making them as impenetrable as possible. On every beach along the coast where a landing was possible, he ordered the erection of a phalanx of obstacles to impale or destroy landing craft—jagged steel triangles, sawtooth iron gates, and millions of metal stakes and concrete cones to which mines were attached. And if not destroyed by such obstacles, the craft would be held up long enough for them to be annihilated by shore batteries. And if some happened to get ashore and to offload troops and vehicles, they would emerge into preset zones of crossfire coming from concrete bunkers and pillboxes. And if some managed to advance up the beaches, Rommel had minefields laid at their paths of egress.

Rommel also expected that an airborne assault would try to land gliders or parachutists behind the beaches. He ordered the flooding of fields for several miles inland, and had other open fields planted with "asparagus" stakes—steel poles that were strung with barbed wire that would shred a glider—or trip wires that would detonate when pulled.

After several months of preparations, Rommel had gained confidence that his fixed defenses could hold up an invasion. The second key element in his view was to then counterattack while the Allies were bogged down on the beaches. On April 23, he wrote to Gen. Alfred Jodl, chief of the operations staff of the German armed forces

High Command: "If, in spite of the enemy's air superiority, we succeed in getting a large part of our mobile force into action in the threatened coast defense sectors in the first few hours, I am convinced that the enemy attack on the coast will collapse completely on its first day."

But across the Channel, American and British commanders were also confident that they had a plan that would succeed. They had arrived at the same conclusion as Rommel: that the battle would be won or lost on the beaches, and that is why they were planning the largest amphibious assault in history. They also thought that the Germans' best chance to knock them back into the sea was to hurl their best Panzer divisions in a counterattack against the vulnerable beachheads as soon after the initial assault as possible, and that is why the Allies had developed extensive plans for the Resistance to thwart the movement of German units toward and near the beachhead. And unlike Rommel, who had to be prepared to defend every stretch of coastline along the entire western Atlantic coast from Denmark to the Spanish border at any time, the Allies had the decided advantage of knowing where, and controlling when, the landings would take place.

The original target date, or Y-Day, for the landings, dubbed Operation Overlord, was set for May 1, 1944, at a summit attended by Churchill, Roosevelt, and Stalin in Tehran in November 1943. The exact date, D-Day, was to be determined by landing conditions. In selecting where to land, the Pas-de-Calais area in northwest France was the most obvious, as it offered the shortest crossing, but it was so obvious that the Germans had built their strongest fortifications there. After investigating other options—Le Havre, Brittany, and the Cotentin Peninsula—and weighing the advantages and disadvantages, planners turned to the Calvados coast of Normandy. It offered thirty kilometers of gradually sloping open-sand beaches for landing craft; it had access to a road network inland; and, perhaps most important, it was not obvious. The main disadvantage was that it did not include a harbor where supplies could be offloaded quickly. Allied engineers came up with the idea of erecting two artificial ports ("mulberries") that would be towed across the Channel and protected by sunken breakwaters. To keep the Germans guessing, elaborate deceptions were put in place to lead them to think that the landings were occurring elsewhere, as far away as Norway, or at the Pas-de-Calais. For the latter, deliberate leaks were made that indicated that Lt. Gen. George S. Patton, whom the Germans believed from his performance in North Africa to be the Allies' best field commander, was in charge of an enormous buildup around Dover, across from the Pas-de-Calais. In fact, Eisenhower was holding Patton back for the land campaign to follow. The Free French were also to be held back until after the initial assault; indeed, they first had to be brought to England from Africa for outfitting and training.

The initial landings were then to be an Anglo-American operation involving the British, Canadians, and Americans, but they would not take place in May. There was a shortage in landing craft. In his plans Eisenhower had doubled the number of divisions to participate in the initial assault on the first day. It became clear that there would not be enough boats available in May to handle the number of men and vehicles going ashore, so Y-Day was pushed back to June 1. On May 8, Eisenhower set D-Day to Y + 4: June 5, 1944.

More Arrests, Another Promotion, and More Waiting

The month of May passed without any specific signal to the Resistance. Meanwhile, yet more of its leaders were arrested. On May 2, the head of the entire FFI in France, General Dejussieu, was captured by the Gestapo in Paris. The chief of the Paris region and Jacques Monod's immediate superior, Pierre Pène ("Périco"), had also been arrested. The arrests necessitated more replacements and promotions: Gen. Alfred Malleret ("Joinville") took Dejussieu's place as head of the FFI in France; Henri Tanguy ("Rol"), former head of the FTP, replaced Périco; and Monod ("Malivert") was promoted to chief adjutant to the general staff of operations (Troisième Bureau) for the entire country.

Monod asked Noufflard to become his secretary. It was a more interesting job than liaison agent, as she was able to read intelligence reports on factories, railroads, navigation canals, and enemy troop positions, but it also meant that she had even more work to do. She still made morning rounds, then spent the afternoon with "Malivert" at her house going over the information she had gathered. After more street meetings, she worked through the evening typing up orders and reports, burned the carbons and drafts, and tried to get to bed before two or three. She changed her alias from "Alix" to "Catherine Vernier."

Since Noufflard's home had been frequented by so many people, she knew that it could be raided at any time. She and Monod worked out a simple signal so that he never went up to the door when she was not home. That way, if she was ever arrested, the Gestapo could not use her house to set a trap for others. She had a flowerpot that she put on the windowsill only when she was home. If someone came unexpectedly to the door, she moved it to the right, which would signal Monod to wait outside. If the situation was all clear, she moved it to the left.

The streets of the capital became even more tense and dangerous. Fewer Parisians walked the boulevards, and those who did moved quickly. Métro stations were often closed, elevators did not work, and theaters were closed during the day. The Germans had installed machine-gun nests in basements of corner buildings so that they could sweep their fire across the streets and sidewalks. Rolls of barbed wire were deployed along the Tuileries and in the Place de la Concorde. Antitank barriers were installed on many streets. For Noufflard and her contacts, bicycling or walking through the city had become more perilous due to endless police controls and roundups. Streets and squares would be suddenly blocked off, and each pedestrian or bicyclist would be stopped; made to show identification, ration cards, and proof of employment; and searched. Those who did not pass inspection were hauled away to the Vel d'hiv, or worse. When carrying incriminating documents, Noufflard had to be constantly on the lookout for new control points.

Monod also took extra precautions; he went completely underground. He told everyone he knew—his friends, those in the laboratory, and the concierge of his apartment building—that he was sick and had to leave the city to recuperate. In order to walk the streets without being recognized, he made up a disguise. He changed the color of his curly black hair and cut it very short. Instead of wearing his typical sheepskin coat without a hat, he changed to wearing a tailored coat, a black felt hat, gloves, and tinted glasses. He put his new look to the test at a street meeting with another FFI member. Monod was delighted when the man walked right by without recognizing him.

Monod continued going back and forth to Saint-Leu-la-Forêt to visit Odette and the twins, only the train was no longer reliable. All of the damage to the rail system had made train journeys unpredictable. The short trip to Saint-Leu could take twenty minutes or four hours. So Monod bicycled home to the relative safety of the suburb. There were German troops stationed around Saint-Leu, air-raid alerts were very frequent, and shrapnel from flak and dogfights overhead sometimes fell on the roof, but there were no significant military targets close by that put the family at great risk. Nonetheless, Odette confessed to Jacques's parents that she was having "apocalyptic visions" of what the invasion might bring, and that Jacques was "very tired."

He was tired, yes, but progress was being made. Arms were being dropped into his region and around France in anticipation of the landings. In the first part of 1944, many thousands of machine guns, pistols, and rifles, as well as bazookas, mortars, ammunition, and high explosives were dropped and distributed to fighters and saboteurs. Plan Vert had identified 571 rail targets near the coast and across the country that were to be demolished, and Plan Tortue marked thirty road cuts to be made once the invasion was under way. Monod and other commanders were awaiting the coded message to get their teams ready.

Radio London frequently spoke openly about the anticipated landing (*le débarquement*). On May 10, listeners were told that "major events were preparing themselves." On May 12, commentators announced a "state of alert" and declared that the watchwords were "discipline and preparation." Citizens were instructed to make sure that they secured a supply of food. Subsequent broadcasts spoke of preparing for a national insurrection following the landings. On May 20, listeners received the first of several directives from a representative of the Supreme Headquarters of the Allied Expeditionary Force (SHAEF), telling them that the Allies counted on their cooperation at the time of the landing and asking them to begin taking note of enemy positions, supplies, and minefields. On May 27, they were instructed on their security and safety. Several times each day, the BBC also broadcast strings of "personal messages," short phrases, many of which were meaningless and intended to confuse the Germans, but some of which were actual messages to Resistance groups, announcing arms drops, for example.

On Thursday, June 1, a stream of almost two hundred personal messages was read and repeated, including:

Ouvrez l'oeil et le bon.
Je suis encore ingambe.
Ma femme a l'oeil vif.

That was it—*"Ma femme a l'oeil vif"* (My wife has a lively expression)—the specific message that Monod and the Paris region departmental commanders had been waiting for: the invasion was imminent. The scores of other messages were to other FFI regions and action teams. Noufflard heard the news that day when she met with a man she knew as "Pêchery."

It was time to make final preparations and to await the confirmatory signal that the invasion had been launched. On Saturday, June 3, Monod told Noufflard he was leaving Paris for Saint-Leu and would not be reachable again until Monday, when he had an FFI staff meeting to review preparations. On Sunday, Noufflard went to meet with a liaison agent for "Gildas" (Pierre Lefaucheux), who was Pêchery's boss and the commander of the Department of the Seine, one of the four departments within the FFI's greater Paris region. The young woman, Françoise, whom Noufflard had known since childhood, looked very upset.

"Isn't it terrible," she said. "Know anything about Leroux?" Leroux was a mutual acquaintance in the organization.

"No. What happened?" Noufflard replied.

"I thought you knew, when you said you wanted to meet me," Françoise said, and proceeded to tell Noufflard about a meeting the day before of chiefs of various sectors of the Paris region that had been raided by the Gestapo. Gildas's wife had called Françoise when he did not return home, and they both had to move out of their homes immediately. Françoise did not know the names of all who were there and had been arrested, but it included at least Gildas and Pêchery.

Noufflard was frantic. She knew that the Gestapo often recovered names and addresses, or the location of a meeting, from the papers or homes of those they arrested, or by torture. She had to find out what had happened to Leroux, and to warn him and others of the danger. She bicycled across Paris all day trying to find him or his address, but without success.

By the next morning, June 5, she began to fret about Monod. She was pretty sure that at least one of those arrested knew about the meeting Monod had that day. She did not know how many, but the Gestapo had in fact arrested eleven people at Saturday's meeting on rue Lecourbe. "What if he came back to Paris and went straight to his meeting without seeing me?" she worried. She knew that the meeting, at which Rol-Tanguy was also expected, could be extremely dangerous. The entire FFI organization around Paris was at risk.

Monod appeared at last. He was shocked by the news of the arrests of so many heads. He quickly determined that Leroux was safe, and was in fact totally unaware of what had happened. He and Noufflard then proceeded carefully to the scheduled meeting in Sceaux, a southern suburb, taking extra precautions to make sure that they weren't followed. As Monod approached the street where the safe house was located, a thought occurred to him that he had pondered on the way to several previous meetings of Resistance chiefs: he could turn and walk into the meeting, and risk not seeing Odette and the children again, or just walk on by and go back home.

He turned into the safe house. Then Rol-Tanguy arrived on his bicycle, followed by other officers of his staff. Noufflard was tremendously relieved to see that all had made it to the meeting, and that there was no sign of the police.

Later that night, at 9:15 p.m., a BBC announcer speaking in a monotone broadcast another string of "personal messages" for six minutes, each sentence being repeated twice. Four of which were:

Il est sévère mais juste.
L'acide rougit le tournesol.
Elle restera sur le dos.
Allô, allô, James, quelles nouvelles.

Each sentence was a specific, confirmatory message to the greater Paris FFI to execute Plan Vert (railroad), Plan Guérilla (guerrilla action), Plan Tortue (road), and Plan Violet (communications), respectively.

The invasion was under way.

CHAPTER 15

NORMANDY

The history of warfare knows no other like undertaking from the point of view of its scale, its vast conception, and its masterly execution . . . History will record this deed as an achievement of the highest order.

—JOSEPH STALIN, June 11, 1944,
telegraph to Winston Churchill

THE INVASION TOOK THE GERMANS *ALMOST* COMPLETELY BY SURprise.

German intelligence had learned the meaning of some of the BBC coded messages, and sent out an alert the night of June 5. However, there had been false alarms in the past, and commanders were wary that the intercepted messages could be a diversion. While one of the two main armies in the west was put on alert, neither Rommel nor the 7th Army, which included all units responsible for the landing zones, were alerted.

Rommel did not think an invasion was imminent; he was not even in France at the time. He was expecting the landings to come on a high tide at dawn, so that landing craft would have the shortest possible stretch of beach to cross at first light. He had consulted moon and tide tables on June 1 and concluded that such conditions would not occur until after June 20. On June 3, he had conferred with Gen. Gerd von Runstedt at the latter's headquarters outside Paris; Runstedt also agreed that there was no sign of invasion. On June 4, Rommel left his own headquarters at La Roche-Guyon, about 40 miles west of Paris and 120 miles from the Normandy beaches, for a long drive to Germany to visit his home in Herrlingen and to celebrate his wife's birthday on the sixth. Afterward, he hoped to go to meet with Hitler in person at Berchtesgarden to request additional reinforcements.

The Germans had failed entirely to notice the massive invasion force as it was assembled over the previous days on the south coast of England. There had been no air reconnaissance for the first five days of June, and both air and sea patrols scheduled for the night of June 5–6 were canceled on account of the poor weather and visibility. The weather was in fact so poor—overcast and stormy—that on June 4, Eisenhower had postponed the operation by twenty-four hours, from June 5 to June 6. The invasion fleet was not detected until the lead ships were already in place offshore just after three a.m., and the defenders had no inkling of the scale of the attack until first light revealed a vast armada of 6 battleships, 20 cruisers, 68 destroyers, more than 1,800 landing craft, and another 900 supporting vessels approaching the coast.

The amassed firepower was unleashed on German shore fortifications shortly after five thirty a.m., just before the landing craft were to approach the Normandy beaches.

The initial wave of the assault was launched by five divisions that would land on five beaches, respectively—Utah and Omaha in the American sector, Gold and Sword in the British sector, and Juno in the Canadian sector. The first landings began at six thirty a.m.

THE BBC INTERRUPTED its regular broadcast at nine thirty a.m. to deliver the first official communiqué that the invasion was under way. The first announcement in French followed at ten a.m.: "Under the command of General Eisenhower, the Allied naval forces, supported by powerful air forces, began the landing of the Allied armies this morning on the North coast of France."

Noufflard thought, "Now comes the time of revenge, the time to show the world what you have prepared for and what you are able to do."

She went to a map store in central Paris to try to find a map of the Cherbourg area. There were hundreds of people asking for exactly the same thing. She bought one of the last copies to put on her bedroom wall, so that she could track the locations of the Allies.

She met Monod and two other staff officers near the Sorbonne. It was a fair day in Paris, so they decided to sit briefly on the terrace of a café to absorb the moment. For the first time in four years, there was excitement on Parisians' faces. And it was satisfying for once to look at the Germans, feeling that their days were numbered.

CAMUS ALSO HEARD the news that morning. He and Maria Casarès had been with Sartre and de Beauvoir at an all-night fiesta hosted by the actor/director Charles Dullin in his very nicely appointed Right Bank apartment in the ninth arrondissement. Camus bicycled home through the excited streets with Maria riding on his handlebars.

—ↀ—

ON THE SHORES of the storm-churned English Channel, however, a fierce battle was raging. Men were dying by the thousands, and the question of which forces would carry the day was still very much in doubt. The losses on Omaha Beach were so great, and the ground gained was so meager, that Gen. Omar Bradley almost ordered a withdrawal that morning.

As news of the landings broke, Allied leaders balanced optimism with caution.

President Roosevelt held a press conference that drew more than 180 reporters. He said:

> The whole country is tremendously thrilled . . . but . . .
>
> The war isn't over by any means. This operation isn't over. You don't just land on a beach and walk through—if you land successfully without breaking your leg—walk through to Berlin. And the quicker this country understands it, the better.

Prime Minister Churchill strode into the House of Commons at noon London time. After a lengthy report on the Allies' capture of Rome, which had taken place two days earlier, he addressed the ongoing invasion:

> So far the Commanders who are engaged report that everything is proceeding according to plan. And what a plan! This vast operation

> is undoubtedly the most complicated and difficult that has ever taken place . . . The battle that has now begun will grow constantly in scale and in intensity for many weeks to come, and I shall not attempt to speculate upon its course . . . The enemy will now probably endeavour to concentrate on this area, and in that event heavy fighting will soon begin and will continue without end, as we can push troops in and he can bring other troops up. It is, therefore, a most serious time that we enter upon.

De Gaulle and Pétain resumed their battle over France's airwaves. The marshal addressed the country that afternoon, repeating familiar, predictable notes:

> The German and Anglo-Saxon armies are fighting on our soil. France becomes a battlefield.
>
> Officials, public servants, railway workers, laborers, remain at your posts to maintain the life of the nation and to accomplish the tasks required of you.
>
> Frenchmen, do not aggravate our misfortunes by acts that could risk tragic reprisals. It would be innocent French people who would suffer the consequences.
>
> Do not listen to those who seek to exploit our distress, and would lead the country to disaster.
>
> France will only be saved by observing the strictest discipline.
>
> Obey the orders of the government, that everyone face their duty.
>
> The circumstances of the battle may lead the German Army to make special arrangements in areas of combat. Accept this necessity; this is a recommendation that I make in the interest of your safety.
>
> I charge you, the French people, to think above all of the mortal danger that our country would face if this solemn warning was not heard.

Laval compounded Pétain's warning by declaring that disobedience of the government's instructions was "a crime against the country." He stated that because France had signed the armistice, she was bound to honor its terms. "We are not at war," he declared, adding, "You must not take part in combat."

De Gaulle had been offered the opportunity to speak over the BBC on D-Day morning, right after a message from Eisenhower was read, but he refused because he objected to having an American give instructions to the French people. After some interventions by the British, he was finally persuaded and broadcast his message at 5:30 p.m.

"The supreme battle has begun!" he declared, before offering his own instructions to the French people: "The actions that we take behind enemy lines ought to be closely coordinated with those taken at the front by the Allied and French armies . . . That is to say that the actions of Resistance forces ought to be sustained and to expand right up to the moment of the rout of the Germans."

Eisenhower had composed a statement in advance in case the invasion failed. He did not have to use it. While the Allies suffered a very large number of casualties—at least 4,400 killed and 5,000 wounded on D-Day—the casualties were not as great as was feared by military planners. By the time offloading paused around ten p.m. the night of June 6, 155,000 troops had made it ashore and the Allies controlled about eighty square miles along the coast.

Behind Enemy Lines

The Resistance had indeed done its job: in the first day, action teams made more than 950 of the planned 1,050 rail cuts. By the day after D-Day, for example, twenty-six trunk rail lines were not usable, including lines that connected towns in the landing areas: Avranches–Saint-Lò, Saint-Lò–Cherbourg, and Caen–Saint-Lò. Designed to prevent the movement of enemy divisions toward the battle zone, and thus to buy time for the Allies to build up and begin to break out of their beachheads, the rail cuts and roadblocks thwarted the movement of eight enemy divisions toward Normandy. The feared Reich SS Panzer division, whose 450-mile movement into Normandy would normally have required just three days, took more than two weeks thanks to the sabotage of rail lines and flatcars, and harassment by the *maquis* and Allied air forces. In the meantime, the Allies landed many more combat divisions in Normandy.

As CONVENTIONAL ARMIES battled in Normandy, the Resistance prepared for the next phases of the campaign, including the eventual liberation of Paris. Noufflard and Monod had more operations work to do than ever before, with an even greater sense of urgency and purpose. Precautions were paramount in every facet of their work. One of Noufflard's first tasks was to find a safer base of operations. Too many people had been seen going in and out of her rue de Varenne house over the previous months to invest the next, critical phase of operations there. Instead, she found a painter's studio in Montparnasse and furnished it like an artist's loft so that she would have a cover story in case anyone looked inside or wondered why Monod was there. She even made several charcoal portraits of Monod and left them lying about the studio. She met Monod there every afternoon and many evenings; almost no one in the organization knew the location.

Monod's responsibilities on the FFI national staff entailed planning and coordinating actions in different zones—a challenging task as situations changed rapidly and communications with different regions were difficult. Noufflard and Monod set up a large map of France, on which the status of all of the railroad and canal networks was kept up to date, with the location of rail cuts and blown locks marked by colored pushpins. Blocking rail transport remained a priority. On July 14, Monod issued an order to all regions advocating an almost undetectable technique for crippling trains on the main lines:

> SUBJECT: SABOTAGE OF RAIL LINES
>
> The hoses connecting train air ducts are becoming more and more scarce and because of this their destruction can create delays and disturbances in rail traffic.
>
> In order to prevent detection of the site of the sabotage (train station or rail yard) where the sabotage of equipment has been executed, a simple means is to perforate the hoses with an awl. It is very rare when the brake is tested at the moment of de-

parture that the sabotage is noticeable. It is only after a certain distance of about four to six kilometers that the hose bursts. At this time the train is on the main route, where it stops traffic.

Transmit immediately the necessary orders to the interested teams in order to systematically organize this sabotage.

Report back the results immediately.

. . . *MALIVERT*

Other high-priority targets were munitions depots, gasoline- and oil-storage facilities, and factories involved in the production of war materiel. Those installations that could not be attacked by sabotage were candidates for air raids. Monod set up an intelligence channel involving a "Monsieur de Saligny," whom Noufflard would meet to pass on information about important targets. De Saligny then passed it on

E.M.N 3ème Bureau À TOUTES REGIONS ET S.N.C.F

14/7/44

OBJET : Sabotage des Chemin de Fer .

Les boyaux d'accouplement servant à raccorder la conduite d'air des trains deviennent de plus en plus rares et leur destruction peut de ce fait créer des retards et des perturbations dans la circulation des trains .

Pour éviter de pouvoir situer le lieu (gare ou voie de garage) où le sabotage des appareils a été exécuté, un moyen bien simple consiste à perforer les boyaux au moyen d'un poinçon. Il est très rare qu'à l'essai du frein, effectué au moment du départ, l'avarie puisse être constatée. Ce n'est qu'après un certain parcours, 4 à 6 Kms. que le boyau éclate. A ce moment le train se trouve sur la voie principale où il arrête la circulation et forme bouchon.

Transmettre d'urgence les ordres nécessaires aux équipes intéressées pour organiser systématiquement ces sabotages .

Rendre compte d'urgence des résultats .

Pour le chef d'E.M.N et F.O

Le chef du 3ème Bureau

MALIVERT .

An order from "Malivert" (Jacques Monod) urging and instructing a specific kind of sabotage of trains so as to block movements on rail lines, July 14, 1944. (Courtesy of Geneviève Noufflard)

to the British Intelligence Service, who then shared it with Allied air commanders. One day while visiting her aunt just outside Paris, Noufflard had the satisfaction of witnessing the bombing and burning of some mills that Monod had recently pointed out to M. de Saligny.

In addition to sabotage and intelligence-gathering, Monod's Troisième Bureau was also tasked with organizing a secret radio network so that they could communicate with different regions. Travel across France had become more difficult, and it was always dangerous for Resistance members to meet in person. Setting up a network, however, was also difficult and risky. Equipment was very scarce and outdated, and the Germans had effective ways of detecting clandestine radios. Monod used his previous, albeit truncated, communications training to organize the network. He recruited radio technicians and scientists and established codes, and Noufflard put together an instruction manual for radio operators. Placing their transmitter was another challenge, because it had to be in a high place and entailed a great deal of risk on the part of whoever was also in the building. A well-connected woman offered several rooms in her home on the avenue Victor Hugo. For security, the transmitter-room door was rigged with a grenade whose safety pin would be pulled if opened by an unwitting visitor, such as a Gestapo agent. When Monod told Odette of the device, she was worried that her sometimes absentminded husband might accidentally blow himself up.

Other important precautions concerned the disposition of documents. There was a large drawer in the Montparnasse loft in which Noufflard kept piles of papers—intelligence reports, diagrams of potential targets, action summaries, radio transmissions, and so forth—that Monod needed to review or that had to be distributed. These compromising papers had to be destroyed, which was a great inconvenience because the studio had no fireplace or stove. Instead, at the end of every day's work, Noufflard and Monod both stuffed their pockets with papers and went into the bathrooms at each end of the long corridor outside the studio. They then tore the documents into tiny fragments and, because the plumbing did not work well, flushed them down the toilet in small batches. Noufflard worried each night that the long, tedious ritual would arouse their neighbors' suspicions, or at least make them think that she and Monod had very odd bathroom habits.

Noufflard's and Monod's security measures were duly warranted. The Germans and the French were arresting more than 100 people a week and executing about 150 a month in Paris alone over the first part of 1944, and the Gestapo's and the Milice's campaigns against the Resistance continued unabated after the Normandy landings. One incident of anti-Resistance vengeance claimed one of Camus's dearest friends only a week after D-Day. Poet René Leynaud, whom Camus had befriended during his stay in Le Panelier, had been caught by the Milice in Lyon on May 16 with compromising documents. Leynaud tried to flee but was shot in the legs and wound up in Fort Montluc Prison. On June 13, as the Germans prepared for their eventual evacuation of the city, they selected nineteen prominent prisoners, including Leynaud, took them away to Villeneuve, and shot them in the woods. When Camus later learned of Leynaud's murder, he declared it an "irreparable loss," a "dreadful death" that would affect him more deeply than anyone else's during the war.

FOUR DAYS LATER in Lyon, the Gestapo and the Milice caught up with *Combat*'s printer, André Bollier, for the third time. After his arrest in March, Bollier had also been sent to Fort Montluc, where he was tortured repeatedly. He divulged nothing, managed to escape on May 2, and resumed his work.

After the invasion, he decided to leave Lyon to join the fighting. But before doing so, he convened a meeting on Saturday, June 17, to prepare the printing of the June issue of *Combat*, the layout for which had arrived from Paris the day before. As his four-person team was working at the printer's, the building was surrounded by the Milice, who opened fire. Bollier fired back with his revolver, and then he and the illustrator made a dash for the street. A volley felled them. Both were severely wounded. Bollier could not endure another round of torture. He said, "My God, forgive me," and he turned his revolver on himself.

There would be no June issue of *Combat*.

Despite yet another loss of an important cog in the organization, the *Combat* staff was determined to continue. In fact, Bernard, Camus, and their colleagues in the Combat movement were anticipating the public debut of the newspaper once Paris was liberated. It was expected that all of the collaborationist newspapers would be forced to close. In going public, *Combat* aimed to have a major voice in postwar France. In early July, Bernard, writer Albert Olivier, journalist Marcel Gimont (alias "Paute"), and Camus met to begin planning the first public issue in Camus's one-room studio that he had leased from the writer André Gide. Located at 1 bis rue Vaneau, it was, coincidentally, just a few steps from the courtyard of Noufflard's home.

Camus's role in the new *Combat,* however, was almost preempted. He was walking with Maria Casarès near the Réaumur-Sebastopol Métro station when they were caught in one of the frequent police controls that materialized without warning in the city. French and German police blocked the road at both ends and started searching the men and asking the women for their identity cards. Camus was carrying a layout page with *Combat*'s logo on it. He first put it in his jacket pocket, then slipped it to Maria.

She saw Camus with his hands in the air and thought that he was going to be arrested. But the police did not find the layout, and they were both released. Camus quickly disposed of the layout and decided to move from his studio into an apartment belonging to an Algerian friend.

Camus and Marcel Gimont did manage to put together and publish an issue of *Combat* in July, the first after D-Day. Camus was no doubt thinking of men like Bollier and Leynaud when he wrote an editorial entitled "You Will Be Judged by Your Actions." After condemning Pétain and Laval for their "treason," Camus wrote:

> The time is fast approaching when the people of this country will be judged not by their intentions but by their actions, and by the actions to which their words have committed them. That alone is just . . .
>
> The Resistance is telling you that we are at a stage where every word counts, where every word is a commitment, especially when

> those words ratify the execution of our brothers, insult our courage, and deliver the flesh of France herself to the most implacable of enemies . . .
>
> The Resistance is telling you that you have no government on French soil and you don't need one . . . We don't need Vichy to settle our score with shame . . . We need men of courage . . .
>
> Frenchmen, the French Resistance is issuing the only appeal you need to hear. The war has become total. But a single struggle remains. The flower of the nation is preparing to sacrifice itself . . . Anyone who isn't with us is against us. From this moment on there are only two parties in France: the France that has always been and those who shall soon be annihilated for having attempted to annihilate it.

The Resistance also needed women of courage, and Camus had a candidate for a new courier for the *Combat* staff—Maria Casarès. Her experience in the Spanish Civil War and her action during the police control had demonstrated her courage. And the stagecraft of street meetings and pseudonyms was perhaps a natural role for someone with such superb acting talent; those skills were then on display at the Théâtre des Mathurins in the first run of Camus's play *Le Malentendu*. Camus set up a meeting to introduce Jacqueline Bernard to Casarès near the theater on the evening of July 11.

Bernard, however, did not show.

Earlier that day, she had gone to meet with someone who turned out to be an informant. The Gestapo was waiting for her, and took her to one of their headquarters on the rue de la Pompe. The Germans discovered her address book, in which telephone numbers were written in a very simple code. Paris phone numbers were comprised of three letters, followed by four numbers; in Bernard's code, one just subtracted three from each of the first two numbers and added three to the last two numbers. One of those numbers was Camus's office number at Gallimard. Bernard looked for a way to warn the organization that she had been arrested and that everyone was in danger because of the informant. She told the Gestapo that she would deliver a letter to one of her contacts to set up a meeting of the staff. The Gestapo went along and allowed Bernard to go into the building alone, where she whispered a warning to her contact, who in turn spread the word.

Camus had to leave town quickly. He went to Janine and Pierre Gallimard's apartment, while Pierre and his cousin Michel bicycled over to Camus's apartment to retrieve his belongings. Camus and the three Gallimards then fled Paris together on bicycles, pedaling some fifty-five miles to a dilapidated house in Verdelot that was owned by a Gallimard editor. Jacqueline Bernard was eventually deported to Ravensbrück concentration camp.

Homecoming

Roosevelt had promised de Gaulle that one major French unit would be sent into northern France in the course of the invasion. That was to be the 2nd Armored Division, or "Deuxième Division Blindée," also known as the DB. After the successful Tunisian campaign in North Africa, Colonel Leclerc's force was combined in August 1943 with various other French units that had been fighting in Africa to form the DB, which was to be patterned after a typical US Army armored division. In Morocco, it was outfitted with new American tanks and equipment and then sent to England in April and May for further training. After almost two months in Hull, the order came for the DB to move south to Southampton. The next destination for the division after the British seaport would be France.

Among the more than 16,000 excited troops of the DB was one twenty-four-year-old, battle-tested medical officer, François Jacob. His four-year odyssey had taken him first to England, where he was one of the first few to join de Gaulle's Free French, and hoped to become an artillery gunner but was assigned to the medical corps on account of his limited training. He then went to French Equatorial Africa—Senegal, the French Congo, Gabon, the French Congo again, Cameroon, and Chad before seeing combat in Leclerc's campaign through Libya and Tunisia, and then being sent back to England. His battalion reached the marshaling area in the American camp outside Southampton on July 29 and spent the night in tents. Roger Dreyfus was not with the DB; Jacob's close friend had been killed two years earlier in Chad.

After more than four years of "exile, anguish, solitude, fighting, despair," and of hoping for and dreaming of this moment, Jacob could barely believe the time had finally arrived. The next day, July 30, his 2nd Medical Company was one of the last units in the division to board its flat-bottomed landing craft. Piloted by a large, sodden, red-bearded British seaman, the craft made its way into the harbor to join the great number of boats that were crossing back and forth to the Normandy bridgehead. The boats entered the English Channel under the cover of darkness.

In the middle of the night, Jacob awoke and realized that the boat had stopped. It was bobbing gently in the waves; the pilot of the boat was nowhere to be seen. He hoped that no enemy plane would pick this moment to fly over. Jacob could hardly bear the waiting, knowing that somewhere out in the darkness lay the coast of France. After a couple of hours, the pilot reappeared, scanned the foggy horizon with a telescope, and started up the motors. Jacob soon saw boats appear, and then the dark line of the coast.

"The soil of France!" he thought. Less than a mile away lay the land of his childhood in the form of the sand dunes of Utah Beach. More waiting as hundreds of ships offloaded. Then, finally, on August 1, Jacob and his unit stepped ashore among the blackened bunkers, wrecked landing craft, destroyed vehicles, and the bustling makeshift port the Americans had built.

Leclerc tried to capture the emotion of his troops that day in a broadcast that went out over the BBC:

> People of France, the long-awaited hour is here. We are setting foot once again on our country's soil.
>
> Four years ago we left France, responding to General de Gaulle's appeal, abandoning our families, determined not to lay down our arms before victory.
>
> We return now at the side of our Allies and at the head of French troops after having maintained our struggle, despite Vichy's surrender.
>
> It is difficult to express the emotions of our officers, noncommissioned officers, and soldiers at such a moment. These men

> come from everywhere. Some joined de Gaulle at the start, they fought in Chad, Libya, Tunisia, they saved our national honor. Others of us rallied as soon as the link with North Africa allowed. Still others have joined us recently from other divisions that have since disappeared.
>
> We want to first fight the Boche, the cursed enemy. This time we have the arms and we are going to use them. Next, we want to find the good French people who for years have been leading the battle within the country that we have been leading outside of it.
>
> A salute to those who have already taken up arms. Yes, we constitute the same army.
>
> Finally, we want to see French grandeur restored tomorrow and we grasp the energy and patriotism that this task will require.
>
> All of the country ought to be devoted to that cause. France standing upright, help us, help our Allies in order to shorten this battle for her liberation, on the soil of our Homeland.

It would take a few days to organize and re-form the division. In the meantime, as his unit moved into Normandy, Jacob was thrilled to see the familiar sights of the Norman farms and hedgerows, to smell the summer hay, and to be handed a glass of fresh cider. All along the route, now marked by signs in English that replaced the German placards, he saw the remnants of battle—ruined and smoldering buildings, incinerated tanks and trucks with corpses still lying inside, and columns of German prisoners. As the DB passed through villages, it often took a while for the onlookers to realize their identity and to notice the Cross of Lorraine on their vehicles, to which they reacted with jubilation. It was an intoxicating journey, a mixture of pure joy and the still-seething desire to fight, sprinkled with the occasional rumor of snipers and the ever-present danger of land mines.

Despite the widespread destruction, the Allies had gained significantly less ground than they had planned or hoped over the first eight weeks after the invasion. Caen was not taken until July 9 by British and Canadian forces, a month behind schedule, and those armies remained bottled up in the Cotentin Peninsula by stubborn German defenses. American forces managed to take Cherbourg by June 26 but were bogged down in the *bocage* (thicket) until the last days of July, when they broke out to the southwest through Avranches in Operation Cobra, opening the door to Brittany and the deeper interior of France.

Patton's 3rd Army, which did not become officially operational until August 1 and to which the DB was attached, was tasked with exploiting that breakout. Leclerc's principal mission was to lead the first troops into Paris, but Patton told him that he thought the Germans might soon surrender, and that if he wanted to get into the fight, he might want to start right away, instead of waiting to liberate Paris. Leclerc seized the offer, and Patton assigned the DB to XV Corps, one of his four battle formations, which was commanded by Gen. Wade Haislip. Elements of the DB caught up to Haislip's units on August 6, south of Avranches. Leclerc established his command post on the seventh at Sainte-James.

On August 8, the order came to head east in the direction of Le Mans as part of a new offensive to try to encircle the Germans. Days earlier, Hitler had ordered a large counterattack toward Avranches in order to sever the Allied line into Brittany, but the attack, launched on August 7, was doomed by insufficient forces. Instead, the German offensive merely left major divisions of the German 7th Army stalled deep in Allied territory. General Bradley, Patton's immediate superior, saw a unique chance to encircle the Germans; he told a visitor that he had "an opportunity that comes to a commander not more than once in a century. We're about to destroy an entire hostile army and go all the way from here to the German border." Bradley's idea was to enclose the Germans in a giant pincer action with Patton's and Haislip's units in the south forming the lower jaw and Canadian forces in the north forming the upper jaw.

On the night of the eighth, Jacob's 2nd Medical Company prepared to move out with other units from their encampment south of Avranches. The camouflage was removed from the equipment, and the vehicles were organized into a column. Before they could get under way, however, at around two o'clock in the morning, Jacob heard the characteristic drone of German bombers. The motors grew louder and then explosions erupted in a nearby field. Jacob and his comrades dove out of their truck for a roadside ditch. Bombs whistled down from the sky, more explosions wracked the area around the column, and a car burst into flames. Jacob remained pressed to the ground until the planes had passed.

He then heard the cries of the wounded. There were many men lying on the ground. Jacob went over to tend to one soldier lying near a car. It was Lt. Lucien Benillouz, who had joined the Free French the year before in Tunisia. He and Jacob had become good friends in England after they discovered that they had been courting the same woman. Jacob saw a bloodstain spreading from Benillouz's side, so he tore open his jacket and shirt and applied a bandage. He and another medic tried to lift his friend onto a stretcher, but Benillouz screamed out in pain.

Just then, Jacob heard the sound of German Stukas approaching again. As the drone grew louder, Benillouz tried to get up but could not. Jacob looked around for cover; there was a ditch less than thirty feet away, but the lieutenant could not be moved. Benillouz grabbed Jacob's hand and said, "Don't leave me."

Jacob looked once more at the ditch, then snuggled up against Benillouz as the bombs came whistling down again. He held still, trying to shelter his friend while at the same time making himself as small as possible. As the earth shook and dirt flew in all directions, Jacob felt a violent jolt along his right side. He held still another moment, not wanting to know what had just happened. More than fifty shell fragments had pierced his body, bursting his elbow and breaking his thigh. He saw blood coming from his elbow and tried to raise his arm; it hung limp. A massive wave of pain washed over him, and he passed out.

Jacob, Benillouz, and several other seriously wounded members of the unit were rushed by ambulance to the 104th Evacuation Hospital south of Pontaubault. As Jacob drifted in and out of consciousness, Lieutenant Benillouz passed away.

When Jacob woke up, he found himself encased in two casts, one around his chest and right arm, another around his right leg and pelvis. Only one week after the landing in Normandy and two hundred miles short of Paris, Jacob's war was over.

CHAPTER 16

LES JOURS DE GLOIRE

A people that wants to live does not wait for its freedom to be delivered to it.

—Albert Camus, *Combat,* August 23, 1944

In order to encircle the Germans, speed was of the essence. The Allies had to close the trap before too many German units could escape to the east. However, moving large formations quickly caused them to become spread out, and their long lines of defense to become thin and vulnerable. As the Allies took Alençon and moved toward Argentan, Haislip and Bradley worried that the Germans still had many divisions in the area and were preparing a counterpunch against the extended Allied lines. Bradley ordered Haislip and Patton, who had freed units to protect Haislip's left flank, to halt short of Argentan and to consolidate their positions on August 13.

Patton was furious. There remained only a twenty-five-mile gap between the two Allied pincers, between his positions south of Argentan and the Canadians north of Falaise. Patton urged Bradley to let him surge ahead and close that gap. Bradley refused. Patton was sure that was a huge mistake, for as he noted proudly in his diary that day, his 3rd Army had "advanced farther and faster than any Army in the history of war."

Leclerc, too, was getting impatient. Since well before his arrival in France, Paris was constantly on his mind. Halted along with the others of Haislip's divisions, he asked Haislip on August 14 when his forces would be released for the liberation mission. Haislip brushed him off.

In the meantime, the irrepressible Patton had come up with an alternate plan. Instead of closing the circle at Falaise, he proposed to Bradley that he take a couple of divisions from Argentan and race east toward the Seine to cut off more retreating Germans—a "long encirclement." Bradley gave his OK to Patton's audacious plan. Leclerc, however, objected, as the DB was not one of the divisions to be sent east. On the fifteenth, he went to see Patton in person. He told Patton that if he was not allowed to advance on Paris, he would resign. Patton, using his best French, told Leclerc that he "would not have [his] division commanders tell [him] where they would fight."

Leclerc appealed to Bradley as well, but got nowhere. He was stuck at Argentan. The "Falaise Gap" would not be closed for another week.

The liberation of Paris was much more than a diplomatic problem. The city of more than three million posed potentially enormous difficulties that could bog down the Allied armies as they raced toward Germany. There was the risk of becoming entangled in house-to-house fighting. There were the logistical challenges of supplying fuel, food, and medical supplies to the population, when the armies' own supply lines were already stretched very thin. And there was also the possibility of having to level the city to extricate the Germans. Eisenhower, who prided himself on putting military considerations above politics, had decided that the Allies would go around Paris and deal with the capital once the Germans had been weakened.

People in Paris, however, had other ideas.

"Chacun Son Boche"

On Saturday, August 19, Geneviève Noufflard was riding her bicycle along the boulevard de la Tour Maubourg near Les Invalides, on her way back from a meeting with her liaison agent. The day was sure to be another in a string of hot, clear summer days in Paris. The atmosphere within the capital, however, had changed almost overnight. There were no police to be seen anywhere. The day before, the entire Parisian police force had gone on strike. Posters had appeared on walls in the name of de Gaulle calling for the mobilization of all eligible Parisians, men and women, to join the FFI or the Milice Patriotique (patriotic militia); from unions calling for a general strike; and from Henri Rol-Tanguy, the commander of the FFI for the greater Paris region, urging citizens not only to join the FFI, but to:

Gather yourselves by household, by neighborhood, knock out the Boches so as to snatch their weapons, liberate Greater Paris, the cradle of France.

Avenge your martyred sons and brothers

Avenge the heroes who have fallen for the independence and liberty of the Country.

Hasten, by your action, the end of the war.

Have as a watchword: "CHACUN SON BOCHE" [to each his own German]

No quarter to the murderers, onward so that

VIVE LA FRANCE!

The city had no or little electricity, no gas, no Métro, and no buses, and it was still full of Germans. The resistants were not content to wait for the Allies; they were taking matters into their own hands. That morning, the striking police had taken over the Prefecture of Police Building on the Île de la Cité opposite Notre-Dame and unfurled the French tricolor from its mast—the first time that the flag had been flown over the capital in four years. The bold seizure surprised not only the Germans but Rol-Tanguy as well.

The insurrection had started.

A bicyclist going the other way called out to Noufflard as he passed her, "That way is dangerous," but it was too late. She had turned onto the avenue de Tourville, only a short distance from her home, but had to stop as a column of German trucks sped by. They were loaded with armed soldiers, their guns pointed at the civilians with their fingers on the triggers. They squeezed off several rounds in the air to clear pedestrians from their path. Just as Noufflard decided that the place was too dangerous, one truck ran over a woman pushing her bicycle. The driver sped on without stopping. As passersby stopped to help, another truck with a machine gun mounted over the cab came to a screeching halt in front of the helpless crowd. Noufflard watched as the machine-gunner, with an extremely tense and spiteful expression on his face, pointed his weapon at the unconscious woman. An officer quickly intervened before the gunner fired, and the woman was carried away.

Some of the trucks sweeping through the streets were leaving the city, part of a "strategic withdrawal" of some units; others were racing to the scenes of the major fight that had broken out at the Prefecture of Police, as well as skirmishes on the boulevard Saint-Germain, the boulevard and Place Saint-Michel, at the Place de l'Odéon, on rue Saint-Jacques, at the Palais de Luxembourg, and at Luxembourg Gardens.

EARLY THAT AFTERNOON, Camus and two friends braved the firing nearby and set out from the neighborhood of Les Invalides to make their way across the Seine. Prompted by news of the Allied advance, Camus had only recently bicycled back to Paris after several weeks of lying low in Verdelot following Jacqueline Bernard's arrest. His purpose that day was a reunion of what remained of the *Combat* staff at 100 rue Réaumur, not far from where he and Casarès had had their encounter with the police. Pascal Pia was already there waiting for everyone. Marcel Gimont was also expected, along with Henry Cauquelin, an old friend of Pia and Camus from the days of *Paris-Soir.*

Once home to the newspaper *L'Intransigeant,* the building had for the past four years housed *Paris Zeitung,* the newspaper of the German occupation forces. A plan for the elimination of all German and collaborationist presses had long been decided. The provisionary French government in Algiers had issued a circular in May to ensure the resumption of news services upon liberation. All fifty-six dailies that had continued to be published in the occupied zone fifteen days after the armistice of June 1940, or the fifty-one dailies that continued to be published in the southern zone after November 1942, were to be suspended. In their stead would be the Resistance newspapers or those, like the Communist *L'Humanité,* whose publication had been banned by the Germans or Vichy. The rue Réaumur offices had been abandoned only days earlier. Now *Combat* was assigned three offices and a large room upstairs, above those of *Défense de la France* and *Franc-Tireur.* The place was a bit dingy, but the new tenants discovered a huge stock of canned food, reams of printing paper, and a case of grenades. Little forethought had been given to the security of the building, so the three newspapers' staffs decided to put some of the grenades on a few windowsills and on the roof, just in case they might be needed.

Their authorization to begin publishing was to come from Alexandre Parodi, alias "Quartus," the delegate-general of the Gouvernement provisoire de la République française (GPRF), the provisional government headed by de Gaulle. The GPRF had been created at the time of the landings in anticipation of the liberation of France. Parodi's main tasks were to reestablish the government ministries and to serve as an intermediary between the GPRF and Resistance organizations. As fighting broke out, he was weighing many conflicting concerns: the determination of de Gaulle to establish his authority in France; the independent-minded Communists who held many positions in the FFI and who were determined to liberate Paris without outside help; the desire to preserve Paris as intact as possible and with minimal loss of life; and the uncertainty of the German's military plans and their reactions to the escalating violence. Uneasy about the response the newspapers might provoke, Parodi was holding them back for the moment. In the meantime, each staff was preparing its first issues. With tanks rolling by on the street and reports pouring in from around the city, no one was leaving the building. Camus and the other staff members each found a corner in which to sleep.

THE POLICE AND the FFI at the Prefecture barely held out. The Germans attacked the building with tanks and inflicted heavy casualties. Outgunned and very low on ammunition, the men in the Prefecture appealed to the Swedish general consul Raoul Nordling to intervene. Nordling went to see Gen. Dietrich von Choltitz, who had just replaced Carl-Heinrich von Stülpnagel as commander of German forces in Paris on August 9. A veteran of both fronts, von Choltitz had earned the reputation of a committed Nazi officer, and Hitler wanted a firm hand in control of the 17,000-man Paris garrison. His direct orders from the Führer were to "stamp out without pity" any uprising in the city or any other acts of sabotage. To von Choltitz's surprise, Nordling proposed a temporary cease-fire to allow for removal of the dead and wounded. Seeing the potential of returning the city to calm and avoiding further escalation of the battle, von Choltitz agreed. Nonetheless, it had been a costly day for both sides, with about 150 resistants and 50 Germans killed in the uprising.

JUST BEFORE DARK, Monod and Noufflard retreated to the roof of her house to survey the city. They sat in silence while listening to sporadic gunfire and the rumble of German tanks and vehicles moving in the direction of the Seine. It had been an intense day—a mixture of excitement, terror, rumors, anxiety, and hope—the first of who knew how many days of the battle for Paris.

To the Barricades!

Early the next morning, the Hôtel de Ville, the seat of political power in Paris, was occupied without a fight in the name of the provisional government. The occupation of buildings, however, worried Monod. He was concerned that the FFI could not hold buildings against the Germans' superior firepower, as the battle at the Prefecture would have shown were it not saved by the cease-fire. Instead, he had an alternate strategy for the insurrection in Paris, one as old as the revolutions of 1830, 1848, and 1870. That same morning—Sunday, August 20—he dictated an order to Rol:

> The development of operations in Paris demonstrates once more the grave error of occupying buildings or strong points, so well guarded as they are. In the large population centers, as much and perhaps even more so than in the countryside, the FFI can only have one tactic, that of the mobile guerrilla.
>
> Concerning the actual situation in Paris, the possession of the Cité is an accepted fact of which the moral if not the strategic value is considerable, there should be no question naturally of abandoning this position, which ought to be held at any cost. But the command must look to create strong diversions everywhere and as often as possible. Consequently:

1) Multiply the armed patrols by car over the entire length of Paris and the outskirts.
2) Build, wherever possible, beginning with the large main streets frequented by enemy patrols, barricades that are powerful enough to stop automobiles, trucks, and scout cars with machine guns. These barricades should be built with twists and turns that allow the passage of friendly patrols.
3) They should be defended by armed groups that will have the mission of preventing enemy vehicles from penetrating the barricades.
4) The unarmed Milice Patriotiques and the population should be encouraged by means of posters and loudspeakers mounted on cars to participate in the construction of the barricades.
5) Alarm systems should be organized between the different groups defending successive barricades, in order to announce the arrival of tanks that the enemy will doubtless seek to use to force their passage. The barricade guards should then withdraw to the nearest buildings, and seek to attack the tank with grenades if they have them. If not, they should let them pass, and restore the barricade immediately thereafter.

. . . *MALIVERT*

Monod turned to Noufflard as she finished transcribing the order and said, with an impatient tone in his voice, "Of course, this order won't be carried out. It is a shame, too; it might prove very successful."

Monod was wrong on one count and right on the other. Rol did in fact promptly issue an order to build barricades, and the obstacles were very effective at hampering the Germans' movements around the city. All across Paris, men, women, and children organized themselves into human chains: they dug up paving stones; hauled out furniture, mattresses, and kitchen stoves; rolled out old vehicles; cut off tree limbs or downed whole trees; gathered up sandbags; and stacked everything together into formidable barriers, some more than one story tall. Armed FFI, many of the men in open shirts, the women in shorts or summer dresses, and all wearing armbands with the FFI insignia, took up positions behind the barricades and in buildings overlooking the street, waiting to ambush vehicles that ran into their traps.

Noufflard did not stop to build barricades. She had other urgent duties, including delivering a message from Rol to Alexandre Parodi. The delegate-general was hoping to extend the cease-fire of the night before into a lasting truce with the Germans that would enable the handover of Paris to the French, while allowing the Germans to withdraw. Rol, who was not part of the discussions of the truce, objected, as did other Communists.

Noufflard bicycled over to deliver Rol's message to Parodi at an apartment on avenue de Lowendal near the École Militaire. The delegate-general happened to be a very good friend of the Noufflard family. He was the brother of the magistrate René Parodi, who had stayed with the Noufflards for a time after the collapse in 1940 and died in Fresnes Prison in 1942. Living clandestinely in Paris for the past year, Alexandre had even shown up at Noufflard's house once, sporting a new mustache. Parodi greeted Noufflard warmly and talked frankly with her about the situation in Paris.

After leaving Parodi, Noufflard had to navigate again the nearly deserted streets of the École Militaire district. She was very nervous, as the area had been reinforced by the Germans with their own barricades of barbed wire at the entrances and exits of all of their buildings. Machine guns also pointed down each street. Noufflard was relieved to get back home without incident.

Parodi was not so fortunate. After he and two aides were driven off toward another meeting, the car was stopped at a checkpoint. The men were found to be armed and carrying incriminating documents; they were arrested immediately. They identified themselves as "ministers of de Gaulle," so the military tribunal to which they were taken contacted General von Choltitz for instructions. The troops had orders to shoot civilians carrying weapons. "Should we shoot them?" he was asked.

"Yes, of course. Shoot them!" von Choltitz said.

Before he hung up, he changed his mind. If the men were who they said they were, von Choltitz wanted to talk to them. Parodi and his aides were brought to von Choltitz's headquarters at the Hôtel Meurice late that afternoon. De Gaulle's representative and Hitler's proxy met face-to-face, with Swedish consul Raoul Nordling trying to mediate. Nordling had told von Choltitz beforehand that if Parodi was imprisoned or shot, the Communists would take charge and chaos would ensue. The Kommandant told Parodi that the fighting must stop, and released him to Nordling's custody.

But the fighting did not stop. Rol and his subordinates ignored the cease-fire and repeated the order to keep fighting. German vehicles were trapped and ambushed at the barricades, von Choltitz had more than 75 men killed, and French sacrifices also continued: 106 were killed and 357 wounded in the course of the day. The truce was defeated at a vote of the military action committee that evening.

IT WAS A momentous day for two particular Frenchmen. That morning, German troops removed Pétain from his headquarters at Vichy and moved him to the occupied city of Belfort on the German border. And on a fighter strip near Cherbourg, a twin-engine Lockheed Lodestar named *France* made an unscheduled landing. After four years in exile, de Gaulle had flown back to France from Gibraltar.

Upon arrival, de Gaulle was told, "There has been an uprising in Paris." Visibly upset, he asked to see General Eisenhower in order to convince him to move on Paris. De Gaulle was promptly driven down to Eisenhower's headquarters at Granville, on the coast near Avranches. As the two generals went over maps of the battle zones, de Gaulle warned Eisenhower that an insurrection could derail Allied war plans. Eisenhower was not at all convinced. He saw de Gaulle's concerns as political, not military, and the Supreme Commander's foremost concern remained routing the Germans. He did not want to risk getting entangled in Paris. De Gaulle came away empty-handed. Paris was on its own.

Combat Continues

The next morning, Monod and Noufflard went to see Rol at his command post. They caught up with the FFI commander just as he was descending into his new secret subterranean headquarters far below the Place Denfert-Rochereau. Entering through a trapdoor in the basement of the headquarters of the Paris Water and Sewers Administration, Monod and Noufflard followed Rol, his secretaries, and a guard down 138 steps, into a labyrinth of dark stone and cement corridors lit only by Rol's gas lamp. After passing by several signs marking the streets that ran above them, they reached a great metal door, which opened after a password was given. Once inside, Noufflard marveled at the clean, brightly lit offices that, despite their location deep within the Paris Catacombs, had fresh air. The offices were supplied with detailed maps of Paris and enough food for Rol's entire staff. In addition to the city's electrical supply, the command post also had diesel-powered and pedal-powered generators. There were public telephone lines, as well as a private network that went out to 250 points across Paris and its outskirts.

Part of the water and sewer works that dated back to the eighteenth century, the command post was organized by a Resistance engineer named Tavès who worked for the utility. It was connected to the more

Jacques Monod's identity card for the French Forces of the Interior (FFI), in his *nom de guerre* "Malivert." (Courtesy of Olivier Monod)

than three hundred miles of tunnels and catacombs running underneath the city so that one could cross between districts without going aboveground. Here, from this fortress eighty feet beneath Paris, protected from probing German eyes and ears, Rol intended to conduct the insurrection.

Monod and Noufflard would each return to Rol's lair in the coming days. To gain entrance, they would have to show their newly issued tricolor FFI identification cards. Thousands of cards would eventually be distributed; Monod was issued number 2 and Noufflard number 7.

For two days, the *Combat* staff had been holed up in their much hotter, stuffier aboveground quarters preparing a first issue. Their electricity was unreliable, but the telephone lines around Paris were, fortunately, still working. The small team followed the rapidly evolving battles in the city, kept track of the Allies' advance and the Germans' retreat, and read the communiqués from the FFI, the Allies, and the provisional government. Of course, much news quickly became outdated, so dispatches had to be refreshed while they awaited final permission to publish.

Since their July meeting in his studio, Camus and his colleagues were determined that the editorial voice and identity of the newspaper was more important, at least to them, than the news they would publish. They had formulated a new motto to replace the original: "A Single Leader: De Gaulle; A Single Fight: Our Liberty." The public incarnation of *Combat* would adopt "From Resistance to Revolution" as its creed. Camus had readied a long article bearing the same title and explaining their purpose, which was to look beyond the approaching liberation and to ask what kind of country they wanted to see emerge from "five years of humiliation and sacrifice." Their answer was a France that honored what they saw as "a revolutionary spirit growing out of the resistance," one that they would "define for the world and for ourselves, the image and example of a nation saved from its worst mistakes and emerging . . . with a youthful visage of grandeur regained."

It was a markedly different tone, and a much more introspective and reflective path, than the one some of their competitors would take.

While *Combat* spoke of honor, justice, and the work ahead, *L'Humanité* would declare "Death to the Boches and the Traitors!"

Finally, in the afternoon, under pressure from Pia and others, Parodi granted permission to publish. Operating with only intermittent supplies of electricity, the presses turned out 180,000 copies of *Combat* issue number 59—a single sheet printed on both sides. Street vendors hawked the newspaper for two francs.

One of the lead stories was an hour-by-hour digest of the insurrection, accompanied by a photo of a shirtless, armed FFI man, beret cocked on his head. It also reported that de Gaulle had arrived in France, and it announced that the newspaper would appear every morning. In the far-left column of the front page, Camus had composed a short editorial; it was signed only with an "x" as the liberation was not a fait accompli and the use of any names or aliases could be fatal. He entitled it "Combat Continues . . .":

> Today, August 21, as this newspaper hits the streets, the liberation of Paris is nearing an end. After fifty months of occupation, of struggle and sacrifice, Paris is rediscovering the feeling of freedom, even as bursts of gunfire erupt at street corners around the city.
>
> It would be dangerous, however, to return to the illusion that the freedom that is due of every individual comes without effort or corresponding pain. Freedom has to be earned and has to be won. It is by fighting the invader and the traitors that the Forces Françaises de l'Intérieur are restoring the Republic, which is the indispensable condition of our freedom . . .
>
> The liberation of Paris is but one step in the liberation of France . . .
>
> The combat continues.

"Mouvement Immediat sur Paris!"

Indeed, the combat in the streets would continue. Late in the day on Monday, Parodi had made a second attempt to convince representatives of the various French factions to agree to a truce. He failed.

Tuesday, August 22, would see some of the heaviest fighting yet. At the intersection of boulevard Saint-Michel and boulevard Saint-Germain, the FFI destroyed several German trucks and took a dozen prisoners. But German tanks rolled through the main arteries, crashing through some barricades and attacking FFI positions. Three tanks attacked the central post office, four tanks attacked at the Panthéon, and several tanks fired on the Hôtel de Ville. The defenders' main weapon against the machines were Molotov cocktails that were now being mass-produced daily by teams in several Paris university laboratories, using a recipe developed by Frédéric Joliot-Curie, the 1935 Nobel laureate in Chemistry. The mixture combined gasoline, sulfuric acid, and potassium chlorate, and ignited on impact. Indeed, Joliot-Curie's lab at the Collège de France was one of the main manufacturing sites that were producing several hundred bottles a day.

Rol directed the distribution of the homemade explosives from his underground command post as he followed the movements of tanks around the city. One of the tanks menacing the Hôtel de Ville was knocked out when a young woman crawled up onto it and tossed an explosive Champagne bottle into its open turret. At a barricade in front of the Police Commissariat in the fifth arrondissement, a man ran out and smashed a bottle into one tank's ventilator, setting it on fire. Both attackers were gunned down before they could make their escapes.

How long the FFI could hold out with its few and primitive weapons and little ammunition (except that captured from the Germans), or how long Parisians could manage with little or no electricity and a dwindling food supply, was not certain.

To the west in Normandy, General Eisenhower was having second thoughts about Paris. Only a day after their meeting in person, he had received a letter from de Gaulle trying to convince him once again to send the Allies immediately to Paris. De Gaulle had been receiving repeated appeals from Parodi and his military delegate in Paris. He wrote to Eisenhower: "Information received today from Paris leads me to believe that . . . serious trouble may shortly be expected in the capital." The situation was so critical, de Gaulle urged, that Paris should be taken "even if it should produce fighting and damage in the interior of the city." While Eisenhower pondered de Gaulle's message, Generals Bradley and Sibert arrived at his headquarters with news that a representative of the Resistance in Paris had reached them and was pleading for help, warning of a potential massacre if the Allies didn't come soon.

Eisenhower relented. The Falaise Gap had been closed; the Allies had taken 50,000 prisoners and killed another 10,000 Germans, bringing the overall German losses in Normandy to a staggering 200,000 dead and another 200,000 taken prisoner. There was no reason to hold Leclerc any longer, and good reasons to send him to Paris. While the DB was not the closest Allied division to the city—it was still 120 miles away—Eisenhower would keep his word and send Leclerc's division to Paris, with the support of the US 4th Armored Division. Bradley and Sibert flew back to give Leclerc the news. Leclerc returned just before nightfall to his command post in an orchard in Fleuré and shouted from his Jeep: *"Mouvement immédiat sur Paris!"*

"They Shall Not Pass"

The next morning's issue of *Combat* (August 23) reported on the previous day's battles around the city, and noted that the Allies were rolling toward Paris. One could ask, as Camus did in his editorial, since the Allies were coming, why should Parisians risk the fight? He was again looking beyond the immediate prospect of liberation to the future dividends of the struggle:

> What is an insurrection? It is the people in arms. What is the people? It is that in a nation which refuses ever to bend its knee.
>
> A nation is worth what its people are worth, and if ever we were tempted to doubt our country, the image of its sons on the march, brandishing rifles, should fill us with overwhelming certitude that this nation is equal to the loftiest of destinies and is about to win its resurrection along with its freedom . . .

The enemy ensconced in the city must not be allowed to leave. The retreating enemy must not be allowed to reenter. They shall not pass.

To those Frenchmen bereft of memory and imagination, forgetful of honor, heedless of shame, and cushioned by their own personal comforts who ask, "What good can any of this do?" we feel compelled to respond here and now.

A people that wants to live does not wait for its freedom to be delivered to it. It takes its own. And in so doing it helps itself as it helps those who seek to help it. Every German who is prevented from leaving Paris means one bullet less for the Allied soldiers and our French comrades in the East. Our future, our revolution, depend entirely on the present moment, echoing with cries of anger and with the wrath of liberty.

While Camus wrote for Parisians, Monod and Noufflard prepared official FFI communiqués for the radio and the outside world. That same morning, Noufflard typed and sent a cable to Philo in Switzerland reporting on the state of Paris as of August 22:

Give through Swiss press and foreign correspondents widest diffusion to following informations [*sic*]. Stop. Since August 19 FFI and risen Parisian population are fighting the enemy. Stop. Entirely rallied to FFI, Police holding Prefecture and Palais de Justice. Stop. Furious enemy assaults with tanks repelled. Several tanks destroyed. Stop. Combats all over Paris between FFI patrols and German patrols. Stop. Enemy checked everywhere suffering heavy losses men and material. Stop. All town halls and ministries occupied by new administration. Stop. New newspapers issued. Three radio transmitters broadcasting informations [*sic*] and instructions. Stop. Military command secured by Rol Regional Chief FFI.

Encouraged by posters, communiqués, and perhaps Camus's very words, the fighting continued. As exhilarating as the sight was of Parisians liberating their city, it made getting from place to place very dangerous. Noufflard took daily communiqués to the newly reoccupied Ministry of Information on the rue de Lille, on the Left Bank, close to the Seine and contested areas. A short distance away, smoke was pouring out of the Grand Palais, which had been destroyed in a German attack. Noufflard and Monod arrived early that evening to find Pierre Schaeffer, the engineer in charge of transmission, discussing his concerns with the minister about the safety of the studios. With the sound of gunfire going on outside Monod said, "Why don't we give up tonight's transmission? Tomorrow we will try to provide you with a guard." Schaeffer shrugged and replied, "You know perfectly well that I *shall* transmit, even with no guard at all," and then he left for the studios nearby.

Monod and Noufflard left the Ministry on their bicycles, but as they approached the rue de Grenelle, near the studio building, they heard gunfire. Then, they saw a tank—the Germans were attacking the radio building. They feared for their friend Schaeffer but could do nothing. Soon they saw FFI men with rifles and hand grenades working their way house-by-house up the street to meet the Germans.

Monod and Noufflard waited nervously for the squad to pass out of sight before they crossed the street and hurried back to a friend's house nearby. There, they anxiously turned on the radio. At seven o'clock, they heard Schaeffer's calm voice reading their communiqué, with gunfire in the background.

A Day of Rumors, the Night of Truth

"The Americans are coming." "The Allies are just outside Paris." "Panzers are coming to defend Paris." "The Germans are going to blow up the bridges and monuments." All sorts of rumors were swirling around Paris. Each of these four, however, did have some element of veracity. Hitler had ordered repeatedly that Paris be "defended to the last man" and that all of the bridges across the Seine be mined in preparation for their destruction. He had also personally ordered the movement of two SS Panzer divisions to defend the city. On the night of August 22, he told his commander in the west, "The defense of the Paris bridgehead is of decisive military and political importance . . . In history the loss of Paris always means the loss of France . . . Paris must not fall into the hands of the enemy except as a field of ruins." All of the bridges had indeed been mined, and only the day before, explosive charges had been placed under Les Invalides, the Palais du Luxembourg, the Palais Bourbon, and Notre-Dame.

The Panzers were not yet in position, but the Allies were very close. Camus confirmed that rumor by sending a writer and Pierre Gallimard out of the city to the Allied lines. They met an American major twenty-five miles southwest of Paris who told them, "We are coming." The story ran in that morning's newspaper, but the major probably did not know that the French DB was intended to be the first into Paris. The DB was fighting its way in along three routes. When it would get there, and how the Germans would react once it did, no one knew.

Fierce fighting continued at German strongpoints within Paris, especially near the Place de la République. Monod and Noufflard spent the day bicycling across the city, going from barricade to barricade. They, too, had heard that the Allies were close. That evening, they were at a friend's home on rue Vaneau, listening to the radio as Pierre Schaeffer was reading out a flurry of communiqués, reports, and rumors. At 9:32, they heard correspondent Pierre Crenesse break in to announce that tanks of the DB had arrived at the Hôtel de Ville.

"Parisians, rejoice!" Schaeffer exclaimed. "The Leclerc Division has entered Paris. We are mad with happiness."

He then read three lines from Victor Hugo's *Punishments*:

Awake! Be done with shame!
Become again great France!
Become again great Paris!

Then Schaeffer played "La Marseillaise":

Allons enfants de la Patrie
Le jour de gloire est arrivé! . . .
[Arise, children of the Fatherland, the day of glory has arrived.]

Then he asked that "all the parish priests ring their church bells."

Monod and Noufflard opened the windows and waited . . . They heard a bell in the distance, then another one closer, and soon a chorus of bells across the city—from Notre-Dame to Sacré-Coeur, ringing for the first time in four years.

NOUFFLARD AND MONOD wanted to join the celebration in the streets, but Monod could not leave the house. He had to wait for a phone call from General Joinville, the head of the FFI. There was a preset plan to take over the Ministry of War, and the call could come at any moment. Monod told Noufflard to go on out while he waited.

She went out into the street. People were singing and shouting; a café had opened its front windows, spilling light onto the street. Noufflard was still wearing her FFI armband; the crowd saw it and started chanting: *"Vive les FFI!"*

She wanted to get across the Seine to the French tanks. But when she reached the river, a German Tiger tank was patrolling and firing its large gun. Gunshots were still ringing out in the area, so she thought better of it and turned back.

Just after she returned, the phone rang. Joinville ordered Monod to go to the Ministry of War. "You are coming with me," he said to Noufflard.

The Ministry was not far—about five large blocks away on the rue Saint-Dominique—but it was very dark and they could not see anything, not even each other. The gunfire was still ongoing, so they pressed against the walls as they crept down rue Casimir Périer. Noufflard kept her hand on Monod's back so as not to lose contact with him. As they started to cross the small park across from Sainte-Clotilde church, there was more gunfire very close by. Too close, Noufflard thought. She could not tell whether it was coming from the bell tower or the bushes. "Who are they shooting at?" she wondered, before realizing that it was she and Monod. It was so dark, she figured whoever it was could only aim in the direction of the sound of their footsteps—the only sound, besides gunshots, that could be heard on the street.

When they reached the door to the Ministry, they were stopped by a group of FFI men with tommy guns; they were there to guard Monod and Noufflard. They entered a grand hallway of large black-and-white flagstones, decorated with armor, and were then led into a large salon. Sitting in a circle of elegant, high-backed eighteenth-century armchairs was a group of military men in civilian clothes. They were mostly generals and colonels attached to the Vichy government. They appeared a bit stunned to see Monod and Noufflard, who were disheveled and dirty from pedaling across the city. She was wearing an old skirt, and her legs were black with grease from her bicycle chain; Monod was in a suit that was too small for him and that had been mended at the knees. Nevertheless, the assembled gentlemen were to turn over the Ministry to them.

Noufflard was awestruck by the surroundings. The salon was in the oldest part of the Ministry building that was once the residence of Laetitia Bonaparte, Napoleon's mother. It had very tall windows that were draped with immense blue damask curtains, fine gold-and-ivory-colored paneling, and a massive mahogany desk in the center. A spectacular chandelier hung from the ceiling.

The conversation was cordial but was interrupted from time to time by the din of battles outside. One of the windows was pierced by a bullet, and the louder detonations made the chandelier shake and swing, so everyone was careful not to stand under it. A case of champagne appeared, booty reclaimed from an apartment that had been occupied by Germans. Fine crystal goblets were brought out and filled. Everyone stood as one of the generals made a formal toast to "Victory." The Vichy men then said farewell.

As the time approached two in the morning, the FFI officers insisted that Noufflard get some sleep. She was too excited and resisted at first, until they showed her the bedroom. It was Madame Mère's—Napoleon's mother's bedroom—with a beautiful little square boudoir with ivory-and-gold paneling and embossed yellow velvet. Still too excited to sleep, she decided to write a letter to her parents to reveal, after eight months of secrecy, that she was in the FFI working for her "illustrious boss," Commandant Monod, and to tell them about the historic night in Paris.

—~—

NOUFFLARD WAS NOT the only Parisian up writing in the wee hours that night; Camus and the staff of *Combat* had a newspaper to put together. The morning's headline would read:

AFTER FOUR YEARS OF HOPE AND STRUGGLE FRENCH TROOPS ENTER INTO THE LIBERATED CAPITAL

For the fifth consecutive day, Camus sought the most apt prose for the historic moment, this "Night of Truth":

> As freedom's bullets continue to whistle through city streets, the cannon of liberation are passing through the gates of Paris amid shouts and flowers. On this sultriest and most beautiful of August nights, the permanent stars in the skies mix with tracer rounds, smoke from burning buildings, and multicolored rockets proclaiming the people's joy. This night unlike any other ends four years of a monstrous history and an unspeakable struggle that saw France at grips with its shame and its fury.
>
> Those who never lost hope for themselves or their country are finding their reward tonight. This night is a world unto itself: it is the night of truth. The truth in arms, the truth in battle, the truth in power after languishing for so many years empty-handed and chest bared . . .
>
> Four years ago, a few men rose up amid the ruins and despair and quietly proclaimed that nothing was lost yet. They said that the war must go on and that the forces of good could always triumph over the forces of evil provided the price was paid. They paid that price. And the cost was indeed heavy: it had the weight of blood and the terrible oppressiveness of prison. Many of those men died, while others spent years enclosed within windowless walls. That was the price that had to be paid . . .
>
> Harsh battles still await us. But peace will return to this gutted earth and to hearts tormented by hope and memories . . .
>
> Nothing is given to mankind, and what little men can conquer

must be paid for with unjust deaths. But man's grandeur lies elsewhere, in his decision to rise above his condition. And if his condition is unjust, he has only one way to overcome it, which is to be just himself. Our truth tonight, the truth that hovers in the August sky, is in fact man's consolation. What gives our heart peace, as it gave peace to our dead comrades, is that we can say before the impending victory, without scolding and without pressing any claim of our own, "We did what had to be done."

Friday, August 25

Dawn broke to a perfect clear-blue-sky summer morning. General Joinville arrived at the Ministry of War. With lumps in their throats, Noufflard and the FFI general staff watched as an FFI guard presented arms and the French tricolor was raised over the building. The staff then went to greet General Leclerc while Noufflard stayed behind to handle communications, the sole representative of the FFI in the Ministry.

Early that morning, American columns and more of the DB arrived triumphantly in the city, swarmed by hordes of ecstatic Parisians. Paris, however, was not yet free. As they reached the city center, the liberators began to encounter German strongpoints. Large numbers of German troops held the École Militaire, the Senate (Palais du Luxembourg), the Chamber of Deputies (Palais Bourbon), and the area around von Choltitz's headquarters on the rue de Rivoli. Fresh battles erupted at each location.

The Chamber of Deputies was just behind the Ministry of War. From her elegantly appointed post, Noufflard looked out over a shady garden where FFI men had taken positions behind a high wall and were exchanging fire with Germans on the other side. The Germans had a heavy gun firing from the Chamber of Deputies, and an occasional shell passed over the Ministry and onto buildings across the rue Saint-Dominique. The chandeliers were swinging again, and the racket was deafening. It was very difficult for Noufflard to answer the constantly ringing phones.

Finally, in the late afternoon, the fighting around the Chamber of Deputies died down and the officers returned to the Ministry. Noufflard showed a newly arrived member of the staff around the building and pointed out the flag flying over the main entrance on rue Saint-Dominique. Then, to their astonishment, the front gate opened and a long black Hotchkiss sedan bearing a tricolor flag with the Croix de Lorraine pulled in. A very tall man in a general's uniform stepped out and looked up at the Ministry building. They had never seen a picture of de Gaulle, but they knew that it was *him*—the voice they had listened to over the BBC during four years of occupation. Noufflard ran to alert the general staff.

De Gaulle had just accepted von Choltitz's surrender at Leclerc's headquarters at the Gare Montparnasse. He then came straight to the Ministry where he had once worked.

Monod and the other officers assembled and stood at attention at the top of the stairs. The general entered the great hallway, strode up the steps, and then shook hands with each one of them.

A LITTLE WHILE later, the FFI officers moved to make room for de Gaulle and his staff. Noufflard went back to gather her belongings from Napoleon's mother's bedroom. She noticed that someone else's luggage had already been moved inside; it bore the inscription COLONEL CHARLES DE GAULLE.

AT LAST, MONOD and Noufflard could enjoy a free Paris on a perfect, beautiful, clear summer night. Monod commandeered a car, upon which he and Noufflard fixed a small tricolor flag that had the Croix de Lorraine and the letters "FFI" embroidered on it, and they went for a celebratory drive. The car's exhaust pipe was broken, so it made a very loud racket, but they ignored it as they sped through the streets enjoying their first ride in years. They cruised past burned-out German vehicles, trucks carrying FFI men, and throngs of joyous Parisians. They reached the Place de l'Étoile near sunset, where they saw their first American tanks and soldiers who had gathered around the Arc de Triomphe.

Darkness fell.

Then, suddenly, the city lit up. All over Paris, streetlights and monuments that had been dark for almost six years, since September 3, 1939, flashed back to life. Paris's engineers and electricians had managed to summon enough power to relight Sacré-Coeur, Les Invalides, Notre-Dame, and the Eiffel Tower, from the top of which the French tricolor now flew for all to see.

On this first evening of freedom, the happiest in all of her glorious history, Paris was once again *la plus belle ville du monde.*

APRÈS QUATRE ANS D'ESPOIR ET DE LUTTE

ÉDITION DE 5 HEURES

COMBAT

DE LA RÉSISTANCE A LA RÉVOLUTION

VENDREDI 25

LES TROUPES FRANÇAISES entrent dans la capitale libérée

La nuit de la vérité

DE GAULLE serait aujourd'hui à Paris

UN JOUR DE BATAILLES et une nuit d'enthousiasme dans Paris libéré

Les partisans attendent

DE FONTENAY-AUX-ROSES A PARIS avec une colonne Leclerc

(De notre envoyé spécial)

PAVOISEZ POUR LA LIBÉRATION !

Le nouveau Préfet de la Seine Marcel FLOURET s'adresse aux Parisiens libérateurs de la capitale

SACHA GUITRY Jérôme Carcopino, Paul Chack et quelques autres ont été arrêtés

LYON et BORDEAUX ont été libérés hier

Le Havre et Rouen menacés

Grosse avance russe vers Cracovie

Parisiens n'utilisez plus le gaz

Les S.S. installent Pétain et ses ministres dans des châteaux du Jura

Les villes de l'Ouest acclament le général de Gaulle

ROUMAINS ET ALLEMANDS SE BATTENT DANS BUCAREST

Un officier allemand est fait prisonnier par les F.F.I.

Combat issue number 63, August 25, 1944. The headline heralds the arrival of French troops in the capital. Camus's anonymous editorial begins at the far left: *"La nuit de la verité"* ("The night of truth"). (Author's collection)

Part Three

Secrets of Life

Genius is talent set on fire by courage.

—Henry van Dyke, *The Friendly Year*

Camus at *Combat* shortly after the liberation of Paris. Camus is at left; André Malraux is at right, wearing a beret. (Rue des Archives/The Granger Collection, New York)

CHAPTER 17

THE TALK OF THE NATION

The first thing for a writer to learn is the art of transposing what he feels into what he wants to make others feel.

—Albert Camus, *Notebooks, 1942–1951*

THE DAYS FOLLOWING THE LIBERATION WERE EUPHORIC.

The first full day of Paris's freedom was celebrated by de Gaulle triumphantly parading down the Champs-Élysées from the Arc de Triomphe, with he, Leclerc, and Parodi preceding columns of French and Allied troops. The next morning's issue of *Combat* accurately proclaimed: "All Paris in the Street to Cheer de Gaulle."

It was also a time of revelations, when aliases and anonymity were dropped, and true identities were revealed. The same issue of the newspaper named the men behind *Combat* for the first time. A small notice after the editorial stated: "Albert CAMUS, Henri FREDERIC, Marcel GIMONT, Albert OLIVIER, and Pascal PIA actually write 'Combat,' after having written it in clandestinity."

Camus perceived that it was a critical moment for the press—"a unique opportunity to create a public spirit and to rise to the level of the country at large." Ever since that first meeting in his studio in July, planning for the public issuance of *Combat,* Camus had anticipated the role the newspaper would play in shaping the public discourse about the direction France would take once liberated. On the eve of liberation, he wrote: "The Paris that is fighting tonight wants to assume command tomorrow. Not for the sake of power but for the sake of justice, not for political reasons, but for moral ones, not to dominate their country but to ensure its grandeur."

It was, Camus would say, a time of danger and hope. Danger because France's economy was in ruins, its farms had no machinery, Germany was not yet defeated, and its army had no arms (except those given to it); hope because freedom had been regained, and if it learned from its very painful and costly lessons, there was the prospect of the rebirth of a greater France. In the days following liberation, Camus seized the platform that *Combat* provided to articulate his and his comrades' highest ideals, to raise readers' hopes, to warn them of dangers, and to criticize those who had not or were not living up to their duty. Some of his first remarks were directed at the press, which served both as a statement of his paper's aspirations and a critique of those who were returning to bad habits. In his signed editorial "Critique of the New Press," published in the first week after liberation, Camus expressed the hope that he and his peers, "who had braved mortal dangers for the sake of a few ideas they held dear, would find a way to give their country the press it deserved and no longer had." He dismissed the prewar press as one that had "forfeited its principles and its morals" and whose "hunger for money and indifference to grandeur" had no goal beyond enhancing the power of a few. Such low aims, Camus asserted, prepared the way for the collaborationist press of the Occupation.

The common desire of the underground Resistance press, Camus believed, was to express "a tone and a truth that would allow the public to discover what was best in itself." Camus exhorted his peers to recommit to this "tone," and to create a press "that is clear and virile and written in a decent style." He reminded them: "When you know, as we journalists have known these past four years, that writing an article can land you in prison or get you killed, it is obvious that words have value and need to be weighed carefully."

Camus urged his colleagues "to write carefully without ever losing sight of the urgent need to restore to the country its authoritative voice. If we see to it that voice remains one of vigor rather than hatred, of proud objectivity and not rhetoric, of humanity rather than mediocrity, then much will be saved from ruin, and we will not have forfeited our right to our nation's esteem."

With the debris from the battle of Paris still littering the streets, and while the FFI and civilian losses were still being counted, Camus was exhorting the press to lead the nation away from its most base reactions to four years of oppression and toward the goal of making a new democracy in France. Camus hoped to influence, or at least to admonish, those newspapers that "seek to please when they ought simply to enlighten" or those that "seek to inform their readers quickly rather than to inform them well," for he noted: "The truth is not the beneficiary in this setting of priorities."

Camus, of course, had to practice what he preached. To enrich *Combat*'s tone and style, he commissioned Jean-Paul Sartre to write a series of articles on the liberation of Paris, then sent him on a long tour of the United States as a correspondent. And day after day, Parisians read and discussed Camus's editorials that spoke repeatedly of truth, courage, character, liberty, and democracy. Beyond establishing a nobler press, Camus largely focused on a few dominant issues. Foremost among these was the question of what kind of nation was to be rebuilt. For Camus, the crux of the issue was reconciling justice and freedom. He wrote: "To ensure that life is free for each of us and just for all is the goal we must pursue . . . Indeed, nothing else is worth living and fighting for in today's world."

A second major concern was the matter of who was fit to lead France. He castigated any thought of reinserting "experienced men" from previous regimes (Vichy or the Third Republic), as they had led the country to ruin, or were at the very least guilty of not having done enough to save it. He declared: "The affairs of this country should be managed by those who paid and answered for it."

A particularly sensitive issue was the recognition of de Gaulle's provisional government (GPRF) by the Allies. Britain and the United States were keenly aware that, despite talk of a united France, de Gaulle, the various parties of the Resistance, and the Communists viewed one another with suspicion. British and American leaders harbored reservations about each faction and were tempted to play favorites. In four editorials, Camus pressed the case for recognition of the provisional government as the authority within France, and cautioned against meddling in what he insisted were France's internal affairs. "If our American friends want a strong and united France, dividing France from outside is not a good way to go about it," he wrote. Camus, who had devoted an entire editorial to paying respect to Britain's "inner strength and tranquil courage" and her defense of freedom during the war, made a simple, direct case: "We have lost many things in this war, but not so much that we are willing to resign ourselves to begging for what is rightfully ours . . . We want to be able to love our friends freely and to prove that there is no bitterness in our gratitude. We believe that we are not asking for much. And if that is still not convincing, then we ask that steps be taken to ease our difficult task in view of our long history of teaching the very name of freedom to a world that knew nothing about it."

While elections would have been the preferred means of settling the issue, no vote was initially possible given the country's disarray. Moreover, some argued, including Camus, that it would be unfair and disrespectful to the more than two million POWs, deportees, and workers outside of the country who were unable to vote. Days after his fourth editorial on the matter, the American, British, Soviet, and Canadian governments recognized the GPRF.

Camus had urged the press to provide the country "with a language that will induce it to listen." And listen to Camus they did. The novelist-playwright turned editorialist now had an audience in the hundreds of thousands. Writer Raymond Aron, who had joined de Gaulle's Free French, edited its magazine during the war, and would later join *Combat* as an editorialist, remarked that readers of Camus's editorials had formed the habit of getting their daily thought from him. *Combat* often sold out, and Camus's editorials were the talk of Paris—both on the street as well as among the intelligentsia.

After the liberation, Camus's private life also underwent a transformation. Francine returned to Paris from Algeria in October, and she was pregnant by December—with twins. Maria Casarès chose to exit the scene.

The Amalgamation

Five years after war was first declared, and after more than four years of occupation, it was understandable that in the joy of liberation most Parisians were eager to reunite their families and to return to some semblance of normal life. Indeed, by October, the Monods were all back together in Paris in their apartment on rue Monsieur-le-Prince: Jacques was working out of an office in the Ministry of War, Odette was spending time again at the Musée Guimet, their maid had returned to the household, and the twins were enrolled in school. There was a lot of talk around the house about politics and the military. When one or the other boy did something good, one brother would "promote" the other to "lieutenant" or "colonel," and he would be allowed to wear their father's FFI armband.

The happy reunion was to be short-lived. The war was not over. As Camus had written earlier in *Combat*: "We have enjoyed our victory but have yet to see the victory of all. And we now know that all we have won is the right to go on fighting." Hitler and Germany appeared determined "to end it all in the most tragic and theatrical of suicides" and to "impose a heavy price in blood." Thus, Camus cautioned his fellow Frenchmen, "We must convince ourselves that the war is going to last and accept the sacrifices of victory as courageously as we assumed the burdens of defeat."

If France was to regain her honor, and to earn a place at the victors' table, the French and their army needed to fight, and to fight well, to the end of the war. De Gaulle told Leclerc in early October 1944: "It is essential that we participate in the future battles of 1945 with maximum force. Nothing is more important for the moment than to form new large units." By that time, the combined British, Canadian, and American forces had well over two million men and counting in the European theater, whereas the entire French Army constituted just 250,000 troops that had been drawn from throughout her empire. The largest source of fresh French manpower and fighting spirit was the FFI, which comprised about 200,000 men. The FFI was far from a professional army, however. Many of its members were under twenty years old and had therefore not received any conventional training. Large bands of *maquis* had fought the Germans and liberated some areas on their own, but advancing into well-defended Germany as the armies now had to do—that was an altogether different proposition and would require close coordination among Allied forces. The immediate challenge to the French command was to integrate the untrained, poorly equipped, largely undisciplined, and politically fractious FFI into the French Army. That would be a complex task that required people with excellent knowledge of the Resistance and the FFI, who had the trust of Communist groups, and who could work within the hierarchy and discipline of the French Army.

It was a job for someone like Jacques Monod.

He was reassigned from the FFI to the French 1st Army (La première armée française) and received his orders in November 1944: "To assist . . . in the study and the development of plans concerning the integration of the French Forces of the Interior in the First Army." Monod was to serve in the cabinet of Gen. Jean de Lattre de Tassigny, the charismatic head of the 1st Army and a future marshal of France. Wounded four times in World War I, de Lattre became the youngest French general in 1939, and his 14th Infantry Division performed well during the invasion of 1940. He then commanded French Army Group B, the predecessor to the 1st Army, in the Allied landings in southern France in mid-August 1944 (Operation Dragoon). He took Toulon and Marseille, then fought his way north, pushing the Germans back into the Vosges Mountains.

Passionately devoted to France, de Lattre was an inspiring leader, but also very demanding of his soldiers and staff, and often a stickler for discipline. Nevertheless, he was enthusiastic about incorporating the Resistance into the regular Army—in spite of the skepticism of both parties—and was determined to "win over this vibrant and tumultuous force without distorting it." The effort was to become known as the "Amalgamation."

One major challenge was the shortage of equipment—the FFI needed uniforms and weapons, and it had to rely mostly on captured German arms. A second, more difficult, challenge was the great difference in cultures within the regular Army and the *maquis*. The general strategy was to pair FFI groups with regular units. De Lattre likened the process to a chemist mixing test tubes of chemicals: some reactions went smoothly; others did not and had to be remixed. Moreover, these experiments had to be conducted in the midst of the battle, as most FFI units to be organized and trained were already in the front lines. Monod oversaw several such experiments and spent much of the brutal winter traveling on icy roads to and from the front lines in a jeep, visiting units, managing problems, and consulting with de Lattre as the 1st Army battled its way to and across the Rhine.

De Lattre could be personable, thoughtful, and charming. He was very appreciative of, and even affectionate toward, Monod, calling him his "most precious *auxiliaire*." By February, an impressive 137,000 FFI had been integrated into the 1st Army that invaded southern Germany; took Karlsruhe, Stuttgart, and Ulm; crossed the Danube; and entered Austria in late April 1945.

By the first of May 1945, Monod believed that the Germans were finished. He wrote to Odette:

> *My dear Odette,*
>
> *I am thinking about you, you before anything else this morning, as I just heard that probably today or tomorrow will be the last day of the war. I am thinking about you, about our little ones, about us, about our life about to resume its course, about our happiness . . .*
>
> *There is a big piece of work to be done in the next couple of weeks, because everything has to be settled between the 15th and the 20th. But I will have at least the satisfaction that I would leave only when the situation would be stabilized, and after I would have reached my goals for the most part, and I am sure by now to be able to ask for demobilization the 31st without remorse.*
>
> *I just heard the radio again: the English will be in Lubeck tonight or tomorrow probably. How wonderful to see Étienne [Odette's brother, a POW] again. If I could go get him!*
>
> *My darling, I hold you in my arms. If only we could have been together today.*
>
> *I adore you.*

On May 8, de Lattre flew to Berlin and signed the Germans' instrument of surrender on behalf of France. The war was officially over.

Parisians again poured into the streets to celebrate. Camus summoned his prose once again to mark the day of Victory in Europe. He cast French joy in the broader terms of the universal quest for freedom and his concept of rebellion:

> History is full of military victories, yet never before has a victory been hailed by so many overwhelmed people, perhaps because

> never before has a war posed such a threat to what is essential in man, his rebelliousness and his freedom. Yesterday belonged to everyone, because it was a day of freedom, and freedom belongs either to everyone or to no one . . .
>
> This war was fought to the end so that man could hold on to the right to be what he wants to be and to say what he wants to say. Our generation understood this. We will never again cede this ground.

Camus also acknowledged the losses that virtually everyone had suffered. But as someone who did not believe in eternity, he could not offer that solace. Rather, as one who saw life as a revolt against death, and who advocated living life to the fullest, Camus had wrestled with how to justify the conscious risk and sacrifice of Resistance members. Could he honestly justify their cause as one worth dying for?

In an earlier editorial, Camus had drawn a contrast between the sacrifice made by religious believers, who saw life as a "way station" and who could hope for martyrdom, and nonbelieving Resistance members: "Many of our comrades who are no longer with us went to their death without hope or consolation. Their conviction was that they were dying, utterly, and that their sacrifice would end everything. They were nonetheless willing to make that sacrifice." That willingness Camus defined and praised as "lucid courage." In his V-E Day editorial, he located the meaning of their courage and sacrifice in the greater good of mankind's lasting freedom: "Those of us who are still waiting or still weeping for a loved one can enjoy this victory only if it justifies the cause for which the missing and the dead suffered. Let us keep them near us and not consign them to the definitive solitude of having suffered in vain. Only then, on this overwhelming day, will we have done something for mankind."

The end of the war in the Pacific three months later brought more relief but also new anxieties, as it was precipitated by the dropping of atomic bombs. Camus was understandably alarmed: "Given the terrifying prospects that mankind now faces, we see even more clearly than before that the battle for peace is the only battle worth fighting. This is no longer a prayer but an order that must make its way up from peoples to their governments: it is the order to choose once and for all between hell and reason."

By the end of the war, and by virtue of scores of editorials, the reprinting of *The Stranger* and *The Myth of Sisyphus*, the publication of his *Letters to a German Friend,* the remounting of his play *The Misunderstanding,* and the publication of *Caligula,* Camus was firmly established (along with his friend Sartre) as a leading public intellectual in France. Even his fiercest critics acknowledged that Camus was "the French editorialist who is most widely read in the world."

But Camus aspired to different goals. As the full magnitude of the war's toll came to light—tens of millions of soldiers and civilians killed in battle, millions more exterminated or nearly so in camps, vast populations left homeless and hungry, much of Europe in ruins, and the advent of new weapons capable of leveling entire cities—his audiences struggled to grasp the catastrophe and despaired for the future. Having justified resistance as a reason for dying, the outstanding challenge was to help his readers find good reasons for living.

CHAPTER 18

SECRETS OF LIFE

There is nothing over which a free man ponders less than death; his wisdom is to meditate not on death but on life.

—SPINOZA, *Ethics*

(Quoted by Erwin Schrödinger in the preface to *What Is Life?*)

AS GERMANY WAS DEFEATED, MORE THAN 2.4 MILLION FRENCH prisoners of war, STO workers, political prisoners, and racial deportees began to return from the former Reich—most of the latter in shocking physical condition and with unimaginable, horrifying stories to tell. Camus and *Combat* celebrated Claude Bourdet's rescue from Buchenwald and Jacqueline Bernard's return from Ravensbrück. Camus and Pia went immediately to see Bernard and said, "*Combat* is waiting for you, your office is ready, when you wish." Marcel Prenant, Monod's former FTP chief, returned in early June, weighing just ninety pounds after barely surviving a typhus-induced coma. Odette's brother Étienne returned home as well. However, tens of thousands of Frenchmen and Frenchwomen, among them Bernard's brother Jean-Guy and Monod's friend Raymond Croland, did not return from the camps.

French soldiers also started coming home. After Germany's surrender, the French Army's mission turned from combat to occupation. Monod had no interest in that role, nor any desire to remain in uniform. As six of his first seven years of marriage and all of the twins' first six years of childhood had been occupied by war, he was eager to be demobilized and to return to his family and scientific work. While he had hoped to be released at the end of May, his papers would not be signed until the beginning of July.

After returning to Paris, Monod wanted only to go back to the lab and to think about nothing else. He consciously "drew a curtain over the memory of wartime." He started to pick up where his work had left off a year and a half earlier. He returned to the Sorbonne, but not for long. That fall, Lwoff invited Monod to join his Department of Microbial Physiology at the Pasteur Institute as a laboratory head. Monod accepted, and went about setting up his lab and catching up on what had transpired while he had been away from biology.

He had learned of one major advance while still in the Army. Before his release, he happened across several back issues of the scientific journal *Genetics* in, of all places, an American Army bookmobile. In the November 1943 issue, an article by Salvador Luria and Max Delbrück seized his attention. The two authors were both refugees from Europe working in the United States: Luria had fled Mussolini's Italy, and Delbrück had left Hitler's Germany before the war. The duo showed elegantly and convincingly that bacteria spontaneously acquired heritable mutations. Specifically, they found that bacteria acquired resistance to infection with bacterial viruses called bacteriophage (literally "bacteria-eaters"). The paper confirmed the interpretation that Monod and Alice Audureau had reached in early 1944 but had not yet published: that the appearance of lactose-utilizing *E. coli* colonies in strains that could not utilize lactose were genetic mutants.

Up until the publication of the Luria-Delbrück paper, there was considerable confusion concerning the properties of bacteria and much doubt about their usefulness as genetic models in biology. Indeed, in what would be one of the most influential books in evolutionary biology (*Evolution: The Modern Synthesis,* published in 1942), Julian Huxley (grandson of zoologist and Darwin apostle Thomas Henry Huxley) considered bacteria irrelevant to processes in more complex organisms: "They have no genes in the sense of . . . hereditary substance . . . Occasional 'mutations' occur we know, but there is no ground for supposing that they are similar in nature to those of higher organisms . . . We must, in fact, expect that the processes of variation, heredity, and evolution in bacteria are quite different from the corresponding processes in multicellular organisms." If these statements were true, then Monod's own work on enzyme adaptation would be of no general importance or utility.

Huxley's claims were not the fault of any particular ignorance on his part—he did not work in the field himself. Rather, it was a reflection of the general ignorance in biology concerning the fundamental nature of heredity, and of the substances that endowed life with properties that distinguished it from nonliving matter. Indeed, so little was known at the outset of World War II about the nature of life that notions of vitalism—the idea that life emerges from some substance or force (a "vital principle") beyond the physical and chemical laws that govern inanimate matter—still lurked, at least among some physicists.

The murky state of knowledge prompted the Nobel Prize–winning Austrian physicist Erwin Schrödinger to ask "What is life?" in a short book of that title. Written in 1943–44 in Dublin, where Schrödinger had retreated after Hitler rose to power, *What Is Life?* had one of the celebrated fathers of quantum mechanics asking whether biology was also reducible to the sorts of mechanics that governed inanimate matter. Schrödinger posed one question at the outset: "How can the events *in space and time* which take place within the spatial boundary of a living organism be accounted for by physics and chemistry?"

His "preliminary answer" was: "The obvious inability of present-day physics and chemistry to account for such events is no reason at all for doubting that they can be accounted for by those sciences."

Schrödinger paid particular attention to the mystery of heredity. Thanks especially to the work of T. H. Morgan's group, the very lab Monod had visited at Caltech before the war, genes were understood to be discrete entities on chromosomes, but their physical structure and chemical makeup were unknown. From the point of view of physics, the existence and properties of genes posed several perplexing problems. It was certain that genes were made up of atoms of some kind, but how could the arrangement of atoms in genes specify the characteristics of an organism—eye color, hair texture, bone length, and so on? Moreover, what properties of genes endowed them with the dual behavior of stability (such that traits were passed faithfully from one generation to the next and the next and so on) and of mutability (such that changes in genes occurred, and those were then also passed from generation to generation)? Schrödinger speculated that genes were some kind of "aperiodic solid" that contained some version of an "elaborate code-script" that specified all of the future development of the organism. How genes worked, how information contained within that solid specified an organism, and how that information was transmitted or altered through time were thus the central mysteries of biology and of what would become known as molecular biology. (For further details or an overview of the science in this book, see Appendix.)

Schrödinger's book would help convince a number of physicists to turn to biology, Francis Crick among them, and a number of young biologists—including James Watson—to pursue genetics (instead of ornithology, in Watson's case).

Schrödinger's book was mostly conjecture. Biologists had to find ways to crack open the gene the way physicists had the atom. For Monod, the Luria-Delbrück paper established bacterial genetics, and thus the foundation for asserting that the great mysteries of heredity were accessible in simple organisms that grew quickly and were easy to work with. Moreover, because of the mutants that he and Alice Audureau had isolated, he had a finger hold on a genetically controlled trait—bacterial growth on the sugar lactose—that could offer a path into understanding how genes work.

To get his lab rolling at the Pasteur Institute, he needed to hire some staff. Madeleine Vuillet, a recent chemistry graduate of a vocational school, heard about an opening and went to present herself to Dr. Monod. As she sat on a stool in a tiny laboratory in the attic of the Pasteur, a handsome young man dressed in canvas pants and an open-collared sport shirt came in whistling. She ignored him, until he introduced himself as Jacques Monod. Vuillet was flustered, as she had thought that all scientists were "strict and severe-looking old gentlemen."

As the interview progressed, Vuillet wondered what she was doing there. She admitted that she knew very little organic chemistry, let alone microbial physiology. Monod was not discouraged. He told her, "At any rate, I prefer that you know nothing, because no school could teach you what we are going to need: I am in search of the secret of life."

Beyond Nihilism

As Monod resumed his research, Camus continued on his quest for the philosophical secrets of life. And the secret for Camus, living in a profoundly traumatized world in which life had been devalued and peace appeared very fragile, was how to move beyond the nihilism of the times and find meaning in one's earthbound life span. While his editorials had occasionally touched upon these subjects, such weighty matters required more extensive and artful examination. For almost exactly one year, Camus had been so immersed in *Combat* that he had had little time for his own projects. On September 1, 1945, Camus stepped back from his daily editorial role in order to devote more time to his writing, as well as to his new family life: the Camus twins, Catherine and Jean, were born on September 5.

Two questions dominated Camus's thinking, his notebooks, conversations, interviews, and speeches: How could individuals find some meaning in their existence? And how could another cataclysm be prevented?

It was understandable and easy to be pessimistic in responding to either question. Two world wars in fewer than thirty years was hardly a basis for optimism. The magnitude of the most recent suffering and death shattered faith for many who asked "Where was God?" and received no satisfactory answer. And for those who did not believe in God in the first place, more than fifty million dead was their proof.

Nihilism, the view that life was meaningless and worthless, followed readily from this loss of faith for many people. But not for Camus. In *The Myth of Sisyphus*, he had rejected nihilism and answered affirmatively the question of whether life is worth living. He viewed nihilism as the central philosophical and ethical challenge of the time, indeed of the twentieth century, and sought a variety of means to counter it. As he stepped away from *Combat,* he returned part-time to the publisher Gallimard as a reader. Gallimard put him in charge of a series that aimed to capitalize on his fame to draw new writers to the publishing house. The series, to be called *Espoir* (Hope), was to bear Camus's name as editor and the back cover of each book would carry a common statement of purpose crafted by Camus: "We are living in nihilism . . . We shall not get out of it by pretending to ignore the evil of our time or by deciding to deny it. The only hope is to name it, on the contrary, and to inventory it to discover the cure for the disease . . . Let us thus recognize that this is a time for hope, even if it is a difficult hope."

For Camus, meaning came from struggle, from revolt. He had already put this philosophy into words in *The Myth of Sisyphus*: "Revolt gives life its value." Now it was time to finally put that philosophy into images, for, as Camus had noted years earlier: "feelings and images multiply a philosophy by ten." His unfinished novel *The Plague,* which he had begun three years earlier as the first element of his "cycle" on revolt, would be his vehicle. Camus's challenge was to create the images and characters in his fictional tale that would evoke the images and feelings of the Occupation and the Resistance with which he hoped all of France could identify.

Among the most resonant images of the novel would be the dead rats, signs of a threat to which no one at first paid attention; the closing of the city gates that cut off the town's inhabitants from loved ones and created a profound sense of separation and exile; the voluntary sanitary squads that bravely risked their lives to fight the plague; the isolation camp in a municipal stadium where afflicted persons were quarantined; and, finally, the joyous reopening of the gates, the ringing of church bells, and the relighting of the streets once the plague was suppressed.

The protagonists of the story, through their courage and solidarity in doing what must be done to combat the plague, would gradually reveal Camus's philosophy of revolt. In a phrase that could have been lifted from a pre-liberation *Combat* editorial, after the plague was declared and the gates ordered closed, the narrator states: "From now on, it can be said that plague was the concern of all of us." The principal characters included Dr. Bernard Rieux, who first identifies the disease, warns the authorities, and works to combat it throughout the epidemic; Joseph Grand, a modest town clerk who is "the true embodiment of the quiet courage that inspired the sanitary groups"; Raymond Rambert, a visiting journalist who, like Camus in Le Panelier, becomes trapped in the city, is separated from his wife, and eventually joins the fight; Jean Tarrou, who organizes teams of volunteers; Father Paneloux, a Jesuit priest who interprets the plague as "the flail of God" and tells his fellow citizens, "Calamity has come on you . . . and, my brethren, you deserved it"; and Joseph Cottard, who takes advantage of the plague to become a smuggler and black-marketeer.

It would take Camus all of 1946 to complete and edit the novel. His routine was generally to write at home in the mornings, then to go to his Gallimard office after lunch to review manuscripts, to read and dictate his correspondence, and to receive visitors. In the evenings, he would socialize with friends such as Sartre and de Beauvoir, meeting them for a drink at one of the cafés on Saint-Germain-des-Prés, then perhaps going to dinner at a bistro and sometimes to another café for a nightcap, or more.

At the time, Sartre was in the process of launching a new literary journal, *Les Temps Modernes* (Modern Times). He penned the review's manifesto in its inaugural issue, which was to publish "engaged literature"—writings that were socially relevant and not merely art for art's sake. Several of Camus's comrades joined the editorial board, but Camus declined Sartre's invitation on account of his many commitments. The two men were often seen together in the Saint-Germain neighborhood, and the press took an interest in the lifestyles of Paris's famous thinkers. In the public's mind, Sartre's and Camus's respective works and perspectives became merged under one label: existentialism.

A once obscure and technical philosophical term, existentialism was moving from a small circle of intellectuals into popular culture—so much so, Sartre thought, that it was in danger of losing its meaning. He clarified the principles of existential philosophy in a speech in the fall of 1945, the first being "Man is nothing else but that which he makes of himself." Sartre explained that existentialism "places the entire responsibility for [man's] existence squarely upon his own shoulders." There was no determinism, no such thing as human nature, no divine guidance or intention. Humans came into the world neither evil nor good, neither heroes nor cowards; all people were free to choose, and were defined not by their hopes or wishes but by their actions.

Existentialism's message of individual freedom and self-determination struck a chord in the atmosphere of post-liberation France. After the failures of governments and religious institutions to prevent the war, after years of the empty nationalist clichés of Pétainism, and after the defeat of Fascism, existentialism suggested a way to begin rebuilding society from a clean slate—by rejecting conventions and authority and empowering the individual.

Both Sartre and Camus were constantly in demand for their views, and not only in France. In early 1946, Camus was invited by the French Ministry of Foreign Affairs to undertake an extended visit to New York. It was a perk for prominent French citizens to promote France's image abroad, and the timing happened to coincide with the upcoming publication of the English translation of *The Stranger.* Camus took along his working manuscript of *The Plague.*

The New York press and francophiles all over the East Coast were eager to hear more from this celebrated young author and Resistance journalist whom the *New York Herald Tribune* pronounced "The Boldest Writer in France Today" and the *New York Times* consecrated the "Apostle of Post-Liberation France." The *Times* went on to say:

> After two wars have shattered the civilization of a continent, few Europeans have any conviction of the meaning or the stability of their world. At no time since the fifteenth century . . . has terror walked so nakedly through the land. It is not simply a physical fear, produced by the brutality and the difficulty of daily life . . . It is also the metaphysical fear of the terrible solitude and helplessness of man in a universe that seems increasingly "absurd" and incomprehensible.
>
> No modern French writer has a greater feeling for these problems than 32-year-old Albert Camus. None has expressed them more strikingly and in a greater variety of forms: the philosophical essay, the novel, the drama. He has led, moreover, an active as well as a contemplative life . . .

> If there is a prophet for the students in Paris at the present moment . . . it is probably Camus . . . Camus commands our attention not simply as a writer . . . but especially as the symbol and the spokesman of the whole generation that reached intellectual maturity under the Occupation.

The press coverage stirred interest wherever Camus was to appear. On his fourth evening in the city, an unprecedented crowd of 1,200 people turned out at Columbia University to hear him speak *in French.* He told his audience that rather than discuss French literature or theater, as might be expected, he had chosen the topic "The Human Crisis." Camus had written only a single editorial over the previous seven months, while he continued to fill his notebooks with observations and ideas. His speech, while dwelling on matters that were his stock-in-trade, reflected that he had carried some ideas further, and that his thinking had clarified about certain issues in his months away from *Combat.* The talk was a combination of editorial and philosophy, and it foreshadowed several directions that future work and projects would take.

Camus explained that he would examine the crisis from the point of view of the "spiritual experience of the men of my generation," men who "were born just before or during the first great war, reached adolescence during the world economic crisis, and were twenty the year Hitler took power. To complete their education they were then provided with the war in Spain, Munich, the war of 1939, the defeat, and four years of occupation and secret struggle. I suppose this is what is called an interesting generation." It was this generation, Camus said, that was "led to think that there might be a Human Crisis, for they had to live the most heartbreaking of contradictions," one in which, while loathing war and violence, they had to accept war and exercise violence.

In order to illustrate what he described as a "crisis in human consciousness," Camus shared vignettes from the war. "A time," he told his American audience, "the world is beginning to forget but which still burns in our hearts." His stories reflected ambiguous attitudes toward cruelty, torture, and death. Camus asserted that these anecdotes, by illustrating that "the death or torture of a human being can, in our world, be examined with a feeling of indifference, with friendly or experimental interest, or without response" demonstrated that there was indeed a human crisis.

Camus said that it would be too easy and too simple to blame all of these symptoms on Hitler. Rather, he sought out more general causes that made Hitlerism possible. Among the many contributors to the rise of terror, Camus suggested that one overriding factor was the emphasis on doctrine and bureaucracy over human dignity. The resulting loss of the protection of "a respect for man," Camus asserted, created a situation in which people of his generation were forced to choose between being either "the victim or the executioner." Camus had used this phrase once before in *Combat* the previous June. He would return to it again in the future. He dwelled upon it in his speech: "The only question for us was whether or not to accept a world in which there was no choice possible save whether to be victim or executioner."

Camus explained that people of his generation wanted to be neither, and thus faced a terrible dilemma. He then described how the Resistance revolted against this choice, against "the civilization of death that was being prepared for us." Hence, for Camus, the spirit of the Resistance embodied his philosophy of revolt that at once denied death and affirmed life. Moreover, in their struggle, the resistants discovered that they were affirming something of value in all people, a common good. "The great lesson of these terrible years," Camus said, "was that we were caught in a collective tragedy, the stake of which was a common dignity, a communion of men which it was important to defend and sustain."

He continued: "It took the war, the occupation, the mass-murders, the thousands of prisons, the sight of Europe wracked with grief, for some of us to finally acquire two or three insights that may somewhat diminish our despair." From this painful experience, Camus declared, "we know what we must do in this crisis-torn world." He offered several prescriptions: to "realize that we kill millions of men each time we permit ourselves to think certain thoughts"; to "cleanse the world of the terror congesting it"; to eliminate the death penalty "throughout the universe"; to put politics "back in its rightful place, which is a secondary one"; to create a "universalism" through which "all men of good will may find themselves in touch with one another."

Lest his audience go away drowning in pessimism and despair, Camus closed by stating that his generation held out an immense hope for man and envisioned a world that was not one of policemen, soldiers, and money, but a world for "man and woman, of fruitful work, and reflective leisure"—a vision to which "we should devote our strength, our thought, and if need be, our lives."

Those who met Camus were struck by the contrast between the gravity of his remarks and the lightness of his personality, which seemed almost "unduly cheerful" to some. Camus lived his days in America to the fullest. His smile, his laugh, his sense of humor caught many questioners off guard, and many found the combination of his personality, talent, and Humphrey Bogart–esque looks attractive. That included nineteen-year-old Smith College student Patricia Blake, who, the day after meeting Camus at one of his speaking engagements, became his romantic companion and tour guide for the remainder of his visit, and even typed part of *The Plague* for him.

Words over Bullets

After returning to France, Camus expanded some of the themes from his "Human Crisis" speech into a series of eight articles for *Combat* in November—the only articles he would publish in the newspaper in all of 1946. Prompted by the conviction that tensions between the East (USSR) and West (USA) were putting the world on the path to conflict in which the casualties would dwarf those of World War II, Camus asked his readers to reflect upon a question of murder: Would they under any circumstances commit or condone murder, or would they categorically refuse to do so? Prominently featured on the front page under the overall title "Neither Victims nor Executioners," Camus's article explored the personal, political, and societal ramifications of either a "yes" or "no" answer to the question.

Camus declared his position: "I, for one, am practically certain that I have made my choice. And having chosen, it seemed to me that I ought to speak, to say that I would never count myself among people of whatever stripe who are willing to countenance murder, and I would draw whatever consequence followed from this." One of those consequences was his call for an international code of justice "whose first article would abolish the death penalty everywhere."

Across several articles, Camus made the case that the world was living in terror and that no matter what ideology one subscribed to—Communist, Socialist, or capitalist—in any contest among these ideologies, the end could never justify violent means. The unprecedented pace of weapons development ensured that war would leave little remaining in the world for the survivors to claim. War was thus unacceptable from any perspective, but nothing was being done to prevent it. Camus appealed for a new "civilization based on dialogue," a new "social contract," and a "new way of life" in order to prevent an apocalypse.

He understood, of course, that in the light of history, the odds were stacked against him, but he was willing to state his wager: "Across five continents, an endless struggle between violence and preaching will rage in the years to come. And it is true that the former is a thousand times more likely to succeed than the latter. But I have always believed that if people who placed their hopes in the human condition were mad, those who despaired of events were cowards. Henceforth, there will be only one honorable choice: to wager everything on the belief that in the end words will prove stronger than bullets."

A few weeks after his series in *Combat, The Plague* went to the printer. Threads of "Neither Victims nor Executioners" were woven into the novel. In a nine-page-long monologue Tarrou confesses to Rieux that as a teenager he discovered that his father was a prosecutor who had men sentenced to death. Repulsed by this revelation, he had joined various revolutionary movements, only to discover they, too, involved killing and the passing of death sentences. When he witnessed an execution in person, he was so horrified that he refused thereafter to kill. "There's something lacking in my mental make-up," he tells Rieux, "and its lack prevents me from being a rational murderer . . . All I maintain is that on this earth there are pestilences and there are victims, and it's up to us, so far as possible, not to join forces with the pestilences."

At the end of the novel, it is revealed that Dr. Rieux is the narrator, who, after the gates had been reopened and the celebrations begun, "resolved to compile this chronicle, so that he should not be one of those who held their peace but should bear witness in favor of those plague-stricken people; so that some memorial of the injustice and outrage done to them might endure; and to state quite simply what we learn in time of pestilence: that there are more things to admire in men than to despise."

In closing his tale, Camus balanced this note of optimism with one of caution:

> Nonetheless, he knew that the tale he had to tell could not be one of a final victory. It could be only the record of what had had to be done, and what assuredly would have to be done again in the never ending fight against terror and its relentless onslaughts . . .
>
> And, indeed, as he listened to the cries of joy rising from the town, Rieux remembered that such joy is always imperiled. He knew what those jubilant crowds did not know but could have learned from books: that the plague bacillus never dies or disappears for good . . . and that perhaps the day would come when, for the bane and the enlightening of men, it would rouse up its rats again and send them forth to die in a happy city.

The novel would be a bestseller in France, selling more than 100,000 copies in its first year at a time when book-buying was a luxury, and it would become one of the top-selling books of the decade following World War II. In Stockholm, the Swedish Academy even considered giving thirty-four-year-old Camus the Nobel Prize in Literature, which would have made him the youngest winning author ever.

As was his compulsion, Camus had already begun a new project well before the first copies were printed. Another plague was festering in the world, but once again, few in France had recognized it. It was a plague that would drive Camus and Sartre apart, and bring Camus and Jacques Monod together.

CHAPTER 19

BOURGEOIS GENETICS

Science is the record of dead religions.
—Oscar Wilde, "Phrases and Philosophies for the Use of the Young"

Monod had hurled himself back into his war-interrupted research. He was happy to be once again designing experiments, working at the lab bench, puzzling over new results, and mixing it up with colleagues. Madeleine Vuillet marveled at Monod's enthusiasm, as he would come bounding up the three flights of stairs to the Pasteur's attic each day and stride down the corridor, smiling and whistling Bach cantatas. A font of ideas, Monod always gave Vuillet several experimental projects to pursue at the same time, so that "there is always something that comes out right." He explained, "It keeps up our spirit and in that way we get several results every day."

After hiring Vuillet, the next orders of business for Monod were to complete and publish the work he had left unfinished when he plunged into clandestinity, and to catch up on what was going on in biology elsewhere in the world. In late 1943, he and Alice Audureau had found that a strain of *E. coli* that could not grow on lactose occasionally produced colonies that *could* grow on lactose, and that these lactose-metabolizing colonies were due to inherited genetic mutations. The result was important because it made Monod understand that the capability to metabolize lactose depended upon genes. Understanding those genes was key to enzyme adaptation.

Monod and Audureau's report was a solid contribution to bacterial genetics in 1946, but there was much bigger news coming out of the United States. Oswald Avery's lab at Rockefeller University in New York was working on the identification of a substance that, when extracted from a virulent type of the pneumococcus bacteria and then given to a harmless, avirulent type of the bacteria, could transform the avirulent bacteria into a lethal, virulent strain. During the war, they published their first results indicating that the active ingredient was associated with the deoxyribonucleic acid (DNA) fraction of the extract, although they could not be certain that the activity was not due to some impurity. In early 1946, they reported that the active ingredient was selectively destroyed by treatment with a purified enzyme that broke down DNA. To a good number of biologists, this was persuasive evidence that DNA was the heredity material. However, since the chemical structure of DNA was unknown, it was entirely a mystery how DNA molecules carried specific genetic information. Indeed, despite Avery's results, some biologists still thought that something else had to be involved.

With such exciting developments unfolding, the small community of the world's "molecular biologists" gathered in the summer of 1946 in Cold Spring Harbor, New York, for a two-week-long symposium on "Hereditary Variation in Microorganisms." The former whaling town on Long Island was home to Cold Spring Harbor Laboratory, a biological research station that had hosted ten previous summer gatherings on major topics in biology. These annual meetings had been interrupted by the war, so it was the first time in a long while, particularly for European scientists, to see colleagues, and for new students to meet the leaders in the young field. Most anyone interested in the nature of heredity attended. Monod and Lwoff took advantage of the new transatlantic commercial plane service to fly to New York from Paris, via Ireland and Newfoundland. Monod described the novel experience in a letter to Odette: "Twelve hours above the water, in reality above wonderful skies of clouds. Constant impression of high mountains. Then the black night with several twinkling stars."

The meeting grounds were situated on a scenic inlet; the symposium mixed formal presentations with swimming, canoeing, picnicking, casual conversation, and late-night beer and whiskey drinking. The attendees heard about Avery's work on the evidence for DNA in heredity, and from Luria and Delbrück and Edward Tatum—all future Nobel laureates—on mutations in bacteria. Joshua Lederberg, a twenty-one-year-old graduate student, reported the stunning discovery that bacteria actually mated and exchanged genes—another Nobel-winning breakthrough that would allow Monod and others to map genes and mutations.

Monod found the many important talks and the new results extraordinary. But at the same time, he had to fend off feelings of discouragement. Research in the United States was progressing so quickly that he confessed to Odette "the difficulty in being current and working usefully in the half-asleep French laboratories." That sense of disadvantage was compounded when he visited labs at Rockefeller University and Yale and saw how much better equipped they were than French labs, having all of the latest gadgets.

There were several very promising developments in the course of Monod's two-month-long visit to the United States. He met with officers of the Rockefeller Foundation who pledged to help French science get restarted with generous grants for equipment. He made many new scientific friends, even among potential competitors, who began to correspond with him, to plan visits to Paris, and to mention the Pasteur Institute to their students as a place to work after their PhDs. And Monod saw that enzymatic adaptation was also starting to get more attention.

Monod was asked to review the field in 1947. As would become his habit, he undertook a comprehensive analysis of all available information and ideas, and developed a persuasive case for the general importance of the phenomenon. Monod drew the connection between the specific puzzle of enzyme adaptation in microbes and the wider, fundamental mystery of cell differentiation in more complex creatures. The latter concerned the unsolved problem, as Monod put it, "of understanding how cells with identical genomes [genetic information] may be differentiated." It was believed from various studies that different cells in the body contained the same chromosomes and the same genes. The differences between the properties of blood cells, brain cells, kidney cells, or other cell types were in the molecules that each cell type produced. Therefore, Monod suggested, understanding how the genes of a microbe caused the production of a certain enzyme under specific conditions and not others "may help in understanding the processes of gene action and cellular differentiation." In the course of writing his review, Monod came to recognize more clearly the potential importance of his own work. He later recalled, "Its significance appeared so profound that there could no longer be any question of my not pursuing it."

Others were also convinced of its importance, and Monod's team at the Pasteur started to grow, although there was not much physical space for anyone. Monod's office was so small it barely had room for a desk, a bench, and a table, around which his small group usually had lunch. Food was still rationed in France, and everyone cooked or warmed their own small meal. It was not the cuisine that drew everyone together; it was the lively conversation and sense of camaraderie that Monod engendered that made attendance at the midday ritual almost obligatory. In addition to scientific ideas, fodder for conversation included music, art, religion, de Gaulle, America, the atomic bomb, new books such as Simone de Beauvoir's *The Second Sex*, and the latest editorials in *Combat*.

For Monod, another part of his putting the war behind him was also putting certain politics aside. He had joined the French Communist Party, albeit with reservations, in order to have some voice in the affairs of the FTP. But during the war, Monod had observed that the Communists had "absolute contempt and hatred" of anything and anyone not with the Party. Because of the Communists' prominent role in the Resistance and the USSR's contribution to the Allied victory over Germany, the Party enjoyed a surge in popularity after the war. Monod, however, had hesitated to become a full-fledged, card-carrying member. By the end of 1945, he had quietly quit the Party and disengaged from Communism for good.

Or so he thought.

On August 26, 1948, the banner across the top of the front page of the Communist newspaper *Les Lettres Françaises* announced "A Great Scientific Event," and heralded an article entitled "Heredity Is Not Commanded by Mysterious Factors: The Soviet Scientist Lysenko Delivers a Major Blow to Anti-Darwinian Theories."

Had there been some breakthrough in Soviet genetics about which Monod and Western science did not yet know?

From the Soviet and French Communist Party points of view, the answer was a resounding yes. But as Monod read the article, he was stunned and horrified to realize that quite the opposite was true, that the USSR and PCF were throwing overboard the very foundation of modern genetics, from the work of Gregor Mendel to that of the Nobel laureate and Monod's former Caltech host, T. H. Morgan.

The newspaper's Moscow correspondent reported that a great debate had recently taken place at the Lenin Academy of Agricultural Sciences on "the situation of biological science in the USSR." During the lengthy proceedings, Russian adherents of Mendel and Morgan were accused by the academician Trofim D. Lysenko of having "turned their backs on science, on materialism," and on the very foundations of Marxist and Soviet principles, and of having embraced metaphysics and idealism.

The fundamental flaw in Mendel's, Morgan's, and other Western scientists' theories, the article explained, was the insistence that inheritance was independent of the conditions experienced by animals and plants, such that no characteristics acquired during their lives were passed on to the next generation. The critics at the Lenin Academy objected to the idea that the hereditary substance of animals and plants was not influenced by the conditions experienced by the organism, and that it acted alone in determining inheritance in a "mysterious and unforeseeable fashion." It was claimed that the prevailing theory had been disproven by Lysenko, who had reportedly achieved transmission of acquired characters from generation to generation.

Moreover, the article reported that the teachings of the late Soviet biologist Ivan Michurin were that there was no limit to the effects of the direct action of external conditions on living organisms: "Human intervention makes it possible to coerce each animal or plant form to be modified more rapidly and in whatever way desirable to man." The difference between Western and Soviet theories had profound implications for the urgent, practical needs of Soviet agriculture: adherents of "Mendelist-Morganist" genetics had long maintained that in order to improve crops, one had to patiently wait for fortuitous, random mutations to arise in strains, and then to select among resulting crop varieties for those with desirable characteristics. It was therefore of limited power to change plant or animal forms. In contrast, the "Michurinist-Lysenkoist" view asserted that any changes imposed on plants by humans could be passed immediately from generation to generation. Clearly, the latter was more beneficial to the goals of Soviet agriculture and the progress of the revolution. Mendelist-Morganist genetics was labeled "reactionary," "bourgeois," and "erroneous."

The article further explained that this debate over genetics had much broader implications, as it reflected the ongoing conflict between two opposing political and ideological views:

> As in our world all fits together, the two hostile and irreconcilable concepts that collided on the seemingly specialized terrain of biology, of genetics more precisely, are the same that are opposed and still oppose each other in all of the modern world—in science, philosophy, economics, politics: that which causes people to exterminate one another on battlefields and sterilize the resources of the earth and human intelligence, and that which wants to unite all citizens of the world . . . In large part, very much in large part, the debates in Moscow saw the defeat of those ideas that, in the matters of heredity . . . constitute before and after Hitler, the basis of all racist doctrines.

In sum, "Mendelist" and "Morganist" views of heredity were not only opposed to Soviet biology but also responsible for Hitler's racist doctrines. They thus needed to be replaced by "Michurinism" and "Lysenkoism." Indeed, two days later in *Le Monde,* Monod read that the Soviet Academy of Sciences had addressed a letter to Marshal Stalin in which it recognized its past errors, promised to correct them, and pledged to coordinate the efforts of scientists in the interest of the country and of Communism. They also admitted that some scientists had applied reactionary and antinationalist theories and were thus incapable of assisting the government in the development of Socialist science. There was no mistaking the meaning: scientists would either repent or lose their jobs, or worse.

Monod was utterly dumbfounded.

For him and other Western observers, the news from Moscow raised many questions: What could possibly have led the Soviet Union, once home to many well-known and respected geneticists, to jettison a cornerstone of modern biology? And to do so at this very moment, when molecular biology was on the verge of great new discoveries about genetics! What was "Soviet biology" and how was it any different from the science they practiced? And who in the hell was this Lysenko?

Monod started immediately to read all he could find about Lysenko and Soviet genetics.

Genetics and the Cold War

The scientific coup in Moscow was at the very least another symptom of the deliberate movement of the USSR away from all things Western. In the three years since the end of the war, the Soviet and Western governments had been drawing increasingly sharp geopolitical and ideological boundaries between their spheres of influence. Mutual suspicion was obvious at the Moscow Conference in December 1945, when the Soviet, British, and American foreign ministers met to discuss problems concerning the occupation of European countries. As early as March 1946, Winston Churchill publicly voiced his concerns over the expanding scope of Soviet influence in coining the metaphor of the Iron Curtain:

> A shadow has fallen upon the scenes so lately lighted by the Allied victory. Nobody knows what Soviet Russia and its Communist international organization intends to do in the immediate future, or what are the limits, if any, to their expansive and proselytizing tendencies . . . From Stettin in the Baltic to Trieste in the Adriatic an "Iron Curtain" has descended across the Continent. Behind that line lie all the capitals of the ancient states of Central and Eastern Europe. Warsaw, Berlin, Prague, Vienna, Budapest, Belgrade, Bucharest and Sofia, all these famous cities and the populations around them lie in what I must call the Soviet sphere, and all are subject in one form or another, not only to Soviet influence but to a very high and, in some cases, increasing measure of control from Moscow . . . The Communist parties, which were very small in all these Eastern States of Europe, have been raised to pre-eminence and power far beyond their numbers and are seeking everywhere

to obtain totalitarian control. Police governments are prevailing in nearly every case.

In March 1947, US president Harry Truman announced a new doctrine of aiding people who were resisting "subjugation by armed minorities or outside pressures." He provided economic and military aid to Turkey and Greece to prevent them from falling into the Soviet camp. On June 5, 1947, the Marshall Plan was made public. Secretary of State George Marshall declared: "Our policy is directed neither against a country nor against a doctrine, but against hunger, misery, desperation, and chaos." The Soviet government was invited to participate, but ultimately rejected the plan as a US plot to exert undue influence over Europe, and the Kremlin persuaded all Soviet satellites not to apply for or to accept US funds.

In turn, the USSR convened a meeting of Communist Party leaders from both eastern and western Europe in September 1947. Stalin wanted to strengthen his hold on eastern Europe and to pressure the Western parties to adhere more closely to Moscow and to break away from coalitions with non-Communist parties in their respective countries. The Communist Information Bureau (Cominform) was established by Stalin's representative Andrei Zhdanov to coordinate the activities of international Communist parties with Soviet policy. At the meeting, Zhdanov proclaimed a new posture toward the West, what would become known as the "Zhdanov line"—namely, that the world was now divided into two camps: a democratic and anti-imperialist camp led by the USSR, and the imperialist-capitalist camp led by the United States. Communist parties outside the Soviet bloc, such as the French Communist Party (PCF), were thus to observe strict discipline and to subordinate themselves to Moscow's central control.

The sealing off of the Soviet sphere was not merely political but also cultural. Both before and again after the war, Stalin and Zhdanov sought to regiment Soviet culture by purging Western influences and promoting distinct forms of Soviet literature, music, art, and science that were aligned with Soviet ideology. There was pressure upon artists to depict so-called Socialist realism, which emphasized appropriate heroes and party spirit. In science, theories were classified as "materialistic" (based on reality and consistent with Marxism) or "idealistic" (a spiritual and mental construct). Under Zhdanov, propaganda campaigns began to claim that certain important achievements in science (such as the invention of the lightbulb and the radio) were actually made by Russians and not the Western scientists who had been credited. The sealing off of the Soviet Union from the West was enabled by the imposition of a "cult of secrecy" on all Soviet affairs. In the realm of science, the translation of Russian journals into English was halted, and Soviet scientists were denied access to most foreign journals and strictly forbidden from publishing their results in the West.

Stalin took a keen personal interest in elevating Soviet scientific heroes, one of whom was Lysenko. As a result of the collectivization of agriculture, the Soviet Union suffered disastrous declines in crop yields in the late 1920s and early 1930s. There was great urgency for innovations to reverse the decline. Working at a plant-breeding station in Azerbaijan, Lysenko experimented with a process dubbed "vernalization" in which wheat seeds were chilled before planting in the spring in order to promote earlier flowering. His claims of great increases in yield and his roots as a Ukrainian peasant's son were widely touted in Party publications. After being promoted and given numerous honors, Lysenko advanced the idea that vernalization caused permanent heritable changes in the plants. He started making claims of transforming other plants simply by altering their growing conditions. With Stalin and the Party's advocacy of "practice" over "theory," Lysenko's purported successes in transforming nature were just what Soviet leaders wanted to hear.

Well before the war, Russian geneticists openly challenged Lysenko's claims concerning the inheritance of any acquired traits, but they were stifled, demoted, and even jailed. The internationally known botanist Nikolai Vavilov died in prison during the war. When Lysenko came under criticism again in 1947–48, he appealed directly to Stalin, who then intervened personally by orchestrating the pivotal meeting at the Lenin Academy and even editing Lysenko's speech. Stalin had previously told Lysenko, "The Michurin position is the only correct scientific approach . . . The future belongs to Michurin." Before the meeting at the Lenin Academy, he told Lysenko to make clear to the assembled body that he was speaking with the explicit approval of Marshal Stalin.

For the many French scientists who were members of the Communist Party, the Lysenkoist policy coming from Moscow would present a difficult dilemma: whether to adhere to their political faith and toe the Party line, or to defend a half century of genetics and break from the Party.

Senseless and Monstrous

The first notable French reaction to the reports from Moscow appeared in the September 5 issue of the Socialist daily *Le Populaire,* in which its foreign political editor, Charles Dumas, declared that the ascent of Lysenkoism "is a return to the Middle Ages: science must again be subservient to the doctrine of political ideology."

Soon afterward, *Combat* entered the fray. Camus and most of its editorial leadership had stepped down more than a year earlier when the paper had run into financial problems; the newspaper was now led by Claude Bourdet and others. The paper announced that it would publish a series of guest articles on the question "Mendel or Lysenko?" Its stated aim was to discover how Lysenko's work differed from his predecessors and to what degree it had suffered from partisan distortion. In one of the first articles, a science historian alleged, "The recent Moscow debates take us back to Galileo's time. With them come the same procedures, the same arguments in order to smear ideas and individuals."

Several issues later, the single Communist biologist who was in the greatest position of authority and who might be able to quell the controversy—Monod's former Sorbonne colleague and FTP chief Marcel Prenant—weighed in. Prenant was both a member of the Central Committee of the PCF and a professor of comparative anatomy and histology at the Sorbonne, so he spoke for the Party as well as for academic science. He sought to accommodate simultaneously both the Lysenko and Mendel-Morgan theories. The headline over his article stated: "According to Professor Marcel Prenant, Lysenko respects the basis of classical genetics, but Prenant considers Lysenko to have obtained the genetic fixation of acquired characters." Prenant explained that, contrary to what others had written, Lysenko did not challenge the existence of genes and chromosomes, but he was merely fighting "with vigor" the exaggerations of classical genetics concerning the role of mutations and the independence of germ cells from the conditions experienced by the rest of the body.

Prenant emphasized the novelty and practicality of Lysenko's results: "The really new point seems to be as follows. Whereas up until now the experimental interventions made by geneticists . . . have enabled them to increase the percentage of mutations, but not to obtain transformations of characters, Michurin and Lysenko say that they have . . . obtained in certain cases the hereditary fixation of characters acquired under environmental influences, and therefore knowable in advance. There is nothing absurd in this." Prenant wrote that whatever Lysenko's faults, they do not negate "the fact that a whole people is profiting today from the work of Michurin and Lysenko." He closed: "Which of our vehement critics has obtained comparable results?"

Monod's turn in *Combat* came the very next day, on September 15. He opened by explaining his approach: "The important thing, the difficult thing in this affair is not to decide whether Lysenko is wrong, or if he is right from the scientific point of view. The cause is quickly understood." Monod wrote that he was going to restrict himself to several citations that he had obtained from an article that appeared a year earlier in the *Modern Quarterly,* a British Communist periodical that could "hardly be suspected of malevolence with respect to the USSR."

Monod offered three brief excerpts:

> Lysenko's claim that Mendelism must be incorrect because it contradicts Darwin . . . is, of course, completely absurd and is on the same level as the fundamentalist position in religious disputes.

> To every scientist the truth of a theory lies in the deductive methods used to establish it and the experimental demonstration of its fundamental premises and its consequences . . . Unfortunately,

> deprived of adequate experimental support, the adherents of the "new genetics" [Lysenko and his school] have chosen to uphold their claims and attack the Mendelists by appeals to authority and by dubious interpretations of dialectical materialism.

> The system by which he [Lysenko] proposes to replace genetics is a muddle of propositions which are mutually contradictory wherever they are not meaningless.

Monod then commented: "These judgments (formulated before the Central Committee took their position) are categorical. They can only be shared by every biologist who takes the trouble to study several of Lysenko's texts. (And one does not understand, or too well, why Marcel Prenant, who is ordinarily so clear, embarrassed himself yesterday, in his 'nuanced' and contradictory explanations)."

Monod continued:

> That such scientific aberrations can be born in the brain of a USSR academician is without a doubt surprising . . . What begs to be understood is how Lysenko was able to obtain sufficient power and influence to subjugate his colleagues, to win the support of the radio and the press, the approval of the Central Committee and of Stalin personally, to the extent that today Lysenko's derisory "Truth" is the official truth guaranteed by the state, that everything that deviates from it is "irrevocably outlawed" from Soviet science . . . and that his opponents who defend science, progress, and the interest of the nation against him are expelled, pilloried as "slaves of bourgeois science," and practically accused of treason.
>
> All of this is senseless, monstrous, unbelievable. Yet it is true. What has happened?

Monod could learn only part of the answer from *Modern Quarterly,* as its authors were not privy to all of the intrigues inside the Party. What Monod could discern was that Lysenko had won over the highest-ranking members of the regime by political maneuvering.

COMBAT

DE LA RÉSISTANCE A LA RÉVOLUTION

CONSEIL DES MINISTRES

…ies budgétaires

des tarifs des

…ces publics (SNCF-PTT)

…X DU TABAC SERAIT AUGMENTÉ

…di : dépôt des projets financiers

MENDEL… OU LYSSENKO

"La victoire de Lyssenko n'a aucun caractère scientifique" estime le Dr Jacques Monod

Chef de laboratoire à l'Institut Pasteur

September 15, 1948, issue of *Combat* featuring Jacques Monod's critique of Soviet biologist T. D. Lysenko and Soviet science. (Archives of the Pasteur Institute)

Lysenko's victory was not scientific, but a matter of "doctrinal fanaticism" that was embraced by the authorities because it aligned with their own mode of reasoning and satisfied their way of thinking, which placed ideology over scientific evidence.

Monod drew his conclusions as to what Lysenko's triumph signified: "What emerges most clearly from this lamentable affair, it is the mortal decay into which Socialist thought has fallen in the Soviet Union. It is so obvious that nothing like this would have happened if the most elementary notions of common sense, of rational thinking, of objective truth were not corrupted by the leaders of the regime. There seems no possible alternative to this conclusion, painful as it may be to anyone who has long set all his hopes on the emergence of socialism in Russia as the first stage of its triumph throughout the world."

The headline over Monod's article read LYSENKO'S VICTORY HAS NO SCIENTIFIC CHARACTER WHATSOEVER.

It was a devastating blow. Monod's article was far more than an indictment of some esoteric branch of Soviet biology; it was a condemnation of the entire Soviet system of thinking and of its leadership.

The PCF continued to defend Lysenko and to publish pro-Lysenko propaganda. In October, the *Europe* review published a special issue that included a translation of Lysenko's entire report. Poet Louis Aragon, a devoted Communist, tried to deflect Monod's criticisms in a twenty-eight-page-long introduction. The "debate" raged on for months. Communist scientists praised Lysenko at public meetings that attracted overflow audiences, as well as at scientific forums. While some accused Monod of partiality, no one could undermine his credibility as a scientist and a geneticist. The damage had been done, both to the prestige of Soviet science and to the credibility of French Communists who supported Lysenko and Party ideology.

For Monod, the Lysenko debate was both surreal and pivotal: surreal because he saw that there was no scientific basis to Lysenko's ascent, that it was "a purely theological affair"; pivotal because it started him on the path of thinking about the origin of such ideology, and trying to understand how such an "insane phenomenon" could arise. To begin to find out, he went around to various meetings that were held to discuss Lysenko—just listening at some, and speaking at others. For a time, he spent one Thursday every month at the meeting of the Michurin-Lysenko Society at the Sorbonne, "debating" the facts of genetics.

Not surprisingly, Monod's front-page attack in *Combat* and his prominence in the debate would estrange him permanently from the hard-liners of the PCF, including former comrades in the Resistance. But his visibility also brought great dividends that he could not have possibly foreseen, for he met many new people because of it, several of whom would influence his life in profound ways.

Monod also attended the meetings of organizations such as the newly formed Groupes de Liaison Internationale. The Paris group was the counterpart of a small American circle that was founded to aid European intellectuals with both spiritual and material support. Cofounded by Camus, the organization was to provide the sort of "universalism" and contact between people of different nations that Camus had advocated in his "Human Crisis" speech and his "Neither Victims nor Executioners" articles. Jean Bloch-Michel, Camus's friend and former colleague at *Combat,* was another cofounder. Bloch-Michel was also a friend of Monod's from the days of La Cantate in the late 1930s. He brought the scientist to a meeting and introduced him to the writer.

Camus had co-written the manifesto for the Paris group:

> We are a group of men who, in liaison with friends of America, Italy, Africa, and other countries, have decided to unite our efforts and our thoughts in order to preserve some of our reasons for living.
>
> These reasons are menaced today by many monstrous idols, but most of all by totalitarian techniques. Those who are prejudiced by blind reason, served by techniques that have gone mad, have led straight to cruel ideologies of domination . . . and, by technical and psychological repression, put the individual at the mercy of the State.
>
> These reasons are especially menaced by Stalinist ideology.

He and Monod had a lot to talk about.

CHAPTER 20

ON THE SAME PATH

A man's growth is seen in the successive choirs of his friends.
—RALPH WALDO EMERSON, "Circles"

MONOD AND CAMUS HIT IT OFF RIGHT AWAY. CAMUS WANTED TO hear about Monod's experiences as a Communist during the war and what Monod had learned about Lysenko. Monod in turn learned that Camus had been devoting a great deal of study to the situation in the Soviet Union. The two former resistants discovered an immediate bond in their respective condemnation of Stalin's regime. Their friendship was cemented over drinks and conversation at La Closerie des Lilas and other Left Bank watering holes.

Camus had reached a conclusion similar to Monod's about the "mortal decay" of Socialist thought in the Soviet Union, but his verdict was based on different evidence. One catalyst for Camus's decisive turn against Soviet-style Communism was his meeting the ex-Communist writer Arthur Koestler. Born in Hungary, Koestler lived in Vienna, Berlin, Palestine, and Paris between the wars. A Communist activist for several years, Koestler quit the Party in 1938 over what he described as its "moral degeneration," which was marked by Stalin's purges of loyal Party members and Moscow show trials. From 1936 to 1938, hundreds of thousands of Russians were executed and millions were sent to labor camps. Even former revolutionary leaders were tried and forced to confess to absurd, fabricated charges. Koestler knew some of the defendants. His disgust and disillusionment gave birth to his novel *Darkness at Noon,* in which the lead character, Rubashov, eventually confesses to false charges and is executed. The book was a sensation in France after the war, selling more than 300,000 copies in less than two years.

Koestler visited Paris in October 1946, and met Camus simply by walking into his office at Gallimard and introducing himself. Koestler tracked down Sartre the next day. He was welcomed immediately into the Camus–Sartre–de Beauvoir social circle. At the time, Camus was in the process of writing "Neither Victims nor Executioners" for *Combat*. Camus paid close attention to what Koestler had to say, especially when he chided Camus for having been too lenient on the Soviet Union. Koestler's break from his Communist past was complete. At a meeting with Sartre and others at André Malraux's home, Camus carefully noted what was said among his famous friends. Koestler said that he hated the Stalin regime as much as he hated the Hitler regime, and for the same reasons. He had once lied for Stalin, he admitted, but now he was certain there was no hope for the regime. Koestler added, "It must be said that as writers we are guilty in the eyes of history if we do not denounce what deserves to be denounced. The conspiracy of silence is our condemnation in the eyes of those who come after us."

Camus would not be silent. His encounter with Koestler hardened his conviction that Communism and Stalinism were the greatest immediate threats to peace in Europe—they were the new plague. In Camus's view, Stalin was indeed akin to Hitler. The terror that he had unleashed on his own people was proof. His government was a dictatorship, not a Socialist state. With his "Neither Victims nor Executioners" articles, Camus had taken an openly critical stance toward Communists, the Soviet Union, and their "ends justifies the means" mentality concerning violence.

And, like Monod, Camus had drawn the ire of Communists. By coincidence, they knew each other's most prominent critics from the Resistance: Camus had met Louis Aragon in Lyon in 1943 in the course of making contact with various Resistance writers, and Camus's critic Emmanuel d'Astier de la Vigerie attended the same summit of Resistance groups in Switzerland with Monod in late 1943. Aragon and d'Astier would each receive the International Lenin Prize from the USSR in 1957.

In 1948, d'Astier was the director of the newspaper *Libération* and a Communist deputy in the French National Assembly. He and Camus exchanged two rounds of mutual critiques following the reprinting of the "Neither Victims nor Executioners" series in *Caliban* magazine in November 1947. In two articles, the first in *Caliban* and the second in the pro-Soviet newspaper *Action,* d'Astier pounced on Camus for myriad sins. He wrote that Camus was being unrealistic in promoting a "third option" other than USSR Communism and American capitalism, arguing that by rejecting the first, he was serving capitalism, which d'Astier linked to Fascism. He accused Camus of having abandoned politics and taken refuge in morality. He asked Camus to clarify what he meant by the justification of violence in totalitarian regimes, and even why, given his convictions, he had taken the side of the Resistance during the war.

Camus was angry enough to take the time from his many projects (two plays and a book-length essay) to write two long ripostes. In his rebuttals, he pointed out that most of d'Astier's criticisms were directed at him personally, as though he were a dangerous threat, and were secondary to the real issues of neither legitimizing violence nor endorsing a revolution that would lead to a third world war. As for violence in totalitarian regimes, Camus reminded d'Astier, "The camps were part of the apparatus of the State in Germany. They are part of the apparatus in Soviet Russia, as you cannot help but know . . . No reason in the world . . . can induce me to accept the existence of the concentration camps." He also took exception to those Communist critics who had so often labeled him as the "son of the bourgeoisie" when in fact he was the son of a worker, raised in poverty. Camus hit back: "The majority among you communist intellectuals have no experience whatsoever in the proletariat's condition, and you are in a poor position to treat us as dreamers ignorant of realities."

Camus was fully aware that he was taking a largely solitary stand, one in which he rejected any legitimization of violence in the name of revolution or future progress. Concluding his rebuttal to d'Astier, he defined his purpose as indeed that of an artist and moralist, and not that of a politician: "My role is not to transform the world, nor man . . . But it is, perhaps, to serve in my way the several values without which a world, even transformed, is not worth living."

Camus's increasingly firm and public stance against the Soviet

Union put a strain on his friendships. Maurice Merleau-Ponty, the political editor of *Les Temps Modernes* and a longtime friend of Sartre's, published an article critical of Koestler and *Darkness at Noon*. Upon encountering Merleau-Ponty at a party of the Sartre-de Beauvoir circle, Camus accused him of making excuses for the Moscow trials and Communist violence. Sartre sided with Merleau-Ponty. Camus stomped out, slamming the door behind him. More than two years later, Camus still refused to share a podium with Merleau-Ponty.

While Camus and Sartre continued to see each other socially, the political distance between them continued to widen. In early 1950, Merleau-Ponty and Sartre coauthored an article in *Les Temps Modernes* in which they were critical of the prevalence of labor camps in the USSR, but they asserted nevertheless: "Whatever the nature of the present Soviet society may be, the USSR is on the whole situated, in the balance of powers, on the side of those who are struggling against the forms of exploitation known to us."

While some friendships were fraying over politics, Camus and Monod were seeing eye to eye. With respect to the Lysenko matter, Camus had no independent grasp of the science in question. But with the benefit of Monod's analysis, he hardly needed a degree in genetics to understand that Lysenko's ascent was a symptom of the same disease that had led to the purges and trials. Camus, who so treasured the sense of solidarity that existed among the Resistance, had in Monod a new comrade who shared both the deep bond of that wartime experience and an unqualified opposition to a new common enemy. Moreover, it was a friendship that could not be complicated by literary egos or competition. For Monod, who admired Camus's talent and work long before the two met, the relationship was both intimate and intellectual.

As their friendship grew, Monod invited the famous author of *The Stranger, The Myth of Sisyphus, The Plague,* and *Combat* editorials to dinner with his family at their home. The other guests who enjoyed the lobster feast were Melvin Cohn, an American scientist conducting his postdoctoral work in Monod's lab, and the great blind organist André Marchal. Camus took a keen interest in the *thangkas* that Odette had collected in Tibet. The twins, now adolescents, were expected to hold their own in conversation with the very worldly adults around the dinner table.

Camus took Monod into his confidence. Shortly before the Lysenko matter broke, and while he was working on his play *L'État de siège* (*The State of Siege*), Camus had reconciled with Maria Casarès. In December 1949, Camus sent a note to Monod, asking him for his advice in finding medical help for Maria's father, Santiago Casares y Quiroga, a former minister in the Spanish Republic, who was seriously ill. He wrote:

> *My dear Monod,*
>
> *On reflection, I am writing to you right away about what I would like to ask of you. Even if you cannot inform me directly, you could perhaps ask around. It concerns Maria Casarès's father, former president of the council of the Spanish Republic. And, what is best after all, a man of quality. Sixty years old, for many years he has been bedridden, an invalid (his condition is worsening). He is suffering from pulmonary fibrosis. Now, it seems that Bogomoletz serum may be indicated in his case. His physician wrote to Doctor Berdach[?], at the Pasteur Institute, and pointed out to him that the case is complicated by a right cardiac insufficiency and asked him for advice and help. This letter remains unanswered.*
>
> *What I would like to know from you, or through you, is the following:*
>
> *1) Is the serum indicated in this case, given the cardiac insufficiency?*
>
> *2) If it is, how does one obtain some?*
>
> *Do nothing, of course, if all of this is outside of your "field." But I assume that you can in this case advise me. I would not bother you without serious reasons. I have affection and respect for Mr. Casarès-Quiroga (that is the name of the patient). And his life is threatened. Regarding Maria Casarès, she lost her mother two years ago, she has no other family other than her father, and she*

loves him. If you knew her better, you would understand that one would want to spare her.

Thank you in advance in any case and do not doubt my already faithful friendship.

Albert Camus

THE GREAT IRONY of Camus's request and his hopes for Maria's father was that they concerned a serum that was in fact yet another absurd and bogus claim of Soviet science, exactly like Lysenko's boasts in agriculture. Oleksandr Bogomoletz was a Ukrainian scientist who had claimed during the war that he had developed a horse serum that was useful in treating scarlet fever, typhus, septicemia, hypertension, digestive diseases, lung abscesses, mental illness, and cancer, as well as promoting the healing of bone fractures and longevity (enabling humans to live as long as 125 years). Stalin named Bogomoletz a Hero of Socialist Labor in 1944. Bardach was a Pasteur scientist who followed up on these claims after the war and produced his own serum. The June 1949 issue of *Paris Match* carried an article heralding Bardach's work, and the French press fanned further interest and hopes in a miracle serum.

Monod could not help Camus; no miracle was forthcoming. Maria's father died two months later.

Throughout the first years of their friendship, Camus was working very hard on several projects. He completed a play, *Les Justes* (*The Just Assassins*), which opened in Paris just before Christmas 1949. Set in 1905 in Russia, and with Maria Casarès in the lead female role as a bomb-making revolutionary, the play explored the moral territory of murder, terrorism, and what revolutionaries were willing or unwilling to justify in the name of a cause. This was exactly the same ground he was immersed in with his greatest effort during the period, an essay on revolt entitled *L'Homme révolté* (*The Rebel*) that was already several years in the making.

A third project was to assemble and edit a collection of his various essays and articles. Entitled *Actuelles: Chroniques 1944–1948*

(*Chronicles*), the volume included many pieces from his first four years of public life—*Combat* editorials, some interviews, as well as responses to critics, including d'Astier. It was published in 1950. On the frontispiece of Monod's personal copy, Camus inscribed:

> *A Jacques Monod*
> *sur un même chemin,*
> *fraternellement*
> *Albert Camus*
> [To Jacques Monod,
> on the same path.
> Fraternally,
> Albert Camus]

CHAPTER 21

A NEW BEGINNING

From small beginnings come great things.

—PROVERB

IN THE FALL OF 1950, A NEW FACE JOINED THE ASSORTMENT OF French, European, and American characters toiling away with test tubes and microbes in Lwoff's attic at the Pasteur Institute.

Ever since the end of the war, François Jacob had been searching for his path, for a profession in which to settle. He spent the first months after liberation in the Val-de-Grâce military hospital enduring countless surgeries to remove just some of the bits of shrapnel that had torn into him. The injuries to his arm and hand eliminated any prospect of his becoming a surgeon. Once he was released, he tried various jobs—journalism, government administration, even acting—but nothing worked out. The only credentials he had were his war record: Jacob had been named a "Compagnon de la Libération"—an honor created by de Gaulle in 1940 and reserved for those who distinguished themselves in the liberation of France. It was awarded to just 1,036 individuals.

Jacob eventually decided to return to medical school to finish his degree, even knowing that his injuries would prevent him from practicing medicine. He crammed two years into one, and completed a required thesis by working at the National Penicillin Center trying to produce the antibiotic tyrothricin. After some initial failures in the laboratory, his preparations ultimately worked on patients' infections. Jacob stayed on after medical school to work at the Center, spending his spare time reading books, including Schrödinger's *What Is Life?*, Julian Huxley's *Evolution: The Modern Synthesis*, and Camus's novels. He courted and married a pianist, Lise Bloch.

After the Center was shut down, Jacob decided that biological research would suit him. The big question was: What field to study? With the eruption of the Lysenko controversy, genetics had acquired a special prominence. Rather than turning French scientists away from Mendel and Morgan, as Soviet biologists were forced to do, the controversy attracted some aspiring scientists toward genetics. Jacob became one of them. At one meeting on Lysenko, Jacob spotted and introduced himself to Monod.

Jacob thought it was incredible that Lysenko had obtained the full support of the Soviet government and press to prohibit "the teaching and practice of one of the most solidly established sciences" and "to impose an idiotic theory on biology." He was astonished that free intellectuals like Louis Aragon would enslave themselves to ideology. He resolved to study heredity. For Jacob, "Studying genetics meant refusing to substitute intolerance and fanaticism for reason."

The problem was that at the time of his epiphany, Jacob had no idea how he—a complete novice, and approaching thirty to boot—might get into the field. Naïve in the ways of academia, he started by knocking on doors. He presented himself to the director of life sciences at the National Center for Scientific Research and declared his desire to devote himself to learning genetics. He was politely refused. He then went to the director of the National Hygiene Institute. Same result.

But at the Pasteur Institute, Director Jacques Tréfouël took an interest in Jacob's war experience and let on that the Pasteur was deeply involved in the Resistance. Then he offered Jacob a research fellowship.

Jacob was thrilled. Of all places to learn and work—the Pasteur Institute!

He accepted on the spot. He started in October 1949 by enrolling in the "Great Course"—an overview of microbiology, virology, and immunology taught by the eminent staff of the institute. In order to get to real research, however, he had to find a supervisor willing to take him on. He learned that in the area of genetics, there were two people to consider: Jacques Monod and André Lwoff. He had briefly met both, Monod at the Lysenko meeting, and Lwoff at the Penicillin Center, and decided that Monod was the less intimidating. He went to see Monod, who told him, "I have no space. And anyway, I am not the boss. Go see Lwoff."

Jacob made an appointment to see Lwoff. He navigated his way around the equipment cramping the corridor of the attic of the Chemistry Building and presented himself to the senior scientist, who was sitting eating his lunch. Jacob made his case, telling Lwoff how much he wanted to study genetics. Lwoff pondered Jacob for a while, then said, "Impossible, I haven't got the least space."

Disappointed but undeterred, Jacob went back to Lwoff several times throughout the winter, and received the same answer each time. His hope nearly gone, he made one last appeal in June 1950. This time, before Jacob could recite his request, Lwoff told him excitedly, "We have just found the induction of the prophage."

"Oh," Jacob said. He had no clue what Lwoff was talking about.

"Would you be interested in working on this phage?" Lwoff asked.

"That's just what I'd like to do," Jacob said.

"Then go off on vacation and come back the first of September," Lwoff told him.

He was in. Not only had Lwoff accepted him, but he would be no more than forty feet away from Monod, who occupied the laboratory at the other end of the corridor. Jacob went immediately to a bookstore to look up "prophage," but had no luck. He had no clue what he was getting into.

UPON JOINING THE lab late that summer, Jacob realized right away that Lwoff was not exaggerating about space. The attic was cramped, and it was hot. The bacteriologists did not want to open any windows and have microbes blowing around. He was assigned to share a laboratory with two Americans at Lwoff's end of the corridor.

His immediate instinct was to find his place in the laboratory's and Pasteur's culture—to get to know the people, learn the hierarchy and the rituals, and then start grappling with the science for which he had no academic background whatsoever. The first cultural issue was language. His two American lab mates wanted to speak French, and

Jacob wanted to develop his English, so they compromised: they spoke French in the morning, English in the afternoon.

His boss, *"le patron"* Lwoff, was a tall, refined person who dressed well and knew about wines, art, and language. He spoke and wrote carefully and precisely. He expected and enforced good French and good grammar throughout the floor, regardless of the nationality of the speaker. He was to be addressed by everyone, Jacob included, as "Monsieur." Yet, despite his formality, Jacob found Lwoff to be very warm, kind, and always encouraging.

Monod, however, was a different personality. With his high cheekbones, firm chin, and wavy black hair, Jacob thought Monod looked "like a cross between a Roman emperor and a Hollywood movie star." And Monod did not go for formalities. When Jacob called him "Monsieur," Monod asked him not to. Instead, he said, "Call me what you like, Monod or Jacques or old fart." Jacob chose Jacques.

The workday was divided by the lunch hour that, in the tradition introduced by Monod, was a free-for-all. The banter included ideas, jokes, scientific gossip, insults for Lysenko and his French supporters, and a discussion of Stalin, literature, films, and more. To Jacob's regret, however, the festivities took place around a large table in the middle of his lab, so experiments had to be suspended from one to two o'clock every day. The ritual ended with everyone sipping a strong coffee before going back to experiments.

Another rite with which Jacob was completely unfamiliar was the research seminar. The attic's inhabitants squeezed into Lwoff's office and sat on stools and chairs while one researcher or a visiting scientist stood in front of a movable chalkboard, presenting their latest results and ideas. The exercise was unlike any lecture he had witnessed, because the speaker was interrupted and peppered constantly with questions, comments, and challenges. In the back-and-forth between scientist and audience, alternative interpretations were floated and new experiments were suggested, and then subjected immediately to the group's scrutiny.

The questioning could be withering. The research seminar was a crucible that tested the confidence, creativity, logic, and quick thinking of every participant. Jacob soon learned that Monod was a master of the form—a form that to Jacob appeared akin to a bullfight, with Monod playing the role of toreador and flaunting the cloth of an argument for others to charge before plunging in his sword.

The greatest ritual of all turned out to be an annual affair that took place at the end of each September, when the entire Institute gathered to mark the anniversary of the death of their great founder, Louis Pasteur. Lwoff told Jacob, "You should go, once, just to see." Jacob joined the streams of Pasteur employees leaving the laboratories, the scientists as well as the cleaning personnel, who were amassing in a large hall. A floor mate started to point out the faces of the most eminent at the head of the line. The oldest sported white goatees and black skullcaps that harked back to Pasteur's time and his first pupils. Jacob saw seventy-eight-year-old M. Camille Guérin, who was the codeveloper of the tuberculosis vaccine known as BCG (he was the "G" in Bacille Calmette-Guérin); others behind him were renowned experts on various microbes and diseases, from tetanus to plague. Jacob understood that "every germ had its representative"—one expert in France for each particular disease.

Suddenly, a hush came over the crowd as the director addressed them, recalling the traditions and achievements of the institution. Then, in silence, the procession into Pasteur's tomb began with the director and trustees in front, the most senior and distinguished scientists second, the heads of laboratories after them, and so forth until the maintenance and kitchen personnel descended through great wooden doors and down the marble steps to the mosaic floor of the tomb. Decorated with green branches of laurel and oak, Pasteur's deep-green sarcophagus of Swedish marble lay in the center. Overhead, Jacob saw colorful mosaics depicting various scenes relevant to Pasteur's life—a boy being bitten by a dog, sheep grazing, garlands of hops, and grapevines representing his achievements in rabies, anthrax, and fermentation, respectively. On the walls were more scenes of Pasteur's victories. Stirring vignettes, to be sure, but to Jacob the grandeur of the mausoleum seemed more befitting Napoleon than a scientist.

Leaving the tomb for the gardens above, Jacob reflected not just on Pasteur's science but on the institution he had created and the unique type of scientist he had cultivated—the Pastorian. Here, recruited from all over the world, were doctors with no patients, pharmacists with no drugstore, professors with no classrooms, and chemists with no industry. It was a cathedral of science—a monastery, really—with its monks devoted to the special calling of research.

Jacob began to wonder how many of those around him had achieved or would achieve their goals. How many years does it take to discover something meaningful? How long before one knew whether one was on the right or the wrong track, whether one had the sort of vision and talent that would merit being at the front of the procession he had just witnessed?

Ten years after leaving for England, a lost decade in terms of his profession, he was starting over from scratch. A mere beginner at age thirty. What were his chances of success, of such glory?

He gave himself five years to find out.

Data over Dogma

Pasteur's shrine was not used exclusively for grand ceremonies. Finding that his desk in Monod's laboratory was too poorly lit for him to work at night, Melvin Cohn discovered that the tomb was very well lit. With the collusion of the building's concierge, the mausoleum was left unlocked for Cohn in the evenings. He would then spread out his papers using Pasteur's sarcophagus as a desk and study late into the night. Monod was startled to find Cohn there one evening when he was taking some visitors around on a tour. But he did not reprimand Cohn for such sacrilege; Monod also found the ambiance rather congenial, and the two men spent many evenings together "chatting across the death mask of the Institute's patron saint."

Cohn and Monod had much to discuss. Shortly after Jacob's arrival in the attic, Monod's team hit a streak during which they acquired a series of new and important results. The experiments provided a large step forward in understanding the phenomenon of enzyme adaptation. But they also had the added virtue of demolishing the Lysenkoist slant on nature in the same blow. One of the ironies of Monod's involvement in the Lysenko matter was that the very phenomenon that he had been studying for more than a decade itself gave the outward appearance that organisms could change readily in accordance with their environment: that is, after some delay, bacteria grew when given an alternative to their preferred sugar glucose. The response appeared as though the bacteria had acquired a new ability that was provoked by the presence of the alternative sugar. The phenomenon could be viewed from the Michurin-Lysenko perspective as evidence that organisms were readily adaptable to whatever conditions they encountered. (It was akin to the notion, promoted by the nineteenth-century naturalist Jean-Baptiste Lamarck, that giraffes acquired long necks by reaching for their food.)

In biochemical terms, what Monod's team knew was that the sugar lactose, a disaccharide composed of the two sugars galactose and glucose, could not be utilized by *E. coli* bacteria unless it was broken down by the enzyme ß-galactosidase into those two sugars. Moreover, that enzymatic activity was not detectable unless cells were grown in the presence of lactose. It appeared as though the lactose molecules somehow "instructed" cells to produce the specific enzyme that was needed. How a specific enzyme was made was therefore the central mystery.

Monod had an idea. He proposed that bacteria cells contained a common set of protein building blocks for making a variety of enzymes, and that these blocks were brought together in different combinations by different sugars to make different specific enzymes. The sugars competed for the building blocks so that, for example, if glucose was present, enzymes for using glucose were assembled instead of enzymes for breaking down lactose. The central feature of Monod's model was that the sugar to be acted upon, what is called the substrate of the enzyme, provoked the formation of the correct enzyme.

Cohn and Monod developed a clever test of this idea by synthesizing a large number of chemically modified forms of lactose. The straightforward prediction of Monod's model was that compounds that were good substrates for the ß-galactosidase enzyme would provoke the formation of the enzyme, and those that were not good substrates for the enzyme would not. That was not at all what the results showed. Monod's team found that certain compounds that were not substrates could elicit enzyme activity, and certain other compounds that were good substrates did not elicit enzyme activity. Monod's model was wrong—and he was delighted.

Monod knew that progress was made when previous ideas, even one's own, were vanquished and new ones had to be formulated and tested. And he was very confident in his abilities to come up with new ideas and new experiments.

Moreover, regardless of how enzyme adaptation actually worked, Monod's results punctured the Lysenkoist balloon. In his laboratory in the Pasteur's attic, bacteria made large amounts of an enzyme in response to sugars they could not use, and failed to make an enzyme in response to sugars they *could* use. Contrary to the Michurin-Lysenko theory, these bacteria were not adapting in accordance with their environment. Cohn, who attended meetings of the Michurin-Lysenko Society with Monod, noted with satisfaction how the absurd Soviet theories that had so irked Monod "had been answered with experimental vengeance." The corpse of Lysenko's fantasies had been buried—with sugar.

In light of their results, Monod decided it would be best to banish the Lamarckian and Lysenkoist connotations of the term "enzyme adaptation." The phenomenon was renamed "enzyme *induction,*" and the substances capable of inducing enzymes were called "inducers." Monod's team had shown that inducers were not always substrates for the enzyme and vice versa. The results with the modified sugars started Monod thinking that inducers and substrates may work through different means—that while the enzyme acted on the substrate, the inducer acted through something else. This would turn out to be a productive new line of thought.

In the meantime, Monod looked forward to talking about his new results and insights. He was invited to the United States in 1951 to speak at meetings of the American Chemical Society and the very prestigious Harvey Society of New York.

But he went to neither. Cold War politics caught up with Monod.

Following the usual procedure, Monod went to the American consulate in Paris to obtain a visa for his trip. He learned that, due to his previous membership in the Communist Party, he was an "inadmissible alien." The year before, in the wake of the outbreak of the Korean War and rising fears about the influence of Communism, the US Congress passed the Internal Security Act of 1950. Among many measures designed to thwart "un-American and subversive activities" was a provision that aliens who "at any time, shall be or shall have been" members of Communist parties were to be excluded from admission into the United States.

The American consul advised Monod that he could apply to the attorney general of the United States for special permission to enter the country. Monod decided against doing so, and explained himself in a letter to the consul (composed in English):

> *In view especially of your extremely courteous and helpful personal attitude in this matter, I feel that I should explain in some detail the reasons which should have led me to this negative decision. These are twofold.*
>
> *To begin with my proposed trip to the U.S. was planned, you may recall, in answer to invitations extended to me by the American Chemical Society and by the Harvey Society. However much I appreciate the honour entailed in these invitations, as well as the pleasure and fruitfulness of a scientific visit to the U.S., I cannot put these in balance with the extremely distasteful obligation of personally submitting my "case" to the Department of Justice, and of having to ask for permission to enter the U.S. as an exceptional and temporary favor of which I am legally assumed to be unworthy.*
>
> *The second reason is that I am not willing to fill in and swear to any "biographical statement" of the type apparently required for the application. This refusal is not based on abstract principles only, but on a sad and terrible experience: this kind of inquisition was introduced into the French Administration under the occupation. I will not submit myself to it, if I can possibly avoid it . . . You will also realize, I believe, that such statements, should they fall into the wrong hands, might conceivably be used as a source of information. The mere possibility of this would make it impossible for me*

to submit one, even though I knew that mine would be uninteresting. The fact that I have been completely estranged from my former political affiliations makes this even more impossible.

That being said, I should like to add that I did not reach this decision light-heartedly, as I fully realize that it means cutting myself partially away from a country which I love, and to which I am attached by very strong links. Not only am I half American [Monod's mother was from Milwaukee, Wisconsin], but I have many very close friends in your country. I have learned by experience to respect and admire American Science. Indeed I owe much to several American scientific or other institutions, such as the Rockefeller Foundation and I may venture to say that, as a scientist, I have had more recognition in the U.S. than in my own country.

However, all this is strictly personal and I would like to mention another more general aspect of these problems. Scientists themselves are quite unimportant. But Science, its development and welfare are overwhelmingly important, isolation is the worst enemy of scientific progress (if proof of this statement were needed I would point to the strange and profound deterioration of Russian biology in recent years). Measures and laws such as you are now obliged to enforce will contribute in no small extent to erecting barriers between American and European science. I do not pretend to know whether or not such measures are justified in general, and in any case I have no right to express an opinion. But I can say, because it is a plain fact, that such measures represent a rather serious matter to the development of science, and that, to that extent at least, they must be contrary to the best interests of the United States themselves.

Monod's letter was subsequently published in the very prominent American journal *Science* amid a growing outcry over the adverse effects US visa policies were having on American and international scientists.

CHAPTER 22

REBELS WITH A CAUSE

He who dedicates himself to the duration of his life, to the house he builds, to the dignity of mankind, dedicates himself to the earth and reaps from it the harvest that sows its seed and sustains the world again and again.

—Albert Camus, *The Rebel*

Cold War politics were never far from Camus and Monod's conversations. One evening, the two met for dinner at Jean Bloch-Michel's home, on rue de Verneuil on the Left Bank, near the Gallimard offices. Another guest that evening, journalist Jean Daniel, cofounder of the *Caliban* review and future cofounder of *Le Nouvel Observateur,* was struck by the intimate rapport between the scientist and the author. He later recalled his first impression of the two men as one of watching a Hollywood movie, with Camus as Humphrey Bogart and Monod as Henry Fonda. Daniel was touched by how Camus or Monod finished the other's sentences, or even anticipated what the other was going to say. The pleasure each took in the other's company and conversation was obvious. When Monod raged about how his visa was denied even though he had long quit the Communist Party, Camus, who was in the same boat, having once been a member of the Party in Algeria, burst out laughing. The two men were so engaged, and so enjoyed their banter, that Daniel had the sense that he and Bloch-Michel were interrupting a private evening.

That intimacy reflected Monod and Camus's deep engagement over their shared central concern of the Soviet Union's deviations. At the time, Camus was putting the finishing touches on *The Rebel,* a book that was more than nine years in the making. The essay was initially conceived to be a philosophical exploration of what Camus viewed as the "first value of the human race"—rebellion or revolt, the individual saying no to some condition of existence.

Even before the publication of *The Myth of Sisyphus* in 1942, Camus had planned to write a full essay on revolt as part of a second "cycle" of his body of work. But as he jotted notes and fragments over the years, the experiences of the war, and of the Resistance in particular, expanded his thinking toward the collective dimensions of rebellion, of what happens when people stand together and say no. *The Plague* was the novelized version; *The Rebel* was to be its companion essay. While both works were in progress, he wrote and published a short essay in 1945—"La Remarque sur la Révolte" ("Note on Revolt")—that framed the central theme of rebellion and ultimately served as the template for the first chapter of *The Rebel* six years later. Written while the spirit of the Resistance still resonated, and at the height of his involvement in *Combat*, Camus's short essay made the point that individual rebellion asserted the existence of values and was the seed of solidarity among humans. He began:

> What is a rebel? It is first of all, a man who says no . . . What does he mean by saying "no"?
>
> He means, for example, that "things are hard enough," "there are limits beyond which one cannot pass," "up to this point, yes, beyond it no," or "you are going too far."

Camus asserted that by saying no, the rebel is in turn affirming that there are limits beyond which his rights are infringed upon. There is thus something to be preserved, something of value, on one side of the limit. Moreover, these limits and rights belong not just to the rebel but also to others. In the act of refusal, the rebel thereby defines a value, a value that Camus alleged "transcends the individual, which removes him from his solitude" and thus joins him to others, and so establishes "the solidarity of man in the same adventure."

The first philosophical secret of life for Camus was the recognition of the absurd condition. This instinct for positive rebellion—against death, oppression, suffering, or injustice—was the second secret of life, and a path to humanity.

In the postwar years, revelations about the Soviet Union and the growing tensions of the Cold War led Camus to recognize that rebellion could go too far. Rebellions that evolved into revolutions somehow ended up creating oppressive regimes that denied freedom and happiness, imposed suffering and injustice, destroyed solidarity, and legitimized murder—which betrayed the very purpose of revolt. Camus's project then took on the added dimension of examining the path from rebellion to revolution to totalitarianism. As he set forth in his introduction, *The Rebel* sought to "understand the times in which we live," and how "logical crime" was justified, how murder became seen as a legitimate means to the realization of revolutions. Camus sought to understand this deviation by first tracing the history of revolutions over the previous century and a half, from the French Revolution's Reign of Terror to the rise of Nazism and Hitler.

Camus observed: "All modern revolutions have ended in a reinforcement of the power of the state." He laid blame once again upon the ascent of nihilism. In those revolutions in which God had been replaced, nihilistic attitudes recognized no limits in ensuring the continuity of the revolution as represented by the state. "Nihilism," Camus wrote, means "one is justified in using every means at one's disposal." Rivals and enemies of the revolution were to be eliminated, and individual freedom suppressed as citizens became mere cogs in the apparatus. When everything is meaningless, power is everything.

In the last third of the book, Camus turned the spotlight exclusively on Marxism, Stalinism, and Communism, and delivered a scathing, comprehensive indictment of the entire Soviet system. Camus declared: "The greatest revolution that history ever knew" has become "the dictatorship of a revolutionary faction over the rest of the people," a regime in which "all freedom must be crushed in order to preserve the empire" and that "contrives the acceptance of injustice, crime, and falsehood by the promise of a miracle." He further condemned "the concentration camp system of the Russians" that had transitioned "from the government of people to the administration of objects," creating a country in which "dialogue and personal relations have been replaced by propaganda or polemic" and "the ration coupon substituted for bread."

The evidence cited in his indictments included the liquidation of so-called enemies of the revolution, trials based on false charges, labor camps, the repression of literature and art, and the denial of modern science.

Camus asked: How could the ambitions of a so-called scientific socialism conflict to such an extreme with the facts as they were in the USSR? He replied: "The answer is easy: it was not scientific." Camus argued that in positing the end of history with the end of class struggle under Communism, Marxist theory had taken on the character of a prophecy, of a religion. He explained: "There remains of Marx's prophecy . . . only the passionate annunciation of an event that will take place in the very far future. The only recourse of the Marxists consists in saying that the delays are simply longer than was imagined and that one day, far away in the future, the end will justify all . . . This new faith is no more founded on pure reason than were the ancient faiths."

He added, "Prophecy functions on a very long-term basis and has as one of its properties a characteristic that is the very source of strength of all religions: the impossibility of proof. When the predictions failed to come true, the prophecies remained the only hope; with the result that they alone rule over our history."

What Camus could not abide were ideologies that sacrificed life in the present, the one fundamental value above all, for some promise of future justice. Christianity "postpones to a point beyond the span of history the cure of evil and murder," he noted, while Russian Communism justified terror and murder and crushed freedom in the promise of some far-off workers' utopia. For Camus, "real generosity to the future lies in giving all to the present."

With respect to the denial of modern science, Camus had the benefit of Monod's direct help in formulating his accusations and conclusions. Well after his *Combat* article, Monod expended considerable effort in getting to the bottom of the Lysenko matter and the derailment of Soviet science. He wrote a sixty-eight-page-long analysis that, although Monod never published it himself, provided sophisticated arguments for his friend to use.

Monod focused on how Lysenko's lack of understanding of a variety of empirically demonstrated facts of genetics was coupled with rigid ideology. One glaring example upon which Monod seized was the matter of mutations and the role of chance. For almost half a century, geneticists studying plants, fruit flies, or other organisms observed that mutations in genes arose spontaneously at some low but measurable frequency. They could not predict whether any given individual plant or animal would bear a particular new mutation; it was a matter of probability, of chance.

According to Monod, Lysenko strongly objected to such unpredictability and argued that the mutations must then lack a material basis, that they were some sort of miracle. In Lysenko's eyes, genetics was thus a false science, irrational and metaphysical because its laws were based on probabilities. "In ridding our science of Mendelism-Morganism," Lysenko stated, "we rid it of chance." He added, "Sciences such as those of physics and chemistry are rid of chance. It is for this reason that they have become exact sciences."

With that declaration, Lysenko had multiplied his errors. As Monod pointed out, the previous century's physicists and chemists had shown "not only that *all* of their knowledge, *all* of their observations, were statistical in nature, and that nearly all of their laws, even the most rigorous, express in reality not certainties, but probabilities. These are classic ideas, familiar today to every enlightened mind. If one were to adopt Lysenko's attitude, it would be necessary to renounce not only genetics, but modern physics, radioactivity and quantum theory, the gas laws, chemical kinetics, thermodynamics." And indeed, following the Lysenko episode, various Soviet science academies met to debate and to purge Soviet science of key Western ("bourgeois and reactionary") theories in chemistry, physics, and astronomy. Monod attributed the denial of science in the USSR to a plague he labeled "Fundamentalist Bolshevism."

Guided by Monod, Camus explained in *The Rebel* that religious adherence to Marxism, enforced by the Soviet state apparatus, had led to the denial of science:

> To make Marxism scientific and to preserve this fiction . . . it has been a necessary first step to render science Marxist through

> terror. The progress of science, since Marx, has roughly consisted in replacing determinism and the rather crude mechanism of its period by a doctrine of provisional probabilities . . . For Marxism to remain infallible, it has therefore been necessary to deny all biological discoveries made since Darwin. As it happens that all discoveries since the unexpected mutations established by De Vries have consisted in introducing, contrary to the doctrines of determinism, the idea of *chance* [italics added] into biology.

Marxism had also negated other great scientific discoveries. Camus wrote, "Marxism is scientific today in defiance of Heisenberg, Bohr, Einstein, and all the greatest minds of our time."

He further pointed out that Marxism

> has also had to rewrite history, even the most recent and best known, even the history of the Party and of the Revolution. Year by year, sometimes month by month, *Pravda* corrects itself . . .
>
> As in the fairy story, in which all the looms of an entire town wove the empty air to provide clothes for the king, thousands of men, whose strange profession it is, rewrite a presumptuous version of history, which is destroyed the same evening while waiting for the calm voice of a child to proclaim suddenly that the king is naked. This small voice, the voice of rebellion, will then be saying, what all the world can already see, that a revolution which, in order to last . . . is living by false principles.

That voice would be Camus's voice, which in 1951 was one of the best-known voices in France and becoming so in the Western world.

And yet, despite 300 pages of grim history, 150 years of brutal, failed revolutions, and the staggering death toll of nihilistic regimes, Camus was still not pessimistic.

What could possibly rescue humanity from its excesses and tragedies?

Rebellion, was Camus's answer, rebellion against nihilism and the very regimes and ideologies that darkened the globe. "Rebellion indefatigably confronts evil," Camus affirmed. But, he urged, what was needed was rebellion "in moderation," rebellion that recognizes limits, one that is "mastered by intelligence."

It was time to take sides: "When revolution in the name of power and of history becomes a murderous and immoderate mechanism, a new rebellion is consecrated in the name of moderation and of life. We are at that extremity now. At the end of this tunnel of darkness, however, there is inevitably a light, which we already divine and for which we only have to fight to ensure its coming."

DURING THE EVENING at Bloch-Michel's, Daniel observed between Camus and Monod a "complicity so intense . . . that only a shared kindness of heart allowed them not to find unwelcome those who interfered in their privacy."

The two rebels had found their "solidarity in the same adventure"—against the oppressive ideology of Communism. On the frontispiece of Monod's copy of *L'Homme révolté,* Camus inscribed:

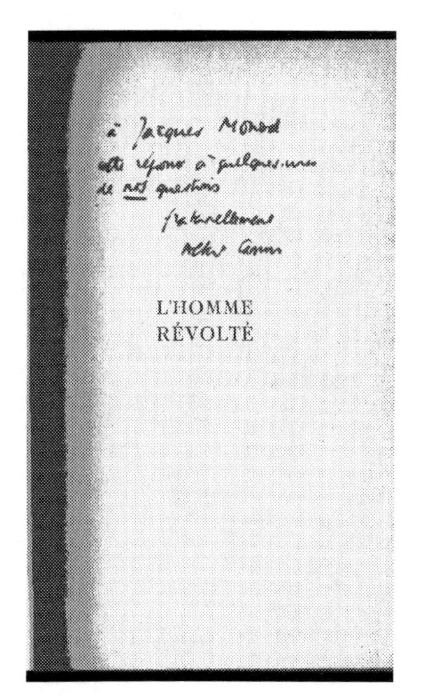

L'HOMME
RÉVOLTÉ

à Jacques Monod
cette réponse à quelques-unes
de nos questions
fraternellement
Albert Camus
[To Jacques Monod,
this answer to a few
of our questions.
Fraternally,
Albert Camus]

Frontispiece of Jacques Monod's copy of Camus's *L'Homme révolté* (*The Rebel*), inscribed to Monod by Camus. (Courtesy of Olivier Monod)

CHAPTER 23

TAKING SIDES

True friends stab you in the front.

—Oscar Wilde

Camus braced himself.

In labeling Marxism a religious prophecy and describing Communist Russia as a delusion ruled by a ruthless dictator and enforced by terror, he declared what no other Frenchman of the left had dared. He knew that there would be strong reactions from both ends of the political spectrum in France. A few days before *The Rebel* was to appear, he had lunch with a friend at the Hôtel Lutétia, on the Left Bank not far from the Luxembourg Gardens. Upon parting, Camus said, "Let's shake hands, because in a few days, not many people will want to shake my hand."

There were the expected vitriolic attacks from Communists, and more polite criticisms from Christian publications. But some reviewers hailed *The Rebel,* especially those of an anti-Communist bent. In *Le Figaro Littéraire,* it was declared to be both Camus's most important book and one of the great books of the postwar era. *Le Monde* agreed that nothing of such value had appeared since the war. Some literary and journalistic colleagues also offered praise. André Malraux, who served as de Gaulle's minister of information immediately after the war, approved. On the non-Communist left, Camus's former *Combat* comrade Claude Bourdet of *L'Observateur* gave it a favorable review that was spread over two successive issues. And a reviewer in *Combat* predicted that the book would change the conversation in France: "More than the coming to awareness of an era by a lucid and courageous spirit, it won't take long to see a reflection of the era on itself, the announcement of a turning-point after which certain problems will be posed differently."

Beyond the critics, the public reception was respectable. The book sold fairly well, 60,000 copies in the first several months, and Camus received a good deal of mail from admiring readers.

Noticeably absent from the raft of reviews that appeared over the first months after publication was that of *Les Temps Modernes* and its editor in chief. It was not because Sartre or anyone else at the magazine had not read the book—all of his team had. The problem was that no one approved of it. The general feeling was that Camus was out of his depth in terms of both history and philosophy. No one was prepared to say so, but with all of the attention being paid to *The Rebel,* the periodical's long silence had to be broken. Given his long-standing relationship with Camus, Sartre was looking for someone to volunteer who would be firm but courteous. Sartre finally tapped Francis Jeanson, a twenty-nine-year-old who had studied at the Sorbonne and risen to manager of *Les Temps Modernes.* Jeanson, like Sartre, was not a Communist but at the same time was opposed to anti-Communism.

While Jeanson was still working on his review in April 1952, Camus met Sartre and de Beauvoir in a small Left Bank café. Camus mocked some of the negative reviews he had received, assuming that Sartre and de Beauvoir approved of the book. The couple did not know how to reply. A short time later, upon meeting in another bar, Sartre warned Camus that the upcoming review of his book would be reserved, and perhaps even severe. Camus was taken aback.

Jeanson's twenty-one-page review appeared the following month, in the May issue of *Les Temps Modernes.* The title, "Albert Camus ou l'âme révoltée," was a pun: *l'âme,* meaning "soul," sounds much like *l'homme*; hence "The Soul in Revolt." Camus did not appreciate the jab, and the article went downhill for him from there. Jeanson mockingly questioned the praise the book had received, noting "the wave of right-wing enthusiasm that has come crashing down from the heights of eternal France." Jeanson sniped, "It seems to me, were I in Camus' place, in spite of everything, I would be worried . . . I have been assured, moreover, that he is worried." Jeanson continued his sarcastic tone: "We can at least attempt to understand those unique virtues of his book that have provided an occasion for such intense delight . . .

What then is this 'good news' that all greet with such joy? What promises does this message contain that each can find in it what he was waiting for, to the exclusion of all others?"

Before addressing the merits of Camus's arguments, or lack thereof, in *The Rebel,* Jeanson critiqued Camus's style. Jeanson suggested that Camus's style in the book was excessive, saying that "his protest is too beautiful." As for the book itself, Jeanson summed up *The Rebel* as "incoherent"—a "pseudophilosophical pseudohistory of 'revolutions.'"

Jeanson's tone was a deliberate choice. He later admitted that he took his approach in order to counter the special privilege that he perceived Camus enjoyed, that was "the privileges of the *sacred*: 'Albert Camus,' in essence the High Priest of absolute Morality."

The gauntlet had been thrown.

Camus was stunned. He felt insulted and disrespected, and he believed that the arguments in his essay had been ignored. Worse, however, Camus knew that, as editor, Sartre was responsible for the content of the magazine and had allowed Jeanson to say what he did. He could ignore such a review from a Communist publication, but not one from his friend's magazine. Sartre's secretary let Camus know that, if he so wished, the magazine would print a reply to Jeanson's review.

Camus hesitated. He was so upset by his treatment in *Les Temps Modernes* that, at first, he had no stomach for a rebuttal. But he decided that he could not remain silent in the face of what he felt as an unjust and misleading review. He adopted a tone similar to Jeanson's in his seventeen-page retort that he sent to the journal at the end of June:

> *Dear Editor,*
>
> *I will use the article that your journal has devoted to me, under an ironic title, as a pretext in order to bring to the attention of your readers a few observations concerning the intellectual method and attitude that this article reveals. This attitude, with which I am sure you are in agreement, really interests me more than the article itself, whose weakness surprised me . . . I apologize for having to be as long-winded as you were. But I will try to be more clear.*

> *I will first attempt to demonstrate what may have been the real intention of your collaborator when he indulges in omissions, makes a travesty of the thesis of the book he sets out to criticize, and fabricates an imaginary biography for its author.*

In addressing his article in the manner that he did, Camus was doing much more than replying to Jeanson. Indeed, in his entire article, he never referred to Jeanson by name, preferring instead to refer to him as "your collaborator" and to the review as "your article." Camus thus redirected responsibility for the article from Jeanson to Sartre, whom he alleged shared the same attitude expressed in the review. He expressed his offense that his longtime comrade had not taken responsibility for the review and addressed the serious subject matter of the book himself: "A loyal and wise critic would have dealt with my true thesis." In Camus's eyes, the review was an act of betrayal: "In it I have found neither generosity nor loyalty toward me."

Camus got his digs in, some aimed at Jeanson, others at Sartre. Referring to an episode during the liberation of Paris when Camus found Sartre sleeping in a seat at the Comédie Française, he wrote: "I am beginning to become a little tired of seeing myself . . . receive endless lessons . . . from critics who have never done anything more than turn their armchair in history's direction."

That summer, Sartre was moving decisively in the opposite political direction from Camus, toward becoming the leading pro-Communist intellectual in France. After several years of tension and criticism, he had decided to draw closer to the Communist Party as a means of countering what he saw as the expanding influence of capitalism and American hegemony in the Cold War. Outraged over press coverage of an aborted workers' strike, Sartre began a series of articles titled "The Communists and the Peace." The first installment, published in the July 1952 issue of *Les Temps Modernes,* stated that "the revolutionary living in our epoch . . . must indissolubly associate the Soviet cause with that of the proletariat." Moreover, Sartre claimed, "the USSR wants peace and proves it every day." When Sartre read Camus's reply to Jeanson's review, he took it as a provocation and decided that he would answer it in the same August issue, and invited Jeanson to do so as well. Sartre was not in a generous or forgiving frame of mind:

> *My Dear Camus:*
>
> *Our friendship was not easy, but I will miss it. If you break it off today, that undoubtedly means that it had to end. Many things drew us together, few separated us. But these few were still too many . . . I would have preferred that our present quarrel went straight to the heart of the matter and that the nasty smell of wounded vanity was not mixed in with it . . . I didn't want to reply. Whom would I convince? Your enemies certainly, perhaps my friends. And you, whom do you hope to convince? Your friends and my enemies.*
>
> *What is certain is that both of us will give our common enemies—and they are legion—a good laugh. Unfortunately, you have implicated me so deliberately, and in such an unpleasant tone of voice, that I cannot keep silent without losing face. Thus, I shall answer you, without anger, but, for the first time since I have known you, I shall speak bluntly.*

Sartre blasted away, telling Camus: "The mixture of dreary complacency and vulnerability that typifies you always discouraged people from telling you unvarnished truths . . . Sooner or later, someone would have told you this; it may just as well be me." Sartre echoed Jeanson's criticism that Camus placed himself above reproach: "You do us the honor of contributing to this issue of *Les Temps Modernes* but you bring a portable pedestal with you . . . But tell me, Camus, for what mysterious reasons may your works not be discussed without taking away humanity's reasons for living? By what miracle are the objections made to you transformed immediately into sacrilege?"

The attack, or counterattack, ran twenty pages. It was brutal and personal. Sartre accused Camus of attempting to terrorize Jeanson by summoning his supporters against the critic, of pomposity ("which comes naturally to you"), and of behaving like a criminal prosecutor by feigning calm so as to make his wrath more dramatic. Sartre twisted in the knife: "Perhaps the Republic of Beautiful Souls should have named you its Chief Prosecutor."

With regard to the criticism of the book, Sartre asked: "Suppose you were wrong? Suppose your book simply attested to your philosophical incompetence. Suppose it consisted of hastily assembled and secondhand bits of knowledge? . . . And suppose you did not reason well? And suppose your thinking was muddied and banal?"

Halfway into his dissection, Sartre shifted his tone, acknowledging Camus's creative work and much of his actual biography: "You have been for us—and tomorrow you could be again—the admirable union of a person, an action, and a work. That was in 1945 . . . You were not far from being exemplary; for in you were summed up the conflicts of our times, and you transcended them through the ardor with which you lived them . . . You united the sense of grandeur with the passionate love of beauty, the joy of life with the sense of death." In 1944, Camus had been the future, but in 1952, Sartre declared him a thing of the past, that his personality had become a "mirage." He closed by saying, "I have said what you meant to me, and what you are to me now. But whatever you do or say in return, I refuse to fight you. I hope that our silence will cause this polemic to be forgotten."

It was not forgotten. The newspaper headlines heralded the "Spat Among the Existentialists" and "Sartre Versus Camus," with one low-brow weekly declaring (accurately): "The Sartre-Camus Break Is Consummated."

For Camus, publishing Jeanson's review was an act of disloyalty, but Sartre's reply to his letter was an act of excommunication. What Camus struggled to digest, what pained him most, was that an important ten-year friendship had been dissolved in an instant. Camus had played a role in its undoing by provoking Sartre, but he did not expect such a violent response. He and Sartre had always had philosophical and political differences, but managed to enjoy each other's company for years. To Camus that was evidence that at least he valued people over ideas, and placed a premium on loyalty. In his notebook, he scrawled: "Sartre, the man and the mind, *disloyal*."

The personal criticism stung. Camus recognized kernels of truth about his weaknesses. He avoided cafés and restaurants where he might run into Sartre. Leaning on those closest to him for support, Camus retreated to his corner to nurse his wounds. He wrote to Francine, who was away from the capital: "I am anguished by Paris . . . *Les Temps Modernes* came out with twenty pages of response from Sartre and thirty from Jeanson . . . As for the replies, one is nasty and the other foolish. Neither answers my questions, except for Sartre at one point, but the fifty pages are deliberately insulting . . . Decidedly, this book has cost me dearly, and today I only have doubts about it, and about myself, who resembles it too much."

As THE WEEKS passed, Camus struggled to rebound from Sartre's blows, the humiliation of their public split, and the waves of commentary it provoked. Camus turned down invitations to participate in a public debate, telling the organizer, "At this point, the least sentence I might say will be used in a way that disgusts me in advance . . . It would be impossible for me in that case to continue expressing myself with academic politeness. I am mistaken for a deliberately polite man whom one may insult in all safety."

Support arrived, however, from many readers who offered encouraging words. Polish artist Józef Czapski, who was taken prisoner by the Soviets at the outbreak of the war and survived to expose Soviet atrocities at Katyn (where some 22,000 Polish officers and intellectuals were massacred on Stalin's orders), wrote to Camus: "Immediately after the attack that was concentrated on you, I wanted to write to tell you why I love you, and why you have more friends than you think."

Camus regrouped. Nothing any critic wrote or said changed his assessment of the problems he had identified. Camus wrote Czapski, "Leftist intellectuals in particular have chosen to be the gravediggers of freedom." He told another correspondent, "The core of the problem remains intact . . . and I feel that they offered nothing serious to oppose to my diagnosis. I therefore consider myself authorized to continue the same road, which I know, besides, is the road of many."

—m—

As WINTER APPROACHED, he left the political and literary battlefields of Paris for the place where he had always gone to restore himself—home to Algeria to see his family, the countryside, and the sea.

Return to Tipasa

Camus arrived from Marseille in early December. After visiting his mother in Algiers, he planned to tour parts of the southern region of the country he had never seen.

But he learned that there had been some unrest in the south and postponed his trip for a few days. He decided to take a day trip to Tipasa, the main attractions of which were its extensive Roman ruins that were perched on several hills overlooking the sea.

It was familiar ground for Camus. Many years earlier he had walked among the ruins as a young man and was inspired to write a short story, "Nuptials at Tipasa," which was published in *Noces* in 1938. He hoped that his pilgrimage in 1952 might rekindle memories of his happy youth, of a time when he once slept "open-eyed under a sky dripping with stars."

The sixty-nine-kilometer road to Tipasa was also a familiar route, one that he had first taken as a teenager by bus to the beaches along the coast, where he would try to impress girls—a time before he contracted TB, before the war and his long exile in France.

Arriving at Tipasa, Camus found an opening in the barbed wire that now surrounded the ruins, and started walking among the stones. He was struck by the silent scene around him—the perfect brown columns of the ruins, the fragrant wormwood plants, the still blue sea below, the Chenoua Mountains looming in the distance against the clear sky, all illuminated by the intense Algerian sun:

> And under the glorious December light, as happens but once or twice in lives which ever after can consider themselves favored to the full, I found exactly what I had come seeking, what, despite the era and the world, was offered me, truly to me alone, in that forsaken nature . . . In this light and this silence, years of wrath and night melted slowly away. I listened to an almost forgotten sound

> within myself as if my heart, long stopped, were calmly beginning to beat again.

Camus lingered at the ruins, listening carefully. Gradually, he began to recognize "one by one the imperceptible sounds of which the silence was made up: the figured bass of the birds, the sea's faint, brief sighs at the foot of the rocks, the vibration of the trees, the blind singing of the columns, the rustling of the wormwood plants, the furtive lizards."

Camus had found the "refuge and harbor" he had come for, and at Tipasa he discovered, or rather rediscovered, perhaps the most vital secret to his creative life:

> *Je redecouvrais à Tipaza qu'il fallait garder intactes en soi une fraîcheur, une source de joie* . . . [I was rediscovering at Tipasa that one must keep intact in oneself a freshness, a cool wellspring of joy,] love the day that escapes injustice, and return to combat having won the light. Here I recaptured the former beauty, a young sky, and I measured my luck, realizing that in the worst years of our madness the memory of that sky had never left me. This was what in the end had kept me from despairing . . .
>
> In the middle of winter I at last discovered that there was in me an invincible summer.

He returned from Algeria to Paris, reporting to friends that he felt "bucked up and calmed down."

Eschewing polemics for more poetic pursuits, Camus polished his short story "Return to Tipasa." One of his most evocative writings, Camus drew on the spirit that was reawakened at Tipasa, and renewed his artistic and political commitments:

> I have returned to Europe and its struggles. But the memory of that day still uplifts me and helps me to welcome equally what delights and what crushes . . .
>
> There is thus a will to live without rejecting anything of life, which is the virtue I honor most in this world . . .

There is beauty and there are the humiliated. Whatever may be the difficulties of the undertaking, I should like never to be unfaithful either to one or to the others.

The story was published in a volume entitled *L'Été* (*Summer*). On the title page of Monod's copy, Camus jotted a new inscription:

à Jacques Monod,
en amicale et fidèle pensée
[To Jacques Monod,
with friendly and loyal thoughts]

CHAPTER 24

THE ATTIC

There is an element of chance to the root of genius.
—ALBERT CAMUS, *Notebooks, 1942–1951*

JACOB LIKED TO START HIS DAY EARLY, BEFORE MOST PARISIANS were out on the street. The walk to the Pasteur was relatively short, about a mile from where he, Lise, and their young son, Pierre, lived near the Carrefour Montparnasse. His route took him along the boulevard Montparnasse, past the train station, left onto the boulevard de Vaugirard, then right onto boulevard Pasteur, before he turned left down rue du Docteur Roux to the wrought-iron gates of the Institute grounds, and then made his way up to Lwoff's attic. The quiet morning journey often reminded him of how much he had missed Paris and longed to return while exiled in Africa during the war.

The walk also built his anticipation of the results of the previous day's experiments. One of the great advantages of working on bacteriophage (or "phage") was that, although invisible to the naked eye or conventional microscopes, the viruses quickly left a mark of their bacteria-killing power. One single phage multiplied so quickly that it made a visible hole, a kill zone on a lawn of bacteria that grew overnight in a laboratory dish. Arriving before most of the other scientists, Jacob delighted in entering "his" laboratory, looking at his bacterial dishes, and then figuring out what experiment he should do that day.

The "induction of the prophage" that Lwoff had proudly mentioned to Jacob at his pivotal last-chance interview was a mysterious property of phages—so mysterious, in fact, that prior reports of the phenomenon had been doubted or discarded. The observation dated back decades: certain bacterial strains, when apparently freed of any bacteriophage in their surroundings, would sometimes burst, or "lyse," spontaneously and release bacteriophages. Such strains were dubbed "lysogenic."

While a few scientists interpreted the phenomenon as evidence that viruses could exist in a hidden, quiescent state in bacteria and then somehow be triggered to reproduce, others attributed the observations to flawed experimental design and the contamination of the medium with phage.

Lwoff, however, was a meticulous and clever experimentalist. He thought that he could settle the issue by studying one large species of bacteria—so large that it could be seen and manipulated under a microscope. Using a micropipette under a microscope, he was able to isolate the bacteria in tiny droplets of sterile media, which ensured that no extraneous bacteriophages were present. He then observed which individual bacteria released phage. He called the quiescent state of the virus in these bacteria the "prophage." He then searched for conditions that affected the efficiency with which lysogenic strains released viruses, and found that ultraviolet light caused almost all such bacteria to lyse and release viruses—this was the "induction of the prophage" he'd mentioned to Jacob.

Thanks to Lwoff's guidance, and the reproductive properties of bacteriophage, Jacob got off to a fast start in the attic. He first studied thirty strains of a different species of bacteria from the one Lwoff used to see which were lysogenic and whether induction with ultraviolet light was a general phenomenon. By taking the medium from the culture of each strain and placing a few drops of it on a lawn of each of the other strains, he quickly found out which strains harbored prophage. Then he examined those strains and determined that ultraviolet induced some viruses and not others. In a strain harboring two types of prophage, he learned that one was inducible and one was not. This result revealed the important fact that inducibility was a genetic property of the virus, not the bacterium. Within just a few months, he produced his first scientific paper, a moment of great pride.

Such progress, however, gave birth to new questions: Where and how did the prophage hide out in the bacterial host? What determined whether it was induced by ultraviolet light or not? Such questions often arose through conversations in the corridor, by bouncing results and ideas off one or more of his floor mates—all of whom had different backgrounds and were of various nationalities and contrasting personalities. While there were relatively few permanent staff members, the attic hosted many scientists from abroad who came to work for some period of time. In 1951, Seymour Benzer, an American, came for a year's sabbatical at Lwoff's invitation and moved into Jacob's crowded lab. Benzer had a good deal of experience working with bacteriophages, as he had previously worked at Caltech with Max Delbrück, a pioneer in the field.

Jacob came to Benzer's rescue early on in his stay. When Benzer and his family first arrived in Paris, they had no permanent place to live and went to a hotel, the same hotel in fact where Sartre was staying, although they never saw the famous existentialist. They eventually found a small artist's studio to lease, with an open-air bathtub and a communal toilet that was just a hole in the floor. They were told that everyone in France lived that way and that as Americans, "you have to lower your standards a little bit." When Jacob came to visit him, he was shocked and asked Benzer, "Why are you living like this?"

"Well, we want to live like the French," Benzer said.

"The French don't live like this," Jacob informed his lab mate. The family broke their lease and eventually found a nice apartment near the Pasteur.

Jacob came to appreciate Benzer's sharp and creative mind, and they eventually teamed up to do some experiments together. He also enjoyed Benzer's vast curiosity and mischievous sense of humor. Accustomed to the more casual atmosphere of American labs, Benzer had some fun with Lwoff's rituals and formalities. Partly out of his sense of adventure, and partly for its shock value, Benzer liked to bring exotic things for lunch—whatever he could find in Parisian markets, such as sea urchin, bull testicles, or dried South African caterpillars. When Lwoff teased him by asking, "Did you ever try *tétine de vache* [cow's udder]?" Benzer tracked down the delicacy at the butcher shop, cooked it for lunch on a laboratory Bunsen burner, and declared it "quite delicious."

While most were not as colorful as Benzer, Jacob discovered that his floor mates were largely just as bright, curious, and helpful. And some carried a more tragic war story than his own. Elie Wollman was another member of Lwoff's group working on bacteriophage. He was not a mere visitor to the Pasteur; he was born into the institution. His parents, Eugène and Elisabeth Wollman, began working at the Institute in 1909, eight years before he was born. His godfather and namesake was Elie Metchnikoff, a great Russian scientist who went to Paris to work with Pasteur in the late 1880s and whose discoveries concerning the cellular basis of immunity earned him a share of the 1908 Nobel Prize. Eugène and Elisabeth Wollman were in fact two of the very few researchers who worked on bacteriophage in the 1920s and 1930s, and they also described the phenomenon of lysogeny. Despite their Jewish backgrounds, and Nazi restrictions that precluded them from publishing their work after 1940, they remained in Paris and at the Institute during the Occupation. In December 1943, despite great efforts by the Pasteur's director to intercede, the couple was arrested by the Germans and died during their deportation to Auschwitz.

Elie had earlier escaped Paris for the south of France, joined the Resistance, and worked as a doctor under an alias. After the war, he returned to Paris and the Pasteur, where Lwoff, who was close to the elder Wollmans and grieved their murder, saw to it that Elie had a scientific home. Jacob found Wollman to be a font of knowledge about both the Pasteur and the world of bacteriophage research. They became good friends.

Surrounded by such enthusiastic, sharp, and rigorously critical people as Wollman, Benzer, Lwoff, Monod, and Cohn, Jacob had the sense that he had found the right place, the right address. And that he had done so at the right time, when so much was beginning to happen in biology. Those feelings were familiar: the atmosphere at the Pasteur, the sense of being where something was happening, stirred him up "as much as had being part of a Free French combat unit in the war."

And while working in the lab gave Jacob a great sense of purpose, his family life gave him boundless joy. In the spring of 1952, fraternal twins Laurent and Odile were born. Every evening, Jacob hurried home to his beautiful wife and growing clan. A fourth child, Henri, would be born in 1954. Each birth was, for Jacob, a rebirth, an affirmation of life. And each moment together with his family helped him replace bitter memories, as if each were "a revenge on the war, on death."

Membership in the Club

In order to obtain bona fide research credentials, to earn the "legal right to practice science," as he put it, Jacob pursued a doctoral degree under Lwoff's supervision. That would require both substantial time—almost four years—and original research. It would also require staying on top of what everyone else was doing that was relevant to his research. And the best ways to do that were to attend gatherings of scientists outside of France, and to hear and meet with the stream of visitors who passed through Paris.

In early 1952, Jacob attended his first international conference, the Society for General Microbiology meeting held in Oxford, England. Lwoff had been invited to give a talk on lysogeny; he brought Jacob and Benzer with him. Jacob discovered that much of the intelligence to be gathered was not so much from the presentations as from the rumors that one picked up in informal chats during breaks, at mealtimes, or in the pub in the evening.

But there were talks that Jacob had anticipated, a chance to see and hear those giants whose names he knew from his reading. Salvador Luria was the most famous on the schedule, but he did not show up. The rumor was that Luria, although living and working in the United States, had been denied a passport for the meeting—another victim, like Monod, of the Internal Security Act. British politesse forbade any blunt announcement of such a controversial fact, so it was handled indirectly. An English scientist in the audience referred a question to Luria, and another audience member asked the chairman, "There is a rumor that Dr. Luria has been prevented from attending by being refused a passport. Are you in a position to scotch that rumor?"

"No sir, I am not," said the chairman.

Luria instead sent his paper to be read by his former student Jim Watson, then a twenty-four-year-old postdoctoral fellow at Cambridge. Watson posed a startling image to Jacob and most attendees—tall and gangly, with wild hair, his shirttails out, socks down around his ankles, he looked like he had literally run over from Cambridge. He had a bewildered look about him, eyes bulging and mouth half open, and spoke in short, choppy bursts. Convinced by Oswald Avery's original evidence that DNA was the genetic material, Watson had recently begun working on the chemical structure of DNA (a change in his research plans that the Merck Fellowship board did not approve, and thereby revoked his fellowship). Luria's paper, however, explained his rationale for thinking that proteins were the genetic material, as they, and not DNA, were the first molecules to be produced after phage infection. But Watson boldly decided not to read it, and instead he explained to the audience that he had just received a long letter from Alfred Hershey, another prominent bacteriophage researcher then working at Cold Spring Harbor, that described new experimental evidence that DNA, and not protein, was the genetic material. Jacob listened intently to this unorthodox American, who was in effect arbitrating the conflicting views of two eminent, but absent, researchers.

Watson explained how Hershey and his collaborator Martha Chase had exploited the fact that phage proteins contain sulfur but that DNA does not, while phage DNA contains phosphorus while phage proteins do not. By growing phage in the presence of precursors containing radioactive phosphorus or radioactive sulfur, Hershey and Chase were able to "label" the DNA and protein components of the phage, respectively, and then follow where each label went in the course of phage infection of bacterial cells. They found that, after infection, DNA quickly wound up inside the bacterial host, while the protein remained outside stuck to the surface of the bacteria. The protein, but not the DNA, could be liberated by stirring the cells in an ordinary kitchen-type Waring blender. The results indicated that the phage, which resembled the shape of a hypodermic needle when seen magnified in the electron microscope, "injected" its DNA into bacteria shortly after infection, while the empty vessel remained outside.

It was a clever but simple experimental design, and it yielded results that clearly distinguished between two alternatives. It was, therefore, the best kind of experiment, and would become a classic in the annals of molecular biology. To Jacob, two implications were clear: DNA *was* the genetic material; and the prophage he was studying must be in the form of DNA inside the bacteria, and not in the form of intact virus particles. Although no one had ever seen a prophage, the mental picture had become clearer.

Jacob was merely a spectator at Oxford; a bigger step was to present his own work, his own discoveries, before a critical audience. And he was making some intriguing progress toward that goal. Working with phage day after day, he was expert at inspecting plates of bacteria in which phage left their telltale pinholes, called plaques. Normally, every plaque was just a bit cloudy due to some bacteria that overgrew the plaque but did not lyse. When those bacteria were isolated and analyzed, they turned out to be lysogenic—they were harboring prophage. One day, Jacob spotted a completely clear plaque. When he isolated bacteriophage from the plaque, they produced only clear plaques. The phage were mutants: able to reproduce in bacteria and lyse them, but unable to hide out in bacteria as prophage, unable to form lysogens.

What was the nature of such a mutation? The question demanded more experiments. When Jacob added the mutant virus to a lawn of lysogenic bacteria, he found that instead of producing clear plaques, it produced no plaques at all. The result demonstrated that lysogenic bacteria were "immune" to infection with other phage. Something in the lysogenic bacteria prevented the clear-type phage from reproducing. That same something could be made by the bacteria or by the prophage. Yet more experiments to do.

Jacob's first opportunity to present his own work came in the summer of 1952, just as the Sartre-Camus split was unfolding. Lwoff hosted a meeting of the First International Congress on the Bacteriophage at the elegant thirteenth-century Cistercian abbey at Royaumont, twenty miles north of Paris. All of the top phage scientists were invited, including Delbrück, Hershey, and Luria, the latter of whom had finally managed to secure a passport. Watson attended as well. Some thought that Watson's attire—shorts and tennis shoes with their laces usually untied—was a deliberate effort to annoy Lwoff. The truth was that his luggage was stolen en route to France and the only clothes he had were a few things he'd bought at an Army PX and those intended for a subsequent hiking trip in the Italian Alps. Watson prudently borrowed a jacket and tie for a garden party held at the country estate of Baron

Edmond de Rothschild. At the end of the evening, Baroness Rothschild told some guests that she regretted that the "mad Englishman from Cambridge" of whom she had heard had not attended. Lwoff had apparently forewarned her that a partially clad eccentric might appear.

Watson had managed to pass his inspection without anyone noticing. Jacob's test came when he gave an account of his work in front of the "jury" of phage luminaries, with Delbrück sitting in the front row. Jacob was surprised when he first saw the former physicist. He'd been expecting a stereotypical paunchy, aging, balding German professor. Instead, he saw a young-looking athletic man, with a full head of hair, wearing steel-rimmed glasses, listening intently to the speakers. Despite the intimidating audience and his inexperience, Jacob's nervousness disappeared once he started speaking. The talk actually went over well.

Even better, Delbrück subsequently invited Jacob to the next Cold Spring Harbor Symposium, a meeting on viruses that was to take place in the summer of 1953. Jacob was thrilled. The invitation was like receiving his "membership card to the club"—the club of insiders, of those in the know, of those who mattered in the very small and select group of international scientists probing the nature of life.

The Double Helix

Membership in that club meant hearing about discoveries before others, and well before they appeared in journals. Some members in Lwoff's attic were made privy to some very big news when Jim Watson came through Paris again in March 1953.

The breakthrough had happened just a couple of weeks earlier.

For more than a year, Watson had been working with Francis Crick in Cambridge, trying to solve the structure of DNA. Watson was certain that the structure was the most important riddle of genetics, indeed of all biology. The central mystery that the correct structure had to solve was how such a molecule was copied, and thus how genetic information was transmitted faithfully generation after generation. Their efforts had stalled repeatedly as they hit various kinds of impasses. Some were caused by a shortage of data; others appeared when they attempted to build models and discovered that they were chemically impossible.

Several things were known about DNA: it was an acid; it contained four different bases—adenine (A), guanine (G), cytosine (C), and thymine (T); and it was a polymer with a sugar phosphate backbone. It had also been reported that within any given species, the ratio of adenine to thymine was always equal, and the ratio of cytosine to guanine was always equal. This chemical analysis, however, had not been sufficient for anyone to decipher the arrangement of atoms within DNA's three-dimensional structure. The critical unknowns were the number of chains in the molecule and the relative arrangement of the bases and the backbone. Crucial clues to those features came from a technique called X-ray crystallography, in which a beam of X-rays was trained on a crystal of DNA, causing the beam to scatter and diffract according to the detailed structure of the crystal. The interpretation of the resulting patterns required a rare expertise.

After struggling for many months, and rejecting one flawed structure after another, the penny finally dropped on the morning of Saturday, February 28, 1953. Tinkering with cardboard-cutout models of the bases in his lab at Cambridge, Watson realized that the A and T bases, and the C and G bases, could form bonded pairs (base pairs) with very similar shapes that fit neatly within and could hold together a two-chain, double helix. Moreover, if A always paired with T, and C with G, then the copying of DNA was easy to explain: the identity of the base on one chain would determine the identity of its complement on the other.

Watson and Crick started building a physical model of DNA right away, and writing up their discovery for publication. A few days later, and despite Crick's objections, Watson decided that it was a good time to make a previously postponed visit to Paris. In addition to the food and fun, the excursion gave him the chance to share the news.

Although Watson did not have a full model of the structure, just of the two base pairs, Monod understood instantly. As the most biochemically inclined of the attic's residents, Monod could best appreciate Watson's chemical and structural reasoning. Moreover, Monod grasped right away that Watson had solved the copying problem.

Jacob, however, did not have the background to weigh Watson's explanations. In fact, he only skimmed Watson and Crick's historic article on the structure of DNA that appeared the following month in *Nature*. The arguments based on the X-ray crystallographs were beyond him. It was not until six weeks later, when he went to Cold Spring Harbor and heard Watson present the complete story, that he understood the full, profound significance of the double helix.

It was Jacob's first trip to the United States. He went by boat with André Lwoff.

At the meeting, Jacob was struck by the contrast with European gatherings. At Cold Spring Harbor, there were no lofty speeches, no formalities whatsoever. The scientists sat where they liked, no matter how eminent their neighbor might be. And students and junior scientists did not hesitate to ask challenging questions of their seniors.

Jacob's talk went fine. He described some new work he had been doing in collaboration with Elie Wollman on a bacteriophage called *lambda* that infected *E. coli*. He reported on the properties of some mutant phage, including different kinds of clear-plaque mutants that could not infect lysogenic bacteria, and another rare mutant that could. The unquestionable star of the meeting, however, was Watson.

Wearing shorts again, and with his shirttail out as usual, Watson gave a detailed explanation of the evidence for the double helix, and the biological consequences that followed from it. The arrangement of the bases implied not only how DNA was copied but also how mutations arose by the substitution of other bases, and that the sequence of bases must somehow encode the characteristics of every organism. The normally feisty and skeptical audience offered no criticisms, no objections. Jacob now appreciated that the structure was both simple and beautiful, and explained so much that he thought, "All this could not be false." Even though the technical background was still beyond him, he understood that the double helix illuminated on the screen above Watson accounted for the fundamental mystery of heredity—"one of the oldest problems posed since antiquity by the living world."

The first secret of life had been revealed.

Before returning to France, Jacob purchased a Waring blender as a gift for Lise.

Fewer than three years into the research game, he could not have possibly imagined that the next big secrets of life would belong to him and his attic neighbor Monod, nor that to get to them, that Waring blender would come in very handy.

Part Four

Nobel Thoughts and Noble Deeds

Where'er a noble deed is wrought,
Where'er is spoken a noble thought,
Our hearts in glad surprise
To higher levels rise.

—Henry Wadsworth Longfellow,
Santa Filomena

CHAPTER 25

THE BLOOD OF THE HUNGARIANS

Rise, Magyar! is the country's call!
The time has come, say one and all:
Shall we be slaves, shall we be free?
This is the question, now agree!
For by the Magyar's God above
We truly swear,
We truly swear the tyrant's yoke
No more to bear!

—SÁNDOR PETŐFI, *"Talpra Magyar"*
(Hungarian national poem, recited by demonstrators in October 1956)

THE SPRING OF 1953 BROUGHT ANOTHER HISTORIC MILESTONE—regime change in the Soviet Union. After ruling the country for nearly three decades, Joseph Stalin died on March 5.

A five-member body of the Presidium assumed leadership of the country at first. After much maneuvering and intrigue, including the arrest and execution of one member who had been head of the secret police, Nikita Khrushchev eventually ascended to first secretary of the Communist Party, the de facto head of government, in September 1953.

After Khrushchev solidified his grasp on power, he signaled some potentially profound shifts in the USSR's posture toward the West—away from the arms race and toward improving relations and "peaceful coexistence." This thaw in the Cold War was made official policy at the ten-day-long Twentieth Soviet Party Congress in February 1956 in which Khrushchev played a starring role. Peaceful coexistence, Khrushchev explained, was "not a tactical move" but now "a fundamental principle of Soviet foreign policy," derived from a position of the growing strength of Socialist states and Communism. He asserted his "certainty of the victory of communism" over capitalism, but also stressed that securing world peace and "the ending of the arms race remains one of mankind's vital tasks." Speaking of matters closer to home, Khrushchev earned prolonged applause when he pledged "to strengthen in every way our fraternal relations" with other states in the Socialist camp. He declared, "The stronger the entire socialist camp, the more reliable will be the guarantee of freedom, independence, and economic and cultural progress of the countries making up this great camp."

After the formal sessions were over, a special "secret" session was called for the early hours of February 25 in the Great Hall of the Kremlin. Khrushchev again took the podium, but this time shocked the assembled delegates by launching into a speech denouncing Stalin for his "use of mass terror," "brutal violence," "cruel repression," and other excesses. Recounting the purges of Party leaders, specific cases of false arrest and execution, and the nearly fatal blunders in preparing for and executing the war with Germany, Khrushchev blamed the "Cult of Personality" that had evolved around Stalin that had made him "akin to a god," and believed him to be infallible. This worship led to "grave perversions of Party principles, of Party democracy, of revolutionary legality" that Khrushchev spelled out for more than four hours in devastating detail. Khrushchev vowed in closing: "Our Party, armed with the historical resolutions of the 20th Congress, will lead the Soviet people along the Leninist path to new successes, to new victories!"

The first serious test of Khrushchev's new path would come just a few months later during a political crisis in Hungary. The plight of Hungarians striving to free themselves from the Soviet sphere would become a cause in which both Camus and Monod would become deeply engaged. Ensuing events would validate Camus's long-running and lonely condemnation of the totalitarian Soviet system, and prompt Monod to plunge into human smuggling.

A Dark History

After fighting on the side of the Axis powers in World War II, and then being invaded, defeated, and occupied by Russian forces at the end of the war, Hungary fell under the firm political, military, and economic control of the Soviet Union. By 1949, all non-Communist opposition had fled, been arrested, or been suppressed.

At the helm of the so-called People's Republic of Hungary was the general secretary of the Communist Party, Mátyás Rákosi. A great admirer of Stalin—he even described himself as "Stalin's greatest pupil"—Rákosi was groomed for the Party boss position while he spent the war years in the Soviet Union. With unlimited power, Rákosi styled his regime after the Stalin model. There were no personal liberties. The regime was the most restrictive of all the Soviet satellites. Anyone straying from Rákosi's dictates, whether the infraction was real or perceived, was dealt with very harshly. The Hungarian secret police, the Államvédelmi Hatóság (AVH), was run by the Party.

Agnes Ullmann, a married, twenty-nine-year-old biochemistry graduate student at Eötvös Loránd University's Faculty of Science in Budapest had witnessed firsthand the excesses of the Rákosi regime, the brutality of Russian occupation, and indeed the entire sequence of calamities that had befallen eastern Europe over the previous seventeen years. A native Romanian, Ullmann was born in northwest Transylvania and raised in Arad, in west Romania near the Hungarian border. Romania first declared itself neutral at the outset of World War II, then joined the Axis powers in November 1940 and fought alongside the Germans against the Soviet Union. The tide of the war turned in 1944 when the Soviets invaded her homeland, pushed back the Germans, and then occupied the country.

Living in Arad during the war, Ullmann had to deal first with the German occupation, then bombings by American and British planes, and finally the Russian invasion. Seventeen years old at the time of the German occupation, she was not hassled by the occupiers. Indeed, with her long blond braids and white socks, she figured that she "looked like a Hitlerjugend [Hitler Youth]." She was able to come and go without attracting interest from the Gestapo, a freedom that her father exploited and that offered Ullmann some wartime adventures.

After the war, Ullmann attended university for two years in Cluj, Romania, before moving to Budapest in 1947. At the university, Ullmann received a Soviet-approved curriculum. "The teaching was Michurin and Lysenko, and the biology courses were about that. That was it," she later recalled.

She had her doubts, however, about Soviet dogma, doubts that were sown by the banning of Hungarian composer Béla Bartók's music as "imperialist," by the debates over "bourgeois science" brewing in the USSR, and by the insistence that all Western ideas were to be discredited.

"It was absolutely unbelievable . . . Nobody believed at that time what this Soviet propaganda said," Ullmann later recalled. The students therefore could not distinguish good Russian science from worthless dogma. Ullmann taught a course in general chemistry and explained that the periodic table of the elements was the creation of Dmitri Mendeleev, a brilliant Russian chemist. Accustomed to fables about Soviet scientists, the students wanted to know who in the West had really developed the table.

It was very dangerous to share opinions with anyone except the most trusted friends. Ullmann had confided her doubts about Lysenko to a friend, Györgi Adám, an economist and committed Communist whose ties to the Party reached back to the days when the Party was illegal in Hungary. Ullmann trusted that he would not betray her.

As chief editor of an official Marxist magazine, Adám was one of the few with access to Western newspapers. One day in 1949, he showed Ullmann the issue of *Combat* in which Jacques Monod had published his critique of Lysenko, saying, "You know, you told me once that you have a bad feeling about [Lysenko]."

Monod's article made an enormous impression on Ullmann: "It was a fabulous discovery for somebody to whom Western information was unavailable."

—w—

LATER THAT VERY summer, Adám failed to show up for work at the editorial office of MTI, the Hungarian news agency. A colleague learned from Adám's landlady that he had been taken away from his apartment in the middle of the night by the AVH. The lifelong Communist had been deemed a threat by the Rákosi regime. Adám's trial was held in secret, and he was sentenced to life in prison.

Adám was just one of the many thousands of victims of Rákosi's purges that swept up writers, intellectuals, and other card-carrying, loyal Party members on fabricated charges, including many who had fought against the Nazis. Ullmann was horrified. "It was absolutely awful . . . People were arrested. People were killed. People were hanged."

Ullmann was working in a laboratory with a young Yugoslavian. At the time, Yugoslavia's prime minister, Josip Tito, had enraged the Kremlin by daring to follow his own course, including accepting support from the West through the Marshall Plan. All Yugoslavians were thus suspected "tools of the imperialists" in the eyes of the Rákosi regime. Ullmann's lab mate was arrested. And because working alongside a Yugoslav apparently made one suspect, she was arrested as well. After being taken at six in the morning from her home and interrogated for a day and a half, she was released.

Her lab mate was executed.

RÁKOSI'S REIGN OF terror was carried out by the AVH, which employed 100,000 police and a vast system of informants in order to root out anyone about whom there might be the slightest doubt regarding loyalty to the regime. These so-called class enemies, Zionist agents, or Party infiltrators were arrested, whether Communists or clergy, and sentenced in secret or in public show trials. Leniency was often promised to those who were willing to denounce others, so the cycle of arrests was perpetuated. Between 1949 and 1953, an estimated 150,000 to 200,000 people were arrested and sent to prison or forced labor camps where torture was commonplace; another 2,000 were executed.

The political witch hunts, however, depleted government ministries of competent management and forced many better-educated

Hungarians into menial labor. The atmosphere of fear intimidated managers and workers from addressing problems at factories. Combined with the forced collectivization of peasant farms, the Hungarian economy crumbled. Food production fell, food shortages were widespread, and rationing was instated. Despite the miserable conditions, there was endless official praise and professions of love for the Soviet Union and Stalin. The populace was cynical and bitter.

Rákosi enjoyed a free hand while Stalin was in power, but after Stalin's death in March 1953, the Soviet leadership moved to curb Rákosi. The lesson from their experience with Stalin was that no individual should hold so much power, so the Soviets moved to dilute the authority of government leaders in the USSR as well as in satellite states. At that time, Rákosi was both prime minister and secretary general of the Party, and held absolute command over Hungary. Rákosi was summoned to Moscow and instructed to replace his one-man rule with collective leadership and to bring into the government some of his most ardent critics, including his minister for farm deliveries and deputy prime minister, Imre Nagy.

Flickers of Hope

Nagy was appointed prime minister (technically the chairman of the Council of Ministers of the People's Republic of Hungary) in July 1953. On his very first day in office, he stunned Rákosi, the Party, and the country by announcing a series of reforms, including putting an end to arbitrary police arrests, dismantling the forced labor camps, and promising to review the cases of all who had been imprisoned on political charges during Rákosi's regime. Nagy proclaimed that he was reversing course from Rákosi's farm and industrial policies, and added, "Intellectuals must be esteemed . . . We must display greater tolerance in religious matters. The foundation of the Socialist . . . state is a strict respect for the rule of law."

Nagy's "New Course" fanned hopes that Hungary was turning away from the past for good. In 1954, Nagy secured permission from Moscow to release thousands of Communist political prisoners, who were then able to tell their stories of the horrors of Rákosi's camps and prisons. When people learned for certain what most suspected had been going on for years, many who had avoided arrest and served the regime felt terrible shame. One writer summed up the guilt of those who had conformed to the regime's dictates: "In sleepless nights it is no sop to one's conscience that one did not directly participate in murder and betrayal. Because responsibility lies not only with the one who bludgeoned but also with those who tolerated evil in whatever fashion—through the thoughtless repetition of dangerous theories, the wordless raising of the right hand, the half-hearted writing of half truths."

In March 1955, spurred by Nagy's New Course and the atmosphere of more open dialogue, intellectuals and leaders of the Communist youth organization (the Dolgozó Ifjúság Szövetsége, or DISZ) formed a discussion group to debate how to go about reorganizing the country. Summoning the spirit and legacy of the 1848 revolution, they named their group the Petöfi Circle, after the famous poet-martyr Sándor Petöfi. Ullmann attended the first meeting and became member number 21 of a movement that would grow into the tens of thousands, nearly all of whom were also Communist Party members who shared the hope of reforming, not overthrowing, the regime.

The wind changed direction, however, yet again, in Moscow. One of Nagy's key supporters resigned and the Soviet Presidium wanted the reformer out. In April 1955, Nagy was removed as premier and excluded from the politburo. In November, despite being a longtime committed Communist, Nagy was kicked out of the Party altogether.

Moscow put Rákosi back in charge again, but the genie of reform had been let out of the bottle and Rákosi would find it very difficult to put it back in. The people now knew about his criminal excesses and economic failures. The press and public discourse had become much more open and critical.

The influence of the Petöfi Circle grew enormously, along with the hope of reform following the disclosure of Khrushchev's "Cult of Personality" speech. Attendance at Circle debates in Budapest swelled into the hundreds and then the thousands in the months following Khrushchev's bombshell speech.

Reports from the Petöfi Circle gathering made Rákosi furious. He condemned the "anti-Soviet nature" of the speakers' statements and forbade further meetings of the Circle. He drew up a list of people to be arrested for their roles in the "anti-Party plot."

But Moscow had Rákosi on a short leash. On July 18, 1956, before he could move on his detractors, Soviet deputy chairman Anastas Mikoyan arrived from Moscow and told Rákosi to resign because his methods were not acceptable to the Soviet leadership. Rákosi was replaced by Ernö Gerö, another hard-liner and one of Rákosi's henchmen.

Gerö spent much of his first weeks in office consulting with the Kremlin and leaders of other Soviet satellites. He was not in the country for the next major act in the Hungarian drama. László Rajk was a foreign minister who had been officially "rehabilitated" the previous spring when Rákosi was forced to admit that the charges that led to his execution were false. Rajk's widow insisted on a public reburial of her husband. Gerö went along with the idea, thinking the event would symbolize his new regime's break with his predecessor.

On October 6, Rajk's body and those of several others who had been executed along with him were dug up from a forest near Budapest, where they had been hidden for seven years, in order to be reburied in Kerepesi Cemetery alongside other Hungarian statesmen.

The weather was lousy—cold, rainy, and windy. Nobody knew how many mourners to expect. The government anticipated perhaps a few thousand.

Despite the grim conditions, Ullmann went to the ceremony. She was astounded to see the throngs of people lining the streets of Budapest—more than 200,000 by most estimates. There had been nothing like this in Budapest since the end of the war. But there was no shouting of slogans, or even banter. The crowd remained quiet and dignified as they followed the funeral procession and listened to the eulogies. The profound irony of the massive turnout was that Rajk had been very unpopular when he was alive and a member of Rákosi's regime. Hungarians had come to bury more than Rajk. One speaker captured the crowd's thoughts: "The hundreds of thousands who are now walking by these coffins not only want to pay the dead their last respects; their passionate hope and immutable decision is to lay an epoch to rest. Lawlessness, arbitrariness, and the moral corpses of those shame-filled years must be buried for good, and the threat emanating from the Hungarian disciples of the law of might and personality cult must be banished forever."

The funeral was a dress rehearsal for putting that passionate hope into action.

Hungarians march for greater freedoms, October 23, 1956. (AP Images)

Day 1: October 23, 1956

It was an unusually warm and sunny fall day in Budapest. The seventy-degree weather was perfect for a stroll around the magnificent capital, with its spectacular medieval Buda Castle and massive gothic parliament building facing opposite banks of the Danube.

It was also a perfect day for a march that would make history.

That afternoon, students were assembling on both the east bank (Pest) and west bank (Buda) of the city to demonstrate their solidarity with Poland. A few months earlier, a massive strike by workers in Poznań was crushed violently by Polish security forces, resulting in the deaths of several dozen workers and protestors. The Polish regime's missteps led to its replacement by a less Soviet-dominated government. In mid-October, the Polish Party leader had managed to stand up against the Soviet Union and extract some concessions sought by the strikers that summer. The events in Poland had nourished Hungarian hopes for greater independence from Soviet influences. The evening before, the students had drafted a sixteen-point list of demands, including the immediate withdrawal of Soviet troops, free elections, return of a multiparty system, and freedom of the press.

As the students made their way toward the statue of a hero of the 1848 revolution, Polish general Józef Bem, their numbers swelled as young workers fell in with the demonstration. Ullmann's laboratory was just a few minutes away from the path of one march. Her husband, Tamás Erdös, a biologist, was in Sweden at the time, so Ullmann joined the ranks by herself. She had no inkling that by the next day, these first steps of peaceful protest would trigger a violent revolution. Nor could she possibly have had any notion that ensuing events would lead her to a life-changing rendezvous with Jacques Monod.

That same morning, Gerö and other Hungarian leaders had returned from a visit to Yugoslavia and were warned by the newspaper's editors that the demands of the masses should be taken seriously. Gerö tried to squelch the demonstrations by having his minister of the interior issue a ban on all public gatherings, which was announced over the radio just before one p.m. But the students did not hear it and were assembling anyway. The regime faced the dilemma of either putting down the demonstration by force or doing nothing and looking impotent. So they reversed the ban shortly after two p.m.

By then the marches were under way.

Emotions were very high. After years of repression, the mass display of unity and purpose was liberating. People watched and cheered from windows and shop doors. The slogans from the marchers grew from simple statements of solidarity—"Poland Shows Us the Way"—to more brazen chants of "Gerö into the Danube." Someone cut the hammer and sickle out of the center of a green, white, and red Hungarian flag. The redesigned flag was passed forward to the front of the marchers.

By four thirty p.m., 50,000 people were crammed into and spilling out of Bem Square. The students had no plans beyond gathering at the square, so they started to disperse and fan out over the city. As they meandered the streets and were joined by workers leaving factories and shops, the crowds swelled again by tens of thousands. Now shouts of *"Russkik haza!"* (Russians, go home!) could be heard. Ullmann returned to the Faculty of Science Building, where some students reassembled in a large auditorium.

Another group of students headed to the government-controlled radio station in order to press for their demands to be broadcast. The bulk of the crowd headed for Parliament Square, a little more than a mile away, in the heart of Pest. By dusk, the square was filling with what would become a throng of 200,000 people. The government tried to disperse the protestors by turning off all the lights in the square. Instead, marchers turned the daily newspaper and students' flyers into torches. The mood was excited but peaceful.

At eight p.m., instead of hearing a broadcast of their demands, they heard Gerö speak by radio. In a twelve-minute, cliché-laden speech, he condemned "those who seek to instill in our youth the poison of chauvinism and to take advantage of democratic liberties that our state guarantees to the workers to organize a national demonstration." Gerö's tone and language angered the crowd, and reminded them exactly why they were protesting.

Not far away, in Heroes' Square, another group knocked down a forty-foot statue of Stalin and dragged the head to an intersection.

The joy of that triumph was short-lived. At the radio station, more than two hundred AVH men had barricaded themselves inside the building in anticipation of some trouble. They were equipped with heavy machine guns and tear gas. Throughout the evening, the students had been negotiating to get their demands broadcast; a delegation had been allowed to enter the building. Now, after hearing Gerö's speech, the crowd was getting more agitated and concerned about the delegation inside. They started hurling bricks and stones at the windows.

The AVH men tried to disperse the throng with tear gas and water hoses. They then tried to clear the street by fixing bayonets and moving the demonstrators back.

Shots rang out. Two demonstrators were hit, then another three. Chaos erupted.

Hungarian Army units arrived on the scene with orders to protect the radio station and crush the demonstration. Instead, seeing the AVH firing on the crowds, they did nothing. A tank regiment arrived, and the commander quickly sized up the situation and declared that he would not attack Hungarian civilians. The crowd asked the soldiers for their weapons so that they could shoot back at the AVH. The soldiers promptly handed them over. Police also opened their armories and provided weapons.

The protest at the radio station erupted into a firefight.

In the auditorium at the university, Ullmann and other students were getting reports from the radio station every fifteen minutes or so. Around eleven p.m., she and some colleagues tried to get close to the radio station but decided that it was too dangerous.

In the course of the battle at the radio station, sixteen protestors were killed and sixty were wounded before the building was taken. Five AVH were killed, and more than eighty were wounded or captured.

Around one a.m., some friends dropped Ullmann off near her home. As she and other residents of Budapest went to sleep that night, many questions loomed: Was the battle at the radio station an isolated incident? How would the government respond?

Day 2: October 24

At two a.m., those questions were answered. Columns of Soviet tanks started to rumble into the city.

Gerö had called his masters for help.

At four thirty a.m., Budapest Radio broadcast from a second transmitter not in the hands of the rebels. After wishing its listeners good morning, it spoke for the government on the previous evening's events: "Fascist and reactionary elements have launched an armed attack against our public buildings and against our forces of law and order. In the interests of re-establishing law and order, all assemblies, meetings, and demonstrations are forbidden. Police units have been instructed to deal severely with troublemakers and to apply the law in all its force."

The Russians sent 700 tanks and 6,000 men into Budapest—an overwhelming force relative to the roaming bands of just a few hundred young, unorganized protestors. Gerö was confident that the streets would be cleared of those troublemakers "without difficulty in a few hours."

But Gerö had miscalculated; the presence of the Russians galvanized the protestors. Now armed, they became freedom fighters. Once the troops were spotted, they were greeted with hails of bullets and Molotov cocktails. The bulky Soviet T34 tanks could not navigate the narrower streets and alleyways within the city, so the rebels engaged in hit-and-run attacks, using their knowledge of the city's interior to stage ambushes and then quickly disappear.

Budapest Radio urged: "The Soviet soldiers are risking their lives to protect the peaceful citizens of Budapest and the tranquility of the nation . . . Workers of Budapest, receive our friends and Allies with affection."

The radio announced shortly thereafter that Imre Nagy had been reappointed prime minister. Gerö remained in charge as first secretary of the Party.

Fighting across the city—at major squares, on main boulevards, and at several intersections—forced the streets to remain empty. Most citizens, including Ullmann, stayed home.

Instead of being put down in a few hours, the rebellion grew. Seeing that the dead rebels in the streets were largely young students, some Hungarian Army troops and their tanks joined the insurgents. In the course of the day, 80 freedom fighters were killed and another 450 were wounded, while the Russians lost 20 men and had 40 wounded.

Day 3: October 25

"The army, the state security forces, the armed workers' guards, and Soviet troops liquidated the counter-revolutionary putsch attempt on the night of October 24," proclaimed Budapest Radio on the morning of the twenty-fifth.

But when people ventured out into the streets they still heard and saw gunfire. The rebellion had not been squelched. The radio was telling lies, as usual. Word spread that there would be another demonstration in front of Parliament. Ullmann and thousands more headed to Parliament Square.

They passed the Russian soldiers and tanks deployed throughout the city that had been reinforced by 14,000 troops overnight. But the Russians did not seem to be bent on the destruction of Budapest or the killing of Hungarians. Their instructions were to maintain order. Marchers engaged the Russians in conversation, explaining that they were workers and university students, not counterrevolutionaries or Fascists. Some Russian soldiers placed Hungarian flags on the turrets of their tanks, then escorted the marchers to Parliament Square.

Despite the ban on assemblies, more than 25,000 demonstrators, including many women and children, soon filled the large square. And despite the events of the previous two days, coupled with the presence of the Russian tanks in front of Parliament and of AVH men on surrounding rooftops, the atmosphere was light and celebratory. With Nagy back in government, many figured that the course of reforms could resume and that the Russians would leave.

Shots rang out.

Ullmann and the rest of the crowd dove, or tried to dive, down to the pavement. It was so crowded, not everyone could lie down. Machine-gun fire burst through the square. When some tried to run for open spaces or into side streets, they were fired upon by the AVH.

When the shooting stopped, Ullmann got up, but, she recalled, "there were people who did not get up. That was absolutely awful. And they didn't let the ambulances in."

It was a massacre. Estimates of those killed varied widely, from seventy-five to several hundred. But regardless of the death toll, the effect of the slaughter was to inflame and expand the ranks of rebels. They went on the offensive in the city, springing more ambushes on Russian tanks.

The wheels were already in motion for Gerö's removal when the debacle unfolded at Parliament Square. Moscow's representatives in

Budapest told Gerö he had to resign. He was replaced by János Kádár, a former minister once jailed by the Rákosi regime. The announcements were made by radio shortly after the killings. Kádár appealed for order and promised that the rebels would not be punished. Moreover, he pledged that Hungarian-Soviet relations would be reviewed. Nagy added that the government would negotiate for withdrawal of the Russians.

But with fierce fighting continuing in the streets, and uncertainty regarding Moscow's stance on the escalating crisis in Hungary, the ability of the new regime to control events was unclear. The Russians imposed a dusk-to-dawn curfew, and the government appealed for calm and the resumption of normal activities.

VICTORY!

The two days of street fighting had left Budapest in shambles. A reporter for the London *Daily Mail* described the scene as he entered the city: "Every street was smashed. Hardly a stretch of tramcar rails was left intact. The jungle of hanging electric cables was denser even than in Buda. Hundreds of yards of paving stones had been torn up, the streets were littered with burned-out cars . . . I counted the carcasses of at least forty Soviet tanks . . . two monster Russian T-54 tanks lumbered past, dragging bodies behind them, a warning to all Hungarians of what happened to the fighters. In another street, three bodies were strung up on a tree."

The bodies belonged to AVH men who had been hunted down and lynched by freedom fighters in reprisal for the massacre at Parliament Square.

After several days of violence, Budapest Radio kept repeating that the Russians were successfully liquidating the counterrevolution. But the freedom fighters continued to hold out.

For the Soviets' part, their commanders recognized that they were in a difficult predicament. They were not prepared for a long urban guerrilla war. Moreover, since Hungary was a friend and ally, the Soviet troops were asking themselves just who were they fighting and why had they become targets? In just a few days, the streets and facades of beautiful Budapest were shot to hell. Would they have to raze the city to wipe out the rebels?

By Saturday, October 27, the streets became calmer as fighting abated and the populace ventured out of their homes to buy food and to get the latest news. While the new Kádár-Nagy government was debating what to do in the rapidly developing situation, and consulting with Moscow's envoys, citizens were taking the next steps in the revolution into their own hands. In the first days of the revolt, in factories across the country, many workers' councils were formed to make their demands known to the government and to take over the management of the factories from Party appointees. And in villages, towns, and cities, revolutionary councils were formed to take over municipal government.

At the university, committees of intellectuals—students, professors, writers, and scientists—were forming. Ullmann, trying to find out how she could be useful, learned that a nascent committee was in the offices of the rector of Eötvös Loránd. There, she met György Adám, who asked her to stay and help what became part of the Revolutionary Committee of Hungarian Intellectuals. Ullmann made contact with workers' committees and other revolutionary committees, and helped distribute weapons that had been stashed in the basement of the building. Her husband, Tamás, hurried back from Sweden and joined the committee as well. By October 28, the Revolutionary Committee had drafted a ten-point appeal to the government. Among its demands were the withdrawal of Soviet forces from Hungary, the management of factories by the workers, complete freedom of speech and the press, and the right of assembly.

Throughout the uprising, the Hungarian Central Committee of the Communist Party had been meeting and debating how to manage the situation. Under the pressure of the revolt, the Central Committee acknowledged the need for wholesale changes in response to the demands of the people and declared: "In consultation with the entire people, we shall prepare the great national program of a democratic and socialist, independent and sovereign Hungary."

That same morning, Nagy and Kádár pressed for a truce with the

Russians. Khrushchev consented. After five days in which more than 1,000 Hungarians and 500 Russians had been killed, a cease-fire was announced over the radio at 1:20 p.m.

Later that afternoon, Nagy acknowledged the revolutionaries' victory on Radio Budapest.

He pledged to dissolve the AVH and promised that the government would abstain from any action against the rebels. On the matter of the Russians, Nagy continued: "The Hungarian Government is initiating negotiations to settle . . . the question of the withdrawal of the Soviet troops stationed in Hungary."

The next day, the Soviets announced that they were withdrawing.

ON TUESDAY, OCTOBER 30, the Soviet tanks guarding Parliament, the bridges over the Danube, and other monuments and strategic points began to leave, taking the bodies of many fallen comrades with them.

Early that afternoon, Nagy announced the end of one-party rule, proclaiming that "the tremendous force of the democratic movement has brought our country to a crossroads." The national radio station, now renamed Free Radio Kossuth (after one of the 1848 revolutionaries), stunned listeners with its candor about "beginning a new chapter in the history of Hungarian radio. For many years our radio has been an instrument of lies; it merely carried out orders. It lied by night and by day, it lied on all wavelengths. Not even in the hour of our country's rebirth did it cease its campaign of lies . . . In the future, we shall tell the truth, the whole truth, and nothing but the truth."

In just one week following the marches to Bem Square, the Hungarian people had managed to replace a hardline regime, stand up to a Soviet invasion, dissolve the AVH and one-party rule, and free its press.

The world had watched with awe and admiration as the Hungarians displayed courage in the face of overwhelming odds. A British diplomat in Budapest declared, "It is nothing short of a miracle that the Hungarian people should have withstood and turned back this diabolical onslaught." Camus, at a gathering to honor a Spanish Republican statesman, spoke of the "heroic and earth-shaking insurrection of the students and workers of Hungary." *Le Figaro* reported that by their "glimmering fires of joy," the citizens of Budapest had "washed away the last traces of communism."

After twelve years of repression and suffering, Hungarians had just cause for hope in a future of their own making. The same day as Nagy's and Free Radio Kossuth's bold announcements, the Kremlin endorsed a pact called "On Friendship and Cooperation between the USSR and Other Socialist States." It repeated the promise that troops would leave Budapest, but most remarkably, it admitted "violations and mistakes which infringe the principles of equality between sovereign states." It further promised to negotiate with Hungary and other members of the Warsaw Pact on the matter of maintaining Soviet troops in their territories.

The Russians' contrition and commitment lasted all of one day.

CHAPTER 26

REPRESSION AND REACTION

If ten or so Hungarian writers had been shot at the right moment, the revolution would never have occurred.

—Nikita Khrushchev

Throughout the crisis in Hungary, the Soviet Union Central Committee Presidium met frequently to decide policy and tactics. Khrushchev had heartily endorsed the generous language used toward the Hungarians on October 30, but by the next morning, he was having second thoughts. He told the Presidium, "We should re-examine our assessment and should not withdraw our troops from Hungary and Budapest. We should take the initiative and restore order in Hungary. If we depart from Hungary, it will give a great boost to the Americans, English, and French—the imperialists. They will perceive it as weakness on our part and will go onto the offensive."

Most of his comrades, several of them hard-liners who were against the initial agreement to withdraw, promptly agreed. Khrushchev asked Marshal Ivan Konev, the commander of all Warsaw Pact forces, how long it would take to crush resistance.

"Three days, no more," the veteran soldier replied.

"We'll do it then," Khrushchev said.

In Budapest, Imre Nagy had been receiving conflicting reports on the Russians' withdrawal. Nagy's advisers sounded the alarm, but Nagy did not believe that the Soviets would renege on their agreement so soon. Nagy warned the Russian ambassador, Yuri Andropov, that if the troops did not retreat, Hungary would declare itself neutral and leave the Warsaw Pact. With no such assurances forthcoming, Nagy made good on his vow; Hungary's withdrawal was announced over the radio that evening.

Nagy still believed, or at least hoped, that disaster could be averted. He was determined to convene the previously planned talks with the Russians about troop withdrawal. At the Parliament, the Russian delegates were received with full honors. Nagy received regular, and encouraging, reports from the talks. The Russians went over their withdrawal plans in detail, which included a festive military parade at which they expected to be cheered. The Hungarians in turn reported that they had ordered all of their tanks back to barracks and called upon all civilians to turn in their arms and ammunition.

In the meantime, however, János Kádár then inexplicably disappeared from Budapest. The Soviet Central Committee had summoned him to Moscow.

Ullmann and Erdös had been camped out at the Revolutionary Committee offices for several days, sleeping on some mattresses under a piano. Ullman had been urging workers' committees to end their strikes because the revolution had succeeded and it was time to go back to work. When she learned of the Russians' demand for a friendly send-off, she also had to explain to revolutionary committees in surrounding villages the need for cooperation. Optimistic about the news trickling out of the talks with the Russians, Ullmann and Erdös decided they could relax. Late in the evening, they climbed into a truck with some armed students and went home to get some rest.

Shortly after midnight, Nagy received word at his office in the Parliament Building that a massive force was approaching Budapest. This time the Russians were much better prepared. Marshal Konev had assembled ten divisions for Operation Whirlwind, comprising 150,000 troops, 2,500 tanks, and accompanying air support that would crush whatever resistance the Hungarians might offer. At four a.m., he gave the code word "thunder," and the shelling erupted.

Nagy knew there was little he could do. The Hungarian forces were no match for the Russians; they would be slaughtered. His deputy encouraged him to make a broadcast while there was still time. Nagy went down to the makeshift studio of Free Radio Kossuth inside Parliament and addressed the nation: "This is Imre Nagy speaking, the President of the Council of Ministers of the Hungarian's People Republic. Today at daybreak Soviet forces started an attack against our capital, obviously with the intention to overthrow the legal Hungarian democratic government. Our troops are fighting. The Government is in its place. I notify the people of our country and the entire world of this fact."

After Nagy's statement and as Soviet tanks rolled up in front of Parliament, Nagy's longtime friend Gyula Háy, a playwright, arrived at the

Russian tanks on the streets of Budapest after the second Soviet invasion. (© Hulton_ Deutsch Collection / Corbis)

Parliament Building and dashed off another appeal to be broadcast: "This is the Hungarian Writers Association speaking to all writers, scientists, writers' associations, academies and scientific organizations of the world. We appeal for help to all intellectuals in all countries. Our time is limited. You all know the facts. There is no need to review them. Help Hungary! Help the writers, scientists, workers, peasants, and all Hungarian intellectuals. Help! Help! Help!"

These were the last words heard over Free Radio Kossuth.

THE HUNGARIANS DID not stand a chance against the Soviet artillery, tanks, and jets. Key Army units were swiftly surrounded while they slept in their barracks, then awakened and persuaded to lay down their arms. The Soviets then concentrated on pummeling several strongholds occupied by the freedom fighters. This time the invaders were not at all concerned about collateral damage. When fired upon, the Russians often leveled buildings without any regard for civilians inside or nearby.

Completely outgunned and outnumbered, the Hungarians knew that resistance was futile, but many vowed to fight regardless. Their sole hope was to hold on long enough for the West to intervene. Desperate to get word of the battle for Budapest to the outside world, one eyewitness started up the Telex connection between the *Szabad Nép* newspaper office and the Vienna bureau of the Associated Press early Sunday morning:

> SOS SOS SOS
>
> YOUNG PEOPLE ARE MAKING MOLOTOV COCKTAILS TO FIGHT THE TANKS. WE ARE QUIET BUT NOT AFRAID. SEND THE NEWS TO THE WORLD.
>
> THE FIGHTING IS VERY CLOSE NOW AND WE HAVEN'T ENOUGH TOMMY GUNS IN THE BUILDING. I DON'T KNOW HOW LONG WE CAN RESIST . . . HEAVY SHELLS ARE EXPLODING NEAR BY. ABOVE JET PLANES ARE ROARING . . .

WE NEED MORE. IT CAN'T BE ALLOWED THAT PEOPLE ATTACK TANKS WITH THEIR BARE HANDS.

WHAT IS THE UNITED NATIONS DOING? GIVE US A LITTLE ENCOURAGEMENT.

THEY'VE JUST BROUGHT A RUMOR THAT AMERICAN TROOPS WILL BE HERE WITHIN ONE OR TWO HOURS.

Americans coming. The same rumors and wishful thinking had circulated in the last hours of free Paris in 1940. All of the news correspondents who had entered Budapest had been asked the same question: "When are the Americans coming?" The rebels believed or had been led to believe, perhaps by broadcasts of Radio Free Europe, that the West would step in to help Hungary.

Of course, the Americans weren't coming. There was no way President Eisenhower was going to risk a direct confrontation with the USSR and World War III over Hungary.

The World Reacts

The Hungarians' hopes for UN intervention were equally in vain. Before dawn on Sunday in New York, when the Russian attack was a few hours old, the United States was able to force an emergency meeting of the Security Council on Hungary. US delegate Henry Cabot Lodge said, "If ever there was a time when the action of the United Nations could literally be a matter of life and death for a whole nation, this is that time."

The Soviets argued that Hungary had fallen into the control of Fascists. After two hours of debate, the USSR vetoed a resolution censuring the attack on Hungary on the grounds that it constituted "interference with the internal affairs of Hungary." That evening, the General Assembly took up an American resolution calling for the USSR to "desist forthwith from all attack on the people of Hungary and from any form of intervention in the internal affairs of Hungary" as well as "to withdraw all of its forces without delay from Hungarian territory." Although the resolution passed 50 to 8, with 15 abstentions, it was merely symbolic, as the UN did not have the means to enforce it.

There was an international outpouring of sympathy for the Hungarians and outrage at the Russians. In an editorial entitled "Repression," *The Times* of London said, "The Russians have given their answer, and the bitter tragedy of Hungary has risen to the height of terror and anguish. Can there be any other end but the extinction of liberty?"

No other capital, however, was more engaged in the Hungarian drama than Paris. Being home to many Hungarian émigrés, as well as a large base of Communist supporters, the revolution and the Soviet response inspired fierce emotions in Paris. Unlike many Communist parties elsewhere in Europe, the French Communist Party (PCF) had not welcomed Khrushchev's February speech and de-Stalinization. Led by the staunch Stalinist Maurice Thorez, the PCF remained completely obedient to Moscow and expected the same of its rank and file. Thorez and the large-circulation Communist daily *L'Humanité* dutifully adopted the Soviet line that the Hungarian revolution was led by Fascists, and had to be crushed. In their statement published in *L'Humanité* the day after the invasion, the PCF applauded the Soviet Army.

The PCF and its leadership immediately became the target of anti-Soviet protests.

On November 7, a demonstration in support of Hungary was announced. By six o'clock that evening, a crowd of 30,000, including numerous members of the National Assembly and elected officials of Paris, had assembled at the Place de l'Étoile. They started up the Champs-Élysées behind French and Hungarians flags, and banners proclaiming "Liberate Budapest," "Outlaw the Communist Party!" and "Thorez to the post!" [firing squad]. By six thirty, the head of the procession reached the Arc de Triomphe, where many laid flowers at the tomb of the Unknown Soldier. Several prominent figures were at the monument, including members of the government such as François Mitterrand, and former presidents of the council such as Paul Reynaud.

A group of young demonstrators broke off from the parade of delegations, and with cries of "Burn the PC," they hustled toward the head office of the PCF. The first wave reached the building at seven thirty and quickly overran the small police force protecting it, but encountered about five hundred Communists in front of the building. The demonstrators and Communists fought on the sidewalk and in the stairwells. The building's defenders showered the attackers with fire hoses. In twenty minutes, the ground floor was taken and the assailants began ransacking it, throwing furniture and documents into the street. A fire broke out on the third floor, and firefighters doused it while the many injured in the melee were attended to in nearby bars.

The protestors abandoned the PCF building and, singing "La Marseillaise," rushed toward the nearby offices of *L'Humanité*. Alerted by telephone, defenders inside the building greeted the mob with fire hoses, bottles of acid, and other projectiles. The protestors plucked bricks from the street, hurled them through the windows, and armed themselves with pickaxes from a nearby work site. A small group reached the roof of the building and started a fire. Others tried but could not penetrate the printing room of the newspaper. A pitched battle unfolded in the street when Communist reinforcements from the Paris suburbs arrived on the scene. By the time calm was finally restored, thirty were wounded and three men, two Communists and one union worker, later died from their injuries.

The clashes further polarized the atmosphere in Paris. In *L'Humanité*, Communist members of the National Assembly characterized the protestors as "fascist arsonists and vandals" and painted the defense of their building in heroic terms akin to the time of the Resistance. In the Assembly itself, insults, threats, and anger were heard throughout days of debate on Hungary. Communist deputies were booed during voting and told: "You are the only communists in the world to insult corpses that are still warm."

It was very difficult for those inside Hungary to understand the United States' inaction or the impotence of the United Nations. Pleas for help continued. On November 8, Camus received a telegram from a Hungarian émigré writers' group. It contained the text of an appeal that was broadcast from one of the last remaining insurgents' radio outposts:

> POETS, WRITERS, SCHOLARS OF THE ENTIRE WORLD. HUNGARIAN WRITERS ARE ADDRESSING YOU. LISTEN TO OUR CALL. WE ARE FIGHTING AT THE BARRICADES FOR LIBERTY OF OUR COUNTRY, FOR THAT OF EUROPE AND FOR HUMAN DIGNITY. WE ARE DYING. BUT OUR SACRIFICE SHOULD NOT BE IN VAIN. AT THIS SUPREME HOUR, IN THE NAME OF A MASSACRED NATION, WE ADDRESS OURSELVES TO YOU, CAMUS, MALRAUX, MAURIAC, RUSSEL [*sic*], JASPERS . . . AND MANY OTHER FIGHTERS OF THE MIND. THE HOUR HAS SOUNDED AND THE TIME FOR SPEECHES IS OVER. ACTS ARE NECESSARY. DO SOMETHING. ACT. THROW OFF THE HORRIBLE INERTIA OF THE OCCIDENT. ACT. ACT. ACT.

The appeal was published on November 9 in the liberal, left-wing newspaper *Franc-Tireur.* Since he was named specifically in the text, Camus felt obliged to respond personally. His reply was published in the next day's issue:

> Our Hungarian brothers, isolated in a fortress of death, most probably do not know of the immense outburst of indignation among all French writers. But they are right to think that speeches are not enough and that it is ridiculous to raise vain laments about a Hungary that has been crucified. The truth is that the international community . . . left Hungary to be assassinated. It has been twenty years already since we allowed the Spanish Republic to be crushed by the troops and arms of a foreign dictatorship. This great courage had its reward: the Second World War. The weakness of and divisions within the United Nations bring us little by little to the third world war, which is now knocking at our door. It knocks and it will enter if international law is not imposed for the protection of nations and individuals everywhere in the world.

Camus then proposed that all of those named in the appeal sign a joint letter to the United Nations General Assembly requesting that it examine "the genocide of which Hungary was now the victim" and that each nation vote for the immediate withdrawal of Soviet troops and the liberation of detainees and deportees. The letter pledged that if the United Nations backed away from its duty, the signers would not only boycott the UN and its cultural organizations but also denounce it at every opportunity.

Camus also suggested that writers throughout Europe gather signatures of other intellectuals and forward those to the Secretary General of the UN. He pled that intellectuals should do all they could to stop the "butchery" in Hungary, and "to demonstrate to the world that besides our cruel and weak governments, and on top of the curtain of dictators, despite the dramatic collapse of the movements and traditional ideals of the left, a true Europe exists, united in justice and liberty, in the face of all the tyranny. The Hungarian fighters are dying today for this Europe. In order that their sacrifice not be in vain, we . . . owe it to them to demonstrate, day after day, our fidelity and our faith and to relay, as far as we can, the appeal of Budapest."

Camus would fulfill that commitment by speaking often throughout the following year about the plight of the Hungarians.

Largely missing from the outpouring of public indignation and condemnation were, not surprisingly, the voices of Communists or their supporters. There were a few important exceptions. Sartre, who began 1956 by publishing a New Year's greeting in *Pravda* to "our Soviet friends" and earlier had stated: "the Party has manifested an extraordinary objective intelligence so much so that it rarely errs," now saw the error of Moscow's ways. In an interview with *L'Express*, he declared, "I condemn absolutely and unconditionally the Soviet aggression" and "the intervention was a crime." "The crime," Sartre added, "was made possible and perhaps necessary (from the Soviet point of view naturally) by twelve years of terror and stupidity."

Sartre announced, "Regretfully but completely, I am breaking my ties with my friends among Russian writers who are not denouncing (or cannot denounce) the Hungarian massacre. It is no longer possible to be friendly toward the ruling faction of the Soviet bureaucracy." He added, "It is not and never will it be possible to re-establish relations with the men who are presently leading the French Communist Party. Each one of their phrases, each one of their gestures, is the end result of thirty years of lying and sclerosis."

Sartre's turnabout was, at last, some public vindication for Camus's long-maintained condemnation of Soviet policies—views for which Camus had paid dearly. But Camus did not stoop to saying "I told you so." He did not have to, and the tragedy unfolding in Hungary was no cause for satisfaction. Moreover, Sartre's analysis of the causes of the Hungarian debacle was quite different from that of Camus. Sartre saw the Soviet action as a consequence of strategic blunders, weaknesses in Hungary, and outside forces rather than the inevitable terror of a totalitarian state, as Camus believed.

Sartre elaborated at length in his *L'Express* interview. He suggested that the "gravest fault was probably Khrushchev's report [in February], for in my opinion the solemn public denunciation, with a detailed list of crimes of a sacred figure who represented the regime for so long, is madness when such frankness is not possible by a prior and considerable rise in the living standards of the population." Sartre believed that Hungarians were as yet too backward for such candor and that it was a mistake "to reveal the truth to the masses which were not prepared to receive it." Sartre also pinned blame on the West and the Marshall Plan.

Camus drew a simpler, unambiguous, conclusion. On November 28, at the grand Salle Wagram, various student organizations held a gathering in tribute to their fellow young Hungarians. In front of two immense French tricolor and Hungarian flags, a message from Camus was read:

> The only thing that I can publicly affirm today, after having participated directly or indirectly in twenty years of our bloody history, is that the one supreme value, the last for which it is worth living and fighting, remains freedom . . . For ten years, we had to fight against Hitler's tyranny and the right wing who supported it. And, for ten more years, we had to combat Stalin's tyranny and the sophisms of its defenders on the left . . . You must know now that when the mind is chained, work is enslaved; when the writer is muzzled, the worker is oppressed; and that when the nation is not free, socialism liberates no one and enslaves everyone.

At the time Camus made his statement, the fighting in Hungary had largely ceased. Condemnations from the West were no deterrent to the Soviets. Nor was all of that French intellectual firepower and sympathy for the freedom fighters of any use against Soviet guns. By November 11, 1956, the last strongholds of resistance in Budapest had crumbled. More than 2,500 Hungarians had been killed since the start of the revolution, the majority of those during the second invasion.

What remained uncertain was the form that daily life would take in Hungary after the invasion, and whether any of the freedoms fought for and secured for a brief moment during the revolution might endure.

The Last March

After the resistance was crushed, Kádár took over. His newly installed regime, dubbed the Revolutionary Workers and Peasants Government, had to deal with the aftermath of the invasion and govern the shattered country. There were several difficult challenges, including getting the people back to work (a general strike that had been called to protest the invasion had brought the country to a halt) and deciding how to deal with the revolutionaries.

Thousands of Hungarians fled the country at the outset of the invasion, and more were continuing to cross into Austria and Yugoslavia. Ullmann, however, felt obliged to remain in the country, as many students she knew had been rounded up and arrested in the course of the invasion. She would not leave while they were in prison. Moreover, she and her compatriots sought to salvage whatever they could of gains made during the revolution.

But within weeks, some all-too-familiar scripts unfolded. Imre Nagy and other officials who had taken refuge in the Yugoslavian embassy since the Russian takeover were duped into leaving and promptly arrested. Budapest workers who tried to organize themselves were blocked from meeting by tank squadrons, and Kádár had about two hundred leading members of the workers' councils arrested.

With virtually all means of protest suppressed, several writers who edited the underground newspaper *Élünk* (We Are Alive)—Gyula

Obersovszky, István Eörsi, and József Gáli (the fiancé of Vera Káldor, one of Ullmann's colleagues at the university)—managed to organize a demonstration for December 4, the one-month anniversary of the second Russian invasion. Figuring that the authorities would be reluctant to use force against women, the writers planned a silent women's march to mourn those lost in the revolution. They made their appeal in the newspaper: "Hungarian mothers! Hungarian women! It is your turn now. Your strength is enormous! Not even bullets can harm you! Your silent, honorable demonstration compels an armistice and calls for respect for our sacred cause."

Several thousand women took the risk and proceeded to Heroes' Square to place flowers and evergreen twigs on the Tomb of the Unknown Soldier. Ullmann was in the third row. The Russians looked on, fully armed. She and the other marchers were convinced that the soldiers would not shoot at a procession of women, some with children in tow.

The march ended peacefully, but all three organizers were arrested within a few days. Obersovszky and Gáli were charged with editing an illegal paper containing antigovernment material, for taking part in the preparation of the women's demonstration, and for distributing leaflets that incited the populace against state order. On December 11, the government banned the Revolutionary Council of Intellectuals and all territorial workers' councils; on December 12, martial law was declared; and on December 13, courts were empowered to imprison citizens without trial and to impose the death sentence for a variety of offenses, including sabotage or concealing weapons and ammunition. The first execution was carried out December 15.

A new security force, many of its members formerly of the AVH, took over for the Soviets in the streets. Workers continued to rebel at factories, so in early January, the government authorized the death penalty for strikers or anyone inciting a strike. On January 17, the writers' union was declared illegal, and on January 20, the journalists' union was similarly banned. One week later, several leading journalists and writers were arrested.

Ullmann was very exposed because of her role in the Revolutionary

Committee, which included the disbursing of weapons and providing papers to people who fled the country. Both were serious offenses that were being prosecuted by the regime. She burned all of the evidence she could. But her name was on a lot of documents. In the spring of 1957, Györgi Adám, chairman of the committee, was arrested. Other friends also were arrested, but they did not give Ullmann's name to the authorities.

The revolution was dead. And each day there was the looming risk of being denounced and arrested. But there was no safe way out of Hungary. The mass exodus of more than 200,000 Hungarians, more than 5,000 per day in the weeks after the invasion, had been checked when the Soviets destroyed several bridges into Austria, and the borders were resealed in December. Attempting to escape the country now incurred the risk of being imprisoned, or shot.

Kádár's Day of Fear

The waves of arrests, the imprisonments without trial, the executions, the suppression of the press, and the crackdowns on workers made the populace bitter, angry, and scornful of the regime. As the March 15 anniversary of the 1848 revolution approached, Kádár was deeply concerned that spontaneous demonstrations would break out and challenge the government. Any public protests would undermine his authority in the eyes of his Soviet bosses. Kádár planned a show of force to deter any gatherings.

Security forces would be posted on street corners; Hungarian Army units were to march through the streets of Budapest while other forces patrolled in trucks. The monuments that were the symbols of the revolution—Parliament Square, the Bem Statue, and the National Museum where Petöfi read his poem in 1848—were to be cordoned off by police. Russian soldiers in armored personnel carriers would also stand by in case they were needed. And to be sure that no rabble-rousers could cause any trouble, Kádár took additional precautions.

Around midnight on March 14, the doorbell rang in Ullmann and Erdös's second-floor apartment on Taragato ut (Taragato Street) in

Buda. Three men belonging to Kádár's security force appeared at the door. They entered the apartment and began to search it room by room. Ullmann followed the men around, even into the bathroom, afraid that they would plant a gun in the toilet, as they were known to do. When the security men began ransacking the bookshelves, Ullmann became furious and asked them to treat the books with more respect. To her surprise, the men complied. But they took papers—even the postcards Erdös had sent Ullmann from Sweden. They found nothing incriminating; all of the important papers had been burned. The men arrested Erdös anyway and would not tell Ullmann where they were taking her husband.

THE ANNIVERSARY WAS marked in Paris with a meeting at the Salle Wagram. Camus appeared in person to express his solidarity with the Hungarian people and to remind the audience and his readers (his speech "Kádár Had His Day of Fear" was published in its entirety three days later in *Franc-Tireur*) what the stakes were for all concerned. Camus's address was also in part a reply to Sartre, who had written at great length in the preceding months about the Hungarian situation and the fate of Communism.

In January, *Les Temps Modernes* published a 487-page triple issue devoted to the Hungarian revolution and Soviet repression. Sartre wrote a 120-page exposition of his views entitled *Le Fantôme de Staline* (*The Ghost of Stalin*) in which he laid most of the blame for the debacle on the reaction of hard-liners in the USSR to the de-Stalinization movement. Sartre also pinned responsibility on the West, whose anti-Communist stance, he alleged, was motivated to preserve the Cold War and the profits of arms manufacturers.

Sartre deemed the USSR's actions in Hungary criminal, but in no way did he indict the Communist system. Camus saw the one-party system as fundamentally bankrupt: "None of the evils that totalitarianism (defined by the single party and the suppression of all opposition) claims to remedy is worse than totalitarianism itself."

Sartre still believed Communism was a path to socialism: "Communism appears to us, in spite of everything that has happened, to be the sole movement which still carries within it the likelihood that it may lead to socialism." Camus underscored where that path had led:

> Foreign tanks, police, twenty-year old girls hanged, committees of workers decapitated and gagged, scaffolds, writers deported and imprisoned, the lying press, camps, censorship, judges arrested, criminals legislating, and the scaffold again—is this socialism, the great celebration of liberty and justice?
>
> No, we have known, we still know this kind of thing; these are the bloody and monotonous rites of the totalitarian religion!

Sartre asserted that "all men of the left" recognized that "the USSR must survive for the cause of communism." Camus rejected this idea completely:

> There is no possible evolution in a totalitarian society. Terror does not evolve except toward a worse terror, the scaffold does not become any more liberal, the gallows are not more tolerant. Nowhere in the world has there been a party or a man with absolute power who did not use it absolutely.
>
> The first thing to define totalitarian society, whether of the Right or the Left, is the single party, and the single party has no reason to destroy itself. This is why the only society capable of evolution and liberalization, the only one that deserves both our critical and our active support is the society that involves a plurality of parties as a part of its structure. It alone allows one to denounce, hence to correct, injustice and crime. It alone today allows one to denounce torture, disgraceful torture, as contemptible in Algiers as in Budapest.

Sartre still rejected the West and claimed that "the USSR is not imperialist, the USSR is peaceful, the USSR is socialist." Camus saw the West as flawed, but also as the only hope: "The defects of the West are innumerable, its crimes and errors very real. But in the end, let's not forget that we are the only ones to have the possibility of improvement and emancipation that lies in free genius."

In closing, Camus paid tribute to the Hungarians and tried to offer some glimmer of hope:

> Our faith is that throughout the world, beside the impulse toward coercion and death that is darkening history, there is a growing impulse toward persuasion and life, a vast emancipatory movement called culture that is made up both of free creation and of free work . . .
>
> The Hungarian workers and intellectuals, beside whom we stand today with so much impotent grief, realized that and made us realize it. This is why, if their suffering is ours, their hope belongs to us too. Despite destitution, their exile, their chains, it took them but a single day to transmit to us the royal legacy of liberty. May we be worthy of it.

Camus and Monod would soon have the personal opportunity to spare Hungarian lives.

CHAPTER 27

A VOICE OF REASON

Whoever saves a life, it is as if he saved an entire world.

—The Talmud, Mishnah Sanhedrin 4:5

Camus's stand on the Hungarian tragedy echoed the feelings of much of France, yet still managed to provide fodder for his many critics. His public statements aggravated his numerous detractors on the left who thought that it was hypocritical of Camus to condemn the Soviets for their actions in Hungary and yet remain silent on the actions of the French government, which was pursuing increasingly harsh policies to repress independence movements in French Algeria. For more than a year, Camus had kept a vow of silence concerning the crisis in his homeland. He had concluded, after extensive public involvement, that he should refrain from commentary "in order not to add to its unhappiness or to the foolishness that is being written on the subject."

The conflict in Algeria was a continuous source of profound anguish for Camus. "Algeria is the cause of my suffering at present as others might say their chest is the cause of their suffering . . . I have been on the verge of despair," he wrote in a letter to a longtime friend that was published in an Algerian newspaper. Since Camus remained closely identified with and attached to Algeria, those pushing for its independence from France expected him to support their cause. However, as an Algerian of European ancestry (Pied-Noir) and Frenchman, he understood that after being part of France for more than 125 years and with a population of 1.2 million Pieds-Noirs, Algeria could not break free without both it and France experiencing a great trauma. He hoped for reforms that would address Arab grievances but preserve Algeria as part of France.

As the violence in Algeria escalated with atrocities committed by both sides, that stand had put him in "the no man's land between two armies and preaching amid the bullets that war is a deception." Unable to support either the militants determined to gain independence or the French government bent on suppressing them, Camus sought to find some principle upon which both sides might find common ground. Horrified at the mounting civilian casualties, and ever mindful that his mother, brother, and many friends still lived in Algeria, in early 1956 Camus conceived of a "civilian truce" in which both sides would agree "for the duration of the fighting the civilian population will on every occasion be respected and protected."

Camus even traveled to Algiers in January 1956 in an attempt to secure support for the idea. It was a dangerous mission. He received death threats and feared that he might be kidnapped. Camus met with French liberals who supported reform, and with members of the militant, pro-independence National Liberation Front (the Front de Libération Nationale, or FLN). Under heavy security, he spoke to an invitation-only audience comprised of both groups at the Cercle de Progrès. As protestors outside chanted "Death to Camus" and pelted the windows with stones, Camus explained that he thought it was his "duty, to come and echo among you a purely humanitarian appeal that might, at least on one point, silence the fury and unite most Algerians." Camus admitted, "My only qualifications for taking a stand are that I have lived through the Algerian calamity as a personal tragedy and that I am incapable of rejoicing over any death whatsoever. For twenty years, with paltry means, I have done all that I could to contribute to the understanding of our two peoples."

Camus said that he could not let "two Algerian populations, each accusing the other of having begun the quarrel . . . to hurl themselves against each other in a sort of xenophobic madness . . . without launching a final appeal to reason." The crux of that appeal was for both sides to recognize that if they could not break the cycle of accusation and violence, they were "condemned to die together, with rage in their hearts." Camus hoped that by agreeing to protect the innocent, one narrow point of recognition might lead to a broader accord. But for the time being, he merely begged that "on a single spot of the globe a handful of innocent victims be spared."

The audience was moved, and there were positive accounts of the speech in the Algiers press. But the proposal languished. It was very soon after the Algiers trip that Camus stopped writing about Algeria. Understanding that he was walking a tightrope—seen as a traitor by the right wing because he did not support the French government's policies in Algeria, but also condemned by the left wing for not supporting the revolutionary violence of the FLN—Camus elected silence.

The shift was tactical, not spiritual. Camus had not abandoned Algeria; he remained committed to acting personally whenever he thought he could make a difference. He found himself intervening often on behalf of Algerians. For example, when he learned that Jean de Maisonseul—a friend from his youth in Algiers and a member of the committee that organized his appearance on behalf of the civilian truce—had been arrested by the French security police, Camus sprang into action. Certain that the liberal architect and painter was not involved in any conspiracy against the state, he wrote letters to Premier Guy Mollet, to the governor general in Algeria, and to *Le Monde* urging de Maisonseul's release.

The newspaper published Camus's appeal, in which he defended de Maisonseul's character and reputation, and explained that he could not remain silent in the face of "such stupid and brutal initiatives." Camus suggested that if de Maisonseul deserved to be arrested, so did Camus, Red Cross workers, and all of those involved in the truce effort. After several days during which no actions were taken, Camus raised his tone and the pressure on the government, demanding that de Maisonseul be released and that reparations be paid, and threatening to appeal for public action. De Maisonseul was freed and the case was later dropped.

But Camus's general silence was not understood, and he found himself explaining, if not defending, his position again and again. That meant being honest and risking the severing of old ties. He replied to a pro-FLN Algerian poet he had known for many years:

> *I owe you . . . the truth about what I think . . . I was painfully shocked by what you wrote on several occasions, about French Algerians in general . . . You have the right to choose the positions of the FLN. For my part, I think of them as murderous in the present,*

> *blind and dangerous in the future . . . I have given up hope on trying to make a voice of reason heard publicly . . . But, in private, I must tell you my reaction, and you should not ignore the shooting, nor justify that they shoot at the French-Algerians* in general, *and thus entangled, shoot at my family, who have always been poor and without hatred and who should not be mixed up in an unjust rebellion. No cause . . . will ever tear me from my mother, who is the greatest cause that I know in the world.*

As requests for his intervention continued to arrive at his office at his publisher, Gallimard, Camus was forced to weigh each case on its merits, which was often difficult given biased accounts of the circumstances, or to try to forge some general rationale to guide his involvement. As French reactions to terror attacks intensified, and as the executions of rebels in Algeria increased, Camus found the rationale for many interventions in his fierce opposition to the death penalty.

Outlawing Death

In the fall of 1956, Camus had been invited to contribute an essay to a book on capital punishment that would also include contributions by Arthur Koestler and Jean Bloch-Michel. Camus researched and composed "Réflexions sur la guillotine" (Reflections on the guillotine) throughout the winter, just as executions were also mounting in Hungary. The lengthy essay was first published in the June and July 1957 issues of the *Nouvelle Revue Française* and in book form the following year.

Camus focused once more on the issue of legitimizing murder that he had raised in his *Combat* editorials "Neither Victims nor Executioners" and in *The Rebel*. Now Camus argued at length that the death penalty must be abolished because it was inhumane, irreversible, ineffective, and in fact "not only useless but definitely harmful." As France remained "one of the last countries this side of the iron curtain to keep capital punishment in its arsenal of repression," Camus hoped to appeal to his readers' reason and sense of French progressivism.

He opened his essay with the story of his father's reactions to attending the public execution of a notorious murderer in Algiers—a story that had been told to him by his mother. Lucien Camus had gotten up early in the morning in order to go to the other end of town to watch the spectacle, the first he had ever attended. He returned home vomiting over what he had witnessed, and never spoke of the experience again. Camus's gory research told him enough to explain his father's reaction to the guillotine.

Camus noted that public executions were stopped in France in 1939 after photographs of one were published. He challenged the logic of moving the operation out of public view because, after all, one alleged purpose of the punishment was the deterrence of crime. Camus suggested that if deterrence were indeed the intended purpose, executions should be performed in broad daylight at the Place de la Concorde, and televised. But he pointed out that the statistics did not, in fact, support a deterrent effect. He cited data from Great Britain that the majority of condemned men there had witnessed at least one execution. Furthermore, in countries that had abolished the death penalty, there was no increase in murder and no correlation with overall criminality.

So why continue an ineffective policy? Camus asked. Because at its core, he answered, "let us recognize it for what it is essentially: a revenge. A punishment that penalizes without forestalling is indeed called revenge. It is a quasi-arithmetical reply made by society to whoever breaks its primordial law . . . whoever has killed must die." But Camus urged, "This is an emotion, and a particularly violent one, not a principle. Retaliation is related to nature and instinct, not to law. Law, by definition, cannot obey the same rules as nature . . . It is intended to correct it." And a law's justification, Camus pointed out, "is in the good it does or fails to do to the society." The lack of effectiveness of capital punishment proved to Camus that its "upholders cannot reasonably defend it" and that it was "a lazy disorder that my reason condemned."

Condemned, Camus thought, because it was also doing great harm. He made his case on two main points: judicial error and its irreversibility, and that it constituted state-sponsored murder that fostered a climate for yet more murder. The former argument rested on the fallibility of the justice system and the probabilities and certainty of putting innocent citizens to death. Once someone was executed, Camus noted, there was no possibility of correcting an injustice.

Even more compelling an argument to Camus was the idea that capital punishment was administrative murder—a murder "no less repulsive than the crime" and furthermore, "this new murder, far from making amends for the harm done to the social body, adds a new blot to the first one." Whereas one might have seen capital punishment as the state protecting individuals, Camus pointed out that in recent history and current events, the number of people killed by the state had assumed "astronomical proportions" that far exceeded private crimes. Moreover, the death penalty was being applied for political crimes, not solely murder. He urged that society must now defend herself more so against the state. Camus asserted that such "bloodthirsty laws" made for "bloodthirsty customs" and the climate for mass murder. Camus then reminded his readers of the executions during the Occupation and elsewhere, and of their consequences: "Without the death penalty Rajk's corpse would not poison Hungary; Germany, with less guilt on her conscience, would be more favorably looked upon by Europe; the Russian Revolution would not be agonizing in shame, and Algerian blood would weigh less heavily on our consciences. Without the death penalty, Europe would not be infected by the corpses accumulated for the last twenty years in its tired soil."

With the contagion spreading into Hungary, Algeria, and elsewhere, Camus declared that "we must call a spectacular halt and proclaim . . . that the individual is above the State" in order to progress toward a "society based on reason" and away from the anarchy created by the excessive powers of the state.

Camus urged that eliminating capital punishment would be a crucial step in that direction and indeed, that "in the unified Europe of the future the solemn abolition of the death penalty ought to be the first article of the European Code we all hope for . . . There will be no lasting peace either in the heart of individuals or in social customs until death is outlawed."

The problems of the Algerian conflict were much broader than the death penalty—too broad for Camus to budge the warring factions.

Therefore, he maintained his commitment to acting personally as appeals continued to stream in on behalf of Algerians and Hungarians. Of special concern to Camus was the prosecution of writers for political activities. In late June 1957, an urgent plea arrived concerning a very prominent case in Hungary, an appeal in which Agnes Ullmann played a part.

Hungarian Justice

Throughout the winter of 1956–57, Kádár filled his prisons with thousands who had participated in the revolution and were deemed a threat to his regime. By the end of January, 148 people had been tried in secret by summary courts, with 29 receiving the death penalty, 14 of whom were promptly executed.

In order to further discredit the revolution, it was decided to hold a public trial open to Western journalists so that the world could see that the rebels were, as the judge in charge put it, "intellectuals, students, and ne'er-do-wells" who had acted "merely out of a spirit of adventure or because of lack of information." Eleven freedom fighters facing various charges were tried together, including two organizers of the December 4 women's demonstration, writer Gyula Obersovszky and playwright József Gáli, medical intern Ilona Tóth, an Army lieutenant, and seven others.

The cases were loosely connected at best. Ilona Tóth, known as Ica to her friends, was a twenty-five-year-old intern in internal medicine at Sándor Péterfy Hospital when the revolution began. As a first-year medical student, she had taken a general chemistry course from Agnes Ullmann. Tóth shared her political concerns with Ullmann at the time, even daring to broach the then-taboo subject of the Rajk trial. Tóth was fully committed to the ideals of the revolution. She joined the Volunteer Rescue Service, which attended to the wounded fighters. After the second Soviet invasion, as the hospital filled, a nearby annex was used for the walking wounded. Tóth volunteered to be the interim leader of the annex, which became a hideout for insurgents whom Tóth helped to protect by listing them as patients.

After the fighting ended, Tóth wanted to preserve the achievements of the revolution, so she helped with the distribution of underground leaflets that were printed in a secret shop in the basement of the hospital. The operation was extremely vulnerable to informants, and the hospital was raided by the police in mid-November. Many involved were arrested, but not Tóth. She and her remaining compatriots were determined to ferret out any informants. Two days after the raid, they became convinced that a stockroom clerk had been a member of the AVH. Fearing that the clerk would inform on them, Tóth and two of her associates decided that the clerk had to be killed quickly.

That same day, Obersovszky and Gáli arrived at the hospital to start their underground newspaper, *Élünk,* in the basement. They met with Tóth in her office while the clerk was being questioned in another room, but did not learn his fate until later. Gáli was told at first that the AVH man had been given a sleeping pill and would later be let go.

Tóth was arrested on November 19 in a raid on the print shop. It was two weeks before the police learned, from Tóth herself, of her involvement in the killing of the clerk. In the meantime, Gáli and Obersovszky had continued printing and distributing their newspaper, and were arrested after the women's march. They would be charged with publishing an illegal journal.

The sight of the pale, thin, blond Tóth on the stand created a great stir. Observers and the press had a hard time believing that she was a murderer, as did Ullmann, and all suspected that she had been coerced into her detailed confession. But her two accomplices also admitted their involvement. Tóth tearfully asked for understanding and mercy. She explained that she was exhausted from her nonstop work, and completely disheartened from the betrayal of the revolution by the Soviets. She testified that she acted out of her conviction: "I thought I had to do everything for the revolution."

Obersovszky was defiant. He told the courtroom, "I want to be a free man, but I do not want mercy or a compromise. I did not fight against the system or the idea, but only against those who besmirched it and discredited it."

On April 8, the judge handed down the sentences: one year for Gáli, three years for Obersovszky, and death for Tóth and her two accomplices. All of the convicted appealed the sentences. Gáli's and

Obersovszky's lawyers wanted a complete acquittal, while the prosecutor sought a more severe sentence.

On June 20, the Council of the People's Court confirmed the death sentences for Tóth and her accomplices but overruled the sentences on the writers and instead imposed the death penalty on each.

The intellectual community, or at least that segment of it that was not yet in prison, was horrified by the extreme sentences imposed on Gáli and Obersovszky. By law, the punishments were to be carried out quickly. So their colleagues had to act quickly if the writers' lives were to be spared. It was hoped that an international appeal might put pressure on the government to reduce the sentences. Letters were drafted by former members of the now-defunct Revolutionary Intellectual Committee to be forwarded to prominent writers and artists around the world via their embassies in Budapest. Ullmann happened to live across the street from, and rode the same bus as, the French consul nearly every day, so she volunteered to deliver the letter. The consul immediately agreed to forward the appeal.

Camus responded immediately, as did Bertrand Russell, Louis Aragon, and Pablo Picasso. On June 25, the prosecutor announced that the sentences were suspended pending reexamination of their cases. On July 4, the court reduced Gáli's and Obersovszky's sentences to fifteen years and life imprisonment, respectively.

The Burden of Responsibility

The sparing of Gáli's and Obersovszky's lives demonstrated that appeals from the international community could affect events in Hungary, but it brought little solace to Camus. After all, the men still faced very long prison terms, and more than fifteen other Hungarian writers were also in Kádár's jails. Some of those awaiting trial were very prominent figures, including novelist Tibor Déry (who was arrested in April 1957), playwright Gyula Háy, and journalist Tibor Tardos. Another twenty or so writers, while still free, had pledged to remain silent as a protest against the regime.

But it was not so much pessimism over the plight of his professional brothers in Hungary that afflicted Camus as it was the heavy burden of his commitments. "Seldom in any country has a writer held so responsible a position as that now occupied in France by Albert Camus," the *New York Times* said. Those responsibilities weighed not only on Camus's conscience but also on his time, and pulled him away from his craft. Camus explained to one correspondent, "The balance between creation . . . and this grip of responsibilities is our only problem." That problem was aggravated by the incessant criticism and second-guessing that went on in the press.

Throughout the summer of 1957, Camus grew increasingly frustrated, and almost despondent, over his lack of progress in his writing. He'd had the plan for a new novel for several years. But he felt incapable of creative work, as if he were just "waiting for inspiration's wing to rustle by me and shake me up."

He retreated to the medieval town of Cordes-sur-Ciel for a vacation and diverted himself with reading the greats Nietzsche, Dostoyevsky, and Emerson. The latter's 1850 essay entitled "Goethe, or the Writer" reminded Camus of the lessons and challenges facing any writer aspiring to have an influence in his own troubled time:

> Goethe teaches courage, and the equivalence of all times; that the disadvantages of any epoch exist only to the fainthearted. Genius hovers with his sunshine and music close by the darkest and deafest eras. No mortgage, no attainder, will hold on men or hours. The world is young, the former great men call to us affectionately. We too must write Bibles, to unite again the heavenly and the earthly world. The secret of genius is to suffer no fiction for us; to realize all that we know; in the high refinement of modern life, in arts, in sciences, in books, in men, to exact good faith, reality, and a purpose; and first, last, midst, and without end, to honour every truth by use.

Despite a restful stay, Camus was still struggling. He shared his frustration with his former teacher Jean Grenier:

> *I am resigned to failing at everything after this summer . . . discouraged to the point that I don't even dare put myself in front of*

> *a blank sheet of paper. Wouldn't it be better for me to drop everything, to give up this sterile effort which has for years prevented me from being totally happy and free anywhere, which takes me away from others, rather guiltily, and in a large part away from myself? . . . These are the kind of thoughts I have been nurturing and that you will undoubtedly recognize.*
>
> *The fact remains that I'm in a strange mood . . . Even a letter seems difficult to write and please forgive my silence this summer.*

Despite his creative crisis, Camus did not forsake his commitment to Algeria. Having articulated his unconditional opposition to state-administrated killing, he felt compelled to intervene in death-penalty cases, but only if the accused had not participated in terrorist acts of the kind that would endanger Camus's family. In late September, he was contacted by a lawyer about a dozen men who had been condemned, most of whom had not actually killed anyone. After reviewing the cases, Camus agreed to intervene. Noting the youth and large families of some of the convicted, he appealed on their behalf to President René Coty to commute their sentences:

> *As a French-Algerian with my entire family in Algiers—conscious of all the dangers that terrorism courts for my family as well as for all the inhabitants of Algeria—the present drama afflicts me every day so strongly that, as a writer and a journalist, I have renounced all public acts, which, despite the best intentions in the world, on the contrary, risk to aggravate the situation. This reserve, authorizes me, perhaps, Mr. President, to ask you to use your right to pardon as many as possible of the condemned whose youth and numerous families deserve your pity. I am convinced, moreover, after long reflection, that your indulgence will, in the end, help preserve the future we all want for Algeria.*

Camus's hope to regain his creative powers and his concerns about restricting his involvement in Algerian matters to private channels would soon be dashed in one stroke—with a stunning message from Stockholm.

The Conscience of a Generation

The Swedish Academy declared that Camus was receiving the Nobel Prize in Literature for "his important literary production, which with clear-sighted earnestness illuminated the problems of the human conscience in our times." The secretary of the Academy noted "a genuine moral pathos" in Camus's writing "which impels him to attack boldly and in his own markedly personal way the great fundamental problems of life."

For Camus, the news of the Prize brought a mixture of pleasure and panic—"a strange feeling of overwhelming pressure and melancholy," he recorded in his notebook. He thought immediately of his mother, and of André Malraux.

He was the youngest writer to receive the honor in fifty years, since Rudyard Kipling. He had, in fact, been nominated several times previously. The successful nomination was submitted by Sylvère Monod, a distant cousin of Jacques's and a professor at the University of Caen.

In a press reception at Gallimard the next day, Camus explained, "I thought that the Nobel Prize should crown an already completed life's work or at least one more advanced than mine. I wish to say that if I had taken part in the voting, I should have chosen André Malraux, for whom I have much admiration and friendship, and who was one of the masters of my youth."

Malraux, almost fifty-six at the time, and one of the acknowledged luminaries of French literature, congratulated Camus, telling him, "Your reply honors both of us." Malraux never would receive the Prize.

Camus's humility and modesty spared him neither the sniping and spite from critics of both his political stands and his literature nor much second-guessing of the Academy. It was to be expected; Camus anticipated the attacks the moment he learned he had won. Perhaps the most outrageous comments came from Lucien Rebatet, a supporter of Nazi Fascism and writer for the collaborationist newspaper *Je Suis Partout* up until the time of the liberation. Rebatet had been sentenced to death in the purge trials after the war. Camus was one of three Resistance writers who, despite finding Rebatet's actions and views despicable, signed a petition for clemency in 1945 as a consequence of his already-resolute opposition to the death penalty. Rebatet was spared and freed in a general amnesty in 1950. Now, ignoring Camus's intervention, though it may well have saved his life, Rebatet claimed that Camus would have liked to have commanded the firing squad. The unrepentant and ungrateful Fascist also played literary critic: "This prize which falls most often to septuagenarians is not at all premature in this case, because since his allegorical *La Peste* [The Plague], Camus has been diagnosed with an arteriosclerosis of style."

But, of course, Camus's loyal friends were elated. Monod sent a note immediately upon hearing the news:

My dear Camus,

My emotion and my joy are profound. There were many times when I felt like thanking you for your friendship, for what you are, for what you managed to express with such purity and strength, and that I had likewise experienced. I wish that this dazzling honor would also appear to you, in some small part, as a token of friendship and of personal, intimate recognition. I would not dare coming to see you right now, but I embrace you fraternally.

Jacques Monod

Former *Combat* companions got together for a round of drinks with the new laureate. Camus told Jacqueline Bernard, Roger Grenier, and other former staff that he had prepared part of his speech for Stockholm: "Remember that shit thou art and into shit thou shalt return."

Camus was deeply touched and grateful for the support of those whom he considered to be his relatively few constant and sincere friends.

Public congratulations were printed in many newspapers. Two previous French laureates in literature, Roger Martin du Gard and François Mauriac, offered their thoughts. Martin Du Gard was generous and effusive. Mauriac, who had sometimes been critical of Camus, acknowledged, "This young man is one of the most listened-to masters of the young generation . . . In a way he is its conscience."

All of the private and public acclaim was not sufficient to quell

Camus's own doubts. Three days after the announcement, he wrote in his notebook:

> Frightened by what happens to me, what I have not asked for. And to make matters worse, attacks so low they pain my heart . . .
>
> No matter what, I must overcome this sort of fear, of incomprehensible panic where this unexpected news has thrown me.

DESPITE THE PARTIES and requests for interviews, and the need to prepare for the days of ceremonies in Stockholm in December, Camus tried to return to business as usual—responding to pleas for intervention and struggling to write anything.

On October 23, just one week after the Nobel, Camus marked the one-year anniversary of the start of the Hungarian Revolution by publishing an open letter entitled "The Blood of the Hungarians"—a moving extract from his March 15, 1957, speech at the Salle Wagram. A week later, he exercised some of the prestige endowed by the Nobel and joined forces with Mauriac to draw attention to the ongoing trials of four Hungarian writers, and even made a personal visit to the Hungarian Legation in Paris.

The Writer's Role and Responsibility

In Stockholm, Camus accepted his Prize from King Gustav VI as "a man almost young, rich only in his doubts and with his work still in progress." In his speech, he deflected the spotlight away from his writing and toward the plight of writers suffering elsewhere, and to the writer's role and responsibilities in the world. He asked rhetorically, "With what feelings could I accept this honor at a time when other writers in Europe, among them the very greatest, are condemned to silence, and even at a time when the country of my birth is going through unending misery?"

Camus admitted that in order to cope with his "too generous fortune" he had to rely on the thing that had supported him throughout his life: "the idea that I have had of my art and of the role of the writer."

Camus then asked to explain to his audience, "in a spirit of gratitude and friendship; as simply as I can, what this idea is."

The author of *The Rebel* and the journalist of the Resistance spoke of the writer's "two tasks that constitute the greatness of his craft: the service of truth and the service of liberty." The witness to the Spanish Civil War, World War II, the Cold War, and the Battle of Algiers said: "For more than twenty years of an insane history, hopelessly lost like all the men of my generation in the convulsions of time, I have been supported by one thing: by the hidden feeling that to write today was an honor because this activity was a commitment—and a commitment not only to write" but "to bear, together with all those who were living through the same history, the misery and the hope we shared."

For Camus, the writer's responsibility was to unite the greatest number of people, and not to compromise his art by serving those in power who make history. Rather, the writer was to be at the service of those who suffer under that power, and to make their suffering "resound by means of his art." Camus asserted, "The nobility of our craft will always be rooted in two commitments, difficult to maintain: the refusal to lie about what one knows and the resistance to oppression." He concluded by accepting his Prize "as an homage rendered to all those who, sharing in the same fight, have not received any privilege, but have on the contrary known misery and persecution."

Four days later, Camus traveled north to the University of Uppsala, where he addressed a rapt audience in the Aula—the grand, ornate auditorium in the main hall of the university. Camus spoke on the subject of "the artist and his time." His main theme concerned how, in contrast to previous times when the artist could remain aloof on the sidelines in history's amphitheater, artists and writers in particular were now *in* the amphitheater. Camus asserted that the era "of the revered master . . . of the armchair genius is over." The question facing all artists in such an age was how, "among the police forces of so many ideologies," to create work of significance.

After examining all of the forces that had threatened and undermined the freedom and respectability of art in recent decades—the submission to tyranny, comfort-seeking superficiality, an indifferent public—Camus suggested that the very convulsions of the times would lead to a rebirth of great art. His conclusion was in many ways an autobiographical statement:

> Let us rejoice . . . at having witnessed the death of a lying and comfort-loving Europe and at being faced with cruel truths. Let us rejoice . . . because a prolonged hoax has collapsed and we see clearly what threatens us. And let us rejoice as artists, torn from our sleep and our deafness, forced to keep our eyes on destitution, prisons, and bloodshed. If, faced with such a vision, we can preserve the memory of days and of faces, and if, conversely, faced with the world's beauty, we manage not to forget the humiliated, then Western art will gradually recover its strength and its sovereignty . . . Danger makes men classical and all greatness, after all, is rooted in risk.

CHAPTER 28

THE LOGIC OF LIFE

In formal logic, a contradiction is the signal of a defeat, but in the evolution of real knowledge it marks the first step in progress towards a victory.

—ALFRED NORTH WHITEHEAD (1861–1947),
Science and the Modern World

AS CAMUS WAS COPING WITH THE GLARE OF THE NOBEL SPOTLIGHT, Jacob and Monod were taking their first steps together that would lead them, too, to Stockholm.

The two men had seen each other nearly every day during the first few years that Jacob worked in Lwoff's group. At the end of 1954, Monod moved out of the attic to the ground floor of the building when he became head of the Department of Cellular Biochemistry, but that did not disrupt their connection. The communal lunches moved to more spacious quarters adjoining Monod's new laboratory. Although their respective research efforts were independent of one another—Jacob was immersed in the genetics of bacteriophage, Monod in the induction of enzymes in bacteria—the two stayed current with each other's work.

By the fall of 1957, Monod realized that if he was going to make further progress in understanding the genetic control of enzyme induction, he needed to apply the latest techniques of genetic analysis. Fortunately for him, the best expert he could have possibly recruited in the world was his former attic mate Jacob.

Over the previous three years, and working closely with Elie Wollman, Jacob had pioneered ways for mapping genes and mutations on bacterial and bacteriophage chromosomes. The advances were due to completely unforeseen properties of bacterial mating. The two men discovered a phenomenon they dubbed "erotic induction," in which the mating of a lysogenic "male" with a nonlysogenic "female" led to the induction of the prophage and the lysis of the nonlysogenic strain. Importantly, Jacob and Wollman determined that the mating in the opposite direction, of a non-lysogenic male with a lysogenic female, did *not* cause induction. The two different results caused them to realize that during erotic induction, the genes of the prophage were being transferred to the recipient, and then being expressed to produce an infectious virus. This notion led Jacob and Wollman to explore how genetic material was transferred from one bacterial cell to another.

It was Wollman who had the idea of breaking off matings to see when and how genes were transferred. They dubbed it the "coitus interruptus" experiment. Madame Jacob had found no use for the Waring blender that her husband had purchased for her in the States. In fact, she hated it and would not use it, so Jacob stored it in the laboratory in case it might be useful. It turned out to be incredibly handy—as a bacterial contraceptive. Upon whirling the bacteria in the blender at different times after they started mating, Jacob and Wollman were astonished to discover that different genes were transmitted at different, but reproducible, times after mating. One gene was detectable after five minutes, another after ten, another after eighteen, and so forth, up to one hundred minutes to transfer the entire chromosome. It was almost unbelievable that a bacterium that took only twenty minutes to divide would engage in mating for a period five times longer than its life cycle. Of course, the fact that sex could last that long was especially impressive, even to Frenchmen.

The discovery was a turning point in bacterial genetics, for it gave the entire field a way to map the relative location of genes as a function of the time when they were transferred during mating. Monod referred to the method as the "spaghetti" approach: the chromosome was taken up by a recipient cell like a string of spaghetti, and then the untransferred portion was broken off at specific points in time. Monod needed such expertise to map mutations in genes involved in enzyme induction. He and Jacob decided to join forces.

It was a pairing of complementary strengths in both scientific expertise and styles. Monod was more quantitative and biochemical, comfortable with enzymes and measurements. Jacob was at home in the more abstract realm of genetics, where the activities of genes were observed indirectly, such as the ability of a bacterium to grow on a plate containing certain nutrients. Monod was a superb and practiced logician who excelled at analyzing all possible interpretations consistent with a set of facts and distilling them into a working theory. Jacob was the more intuitive thinker. Both had demonstrated ingenuity at designing experiments that pushed the boundaries of what had been possible in their respective fields.

Their intense collaboration would, in just three exceptionally creative years, catapult them to the pinnacle of the new world of molecular biology. And during this momentous and most demanding stretch of his scientific career, Monod would undertake another daring project outside of the lab. In a clandestine operation reminiscent of his days in the Resistance, he would attempt to orchestrate Agnes Ullmann's and Tamás Erdös's escape from behind the Iron Curtain.

Black Boxes

In late 1957, there remained gaping holes in biologists' general understanding of how genes worked. It was known that genes were located within DNA, but one crucial mystery was how the information in genes was related to the production of proteins—the molecules, like Monod's enzymes, that did all of the actual work inside cells and bodies.

The first complete structure of a protein, the cow's version of the insulin hormone, was only just solved in 1955. That achievement and related efforts demonstrated that proteins were made of chains of twenty different amino acids, and suggested that each of the thousands of different proteins in a cell was comprised of a unique sequence of amino acids. But it was entirely unclear how any one protein, let alone the thousands in a cell, was manufactured, or what role genes and DNA played. Understanding the relationship between genes and proteins was crucial to understanding how genes determined the characteristics of all organisms—their life, their growth, and their physiology.

In mid-September, Jacob and other biologists broadly concerned with genes and proteins convened for a meeting of the British Society for Experimental Biology at University College, London. At the time, different scientists were working on different models—bacteria, viruses, rat liver, pigeon pancreas—and different proteins, and employing vastly different experimental approaches. As a result, there was a variety of ideas circulating both at the meeting and in the scientific literature about how proteins were manufactured, as well as a great deal of confusion about what had been learned that was particular to the protein, tissue, or species examined, and what might be generally true.

Then Francis Crick stepped up to the podium.

The theoretician had helped solve the structure of DNA using a lot of data that had been obtained by others. Now, without having conducted a single experiment himself, he tried to bring some clarity to the issues swirling around genes and the making of proteins.

In the printed version of his address, "On Protein Synthesis," Crick argued that "in biology proteins are uniquely important" and "they can do almost anything" because they are responsible for all of the chemical reactions in cells and bodies. He asserted that "the main function of the genetic material [DNA] is to control (not necessarily directly) the synthesis of proteins." He acknowledged, "There is a little direct evidence to support this, but to my mind the psychological drive behind this hypothesis is at the moment independent of such evidence. Once the central and unique role of proteins is admitted there seems little point in genes doing anything else."

Crick reviewed the various bits and pieces of what was known at the time, and managed to boil them down into two concrete principles, which he called the "Sequence Hypothesis" and the "Central Dogma." He admitted, "The direct evidence for both of them is negligible, but I have found them to be of great help in getting to grips with these very complex problems. I present them here in the hope that others can make similar use of them. Their speculative nature is emphasized by their names. It is an instructive exercise to attempt to build a useful theory without using them. One generally ends in the wilderness."

Crick explained that in its simplest form, the Sequence Hypothesis "assumes that the specificity of a piece of nucleic acid is expressed solely by the sequence of its bases, and that this sequence is a (simple) *code* [emphasis added] for the amino acid sequence of a particular protein." In other words, the linear strings of just four different bases in DNA was somehow a specific chemical blueprint for assembling the proper amino acids, in order, into the unique chains that made each protein.

Crick then stated his Central Dogma, which suggested that this "information" in DNA, once it has "passed into protein . . . *it cannot get out again* [emphasis in original]." Crick insisted that there was thus a directional flow of information from DNA to protein, but not from protein to DNA or other nucleic acids in the cell.

In both principles, Crick was asserting that there was a simple and general logic underlying the making of all proteins in all organisms. Exactly what that logic could be was entirely unknown:

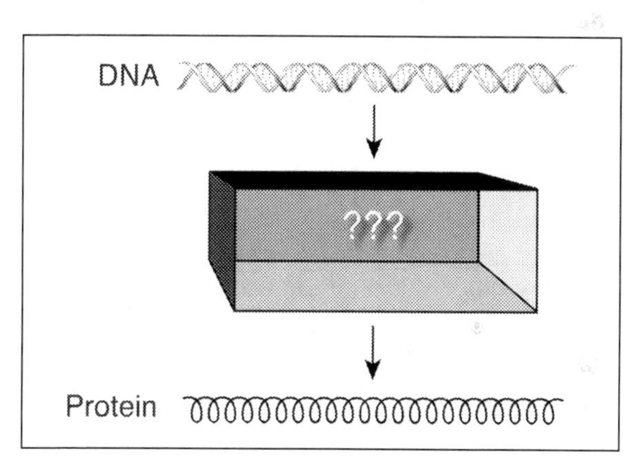

CRICK OFFERED SOME ideas about what might occupy that black box between DNA and protein. There was evidence from a variety of sources that ribonucleic acid (RNA) played a role in protein synthesis, and that protein synthesis took place in particles that would soon be named ribosomes. But the RNA known to be in these particles was uniform, which presented a paradox: In light of the great diversity of proteins made in the cell, how could such uniform particles manufacture different proteins? Crick suggested that there may also be another type of RNA, a "template RNA," in the particles that carried the information for proteins. He also imagined some kind of molecular "adaptors" that shuttled amino acids to the template so that they could be assembled into proteins according to the information in the template RNA.

Crick acknowledged that "it is remarkable that one can formulate so many principles such as the Sequence Hypothesis and the Central Dogma, which explain many striking facts and yet for which proof is completely lacking. This gap between theory and experiment is a great stimulus to the imagination."

Crick stimulated a number of imaginations that day. Jacob later recalled, "In comparison to the confusion, the mess that was the synthesis of proteins, Crick's speech was extraordinarily simplifying. He brought things into perspective. He showed where we needed to look for something and where we shouldn't be looking for anything. And he was very surprising."

For those who were not present to hear Crick's lecture, they would have to wait to learn about his bold ideas by word of mouth, or to read his article when it was published the following year.

AGNES ULLMANN'S DOCTORAL studies were on the very subject of protein synthesis. She had tried very hard to get to England in time to attend the conference at which Crick presented, but she missed the symposium. Despite all that had happened in the previous year, the Kádár regime was allowing some travel outside of Hungary. Ullmann's thesis adviser, F. Bruno Straub, had encouraged her to go to London for the meeting. She had applied for a passport, and was amazed to receive it while Tamás was still in prison. She had used her maiden name when she applied, and the authorities apparently did not make the connection to her husband. Ullmann thought, "My God, they are not very efficient."

The obstacle in getting to England was not the Hungarian government but the British immigration authorities. "The British were awful," Ullmann recalled. "They didn't give me the visa." Despite the well-known plight of Hungarian intellectuals, Ullmann was stonewalled: "They said that maybe they will give it in two months or three months and so on." By the time the meeting was starting, she did not have her visa.

Her boss, Straub, however, did get his visa and made it to the symposium. He presented Ullmann's work on the synthesis of amylase, an enzyme that breaks down starch and that is produced in the pancreas. Their approach was to develop a cell-free chemical extract of pigeon pancreas that was able to produce the enzyme. The idea was that in such an in vitro system, they could treat the extract in various ways to determine which components were necessary for making the enzyme. For example, when Ullmann treated the extract with ribonuclease, an enzyme that destroyed RNA, enzyme synthesis was abolished. This result was consistent with experiments done by many others who found a similar requirement for RNA in protein synthesis. The shortcomings of Ullmann's and others' experiments were pointed out by Crick in his lecture: RNA was found in many places inside cells; the RNA associated with different parts of the cell was produced and destroyed at different rates, and so there was probably more than one type of RNA. Crude experiments such as adding ribonuclease to a pancreatic extract could not resolve what role(s) RNA played in protein synthesis.

When Straub returned to Budapest and Ullmann asked him the news from the meeting, he reported: "There was nothing new." He did not mention Crick's lecture.

FOR JACOB, THE lecture prompted him to be on the alert for results that might bear on understanding what was inside Crick's black box. But as masterful as Crick's presentation was, he did not address all of the fundamental mysteries surrounding genes and the control of protein synthesis. Crick's primary interest, and that of a number of other molecular biologists, was the flow of genetic information from DNA to protein, and the nature of the genetic code. He made no mention of what had occupied Monod for years, which was the *regulation* of protein synthesis, the questions of how and why an enzyme was synthesized by bacteria in response to a nutrient—questions that now also engaged Jacob. Crick did not even cite a single piece of work from either of the Frenchmen.

This was not a slight on Crick's part. His focus was elsewhere. Crick was unquestionably the leader of the most elite circle within molecular biology. His lack of attention to the regulation of protein synthesis could be rationalized as him thinking "first things first." It was reasonable to believe that the fundamentals of DNA, RNA, and proteins needed to be understood before much more complex matters, such as why certain proteins were made at some times and not others, could be tackled successfully.

But science does not progress through central planning. At rare moments, an experiment may cast light where there was before only darkness. And in extremely rare instances, an experiment may fortuitously open the way to multiple discoveries. In one such experiment, Jacob and Monod would both break open the mystery of gene regulation and unexpectedly lead the way to discovering a key ingredient inside Crick's black box.

Geniuses in PaJaMas

The fundamental biological question posed by enzyme induction was: What was the "logic" that a simple, single-celled bacterium went through in order to make something (an enzyme) only when it could be utilized (when a certain sugar was present)? There were two approaches available to the puzzle: biochemistry and genetics. The thrust of the biochemical approach was to identify the molecules involved in the process, but this required one to sift through the thousands of different substances present in a cell and somehow fish out every component involved. This would be nearly impossible to do if one did not know how many and what kinds of molecules participate in a process.

By contrast, in the genetic approach, the experimenter may identify through mutations all genes that affect a process. The power of the genetic approach is that it makes no assumptions about the number or type of substances involved. Each mutation, in effect, breaks or alters a piece of the machinery. The experimenter may then deduce how each component works by observing how each gene mutation affects the overall process or the activity of known components. The logic of the machinery may then be understood without having isolated any component in a test tube.

By 1957, Monod and his team had identified mutations in two

different genes that affected the ability of *E. coli* to synthesize ß-galactosidase. One type of mutation, called *z*–, eliminated the ability of the bacterium to produce an active enzyme. The other type, called *i*–, changed enzyme synthesis from being inducible to being produced all the time—or *constitutively*—regardless of whether an inducer was present. Monod and Jacob needed to better understand these mutations if they were to figure out how enzyme induction worked.

It was a well-established test in genetics to determine whether a given mutation was recessive or dominant to a normal, unmutated version of a gene. Recessive meant that the trait was normal in the presence of one copy of the mutation, whereas dominant meant that the trait was altered by the presence of just one copy of the mutation. For example, albinism in humans and other mammals is recessive to other skin or coat colors. Plants and animals typically carry two copies of each gene and chromosome (except for the sex chromosomes in males). Monod and Jacob knew that they could better understand the effects of the *z*– and *i*– mutations if they tested them in the presence of one copy of normal, unmutated *z* and *i* genes (denoted *z*+ and *i*+). The hurdle for Jacob and Monod was that the *E. coli* bacterium naturally had only one copy of each gene in its single chromosome.

But Jacob and Wollman, in developing their techniques for mating bacteria and mapping genes, knew how to make cells that contained two copies of a portion of a bacterial chromosome. Jacob and Monod drew up a scheme to transfer normal *z*+ and *i*+ genes into recipient bacteria carrying mutant *z*– and *i*– genes, and to determine whether the ß-galactosidase enzyme was produced.

IN SEPTEMBER 1957, Arthur Pardee arrived from the University of California–Berkeley for a year's sabbatical with Monod at the Pasteur. Pardee's own laboratory had been working on the control of the synthesis of other enzymes. He had met Monod several years earlier, and subsequently heard enthusiastic reports about the atmosphere at the Pasteur from others who had visited.

Pardee took on the challenge of carrying out the mating experiments that Jacob and Monod had designed. The novel idea was to measure enzyme synthesis in mated cells that carried various combinations of normal and mutated genes. Monod had constructed the necessary bacterial strains before Pardee arrived, but had hesitated to do the experiments himself because of various technical difficulties that he had anticipated. Monod, Jacob, and Pardee did not know at first whether enough cells would mate, and remain mated over the course of the experiment, such that they would be able to detect enzyme synthesis—if it indeed occurred.

Pardee had to devise a series of tricks to make the experiments work. Some of his challenges were a matter of the differences in techniques and equipment between Berkeley and Paris. For example, Pardee wanted to maximize the amount of mating occurring in a mixture of the cells, but in order for the cells to be healthy over the course of the experiment, the cultures had to be well aerated. Usually, one would swirl the flasks vigorously on a rotating platform, but too much swirling disrupted mating. Pardee solved the aeration problem by putting only a small volume of liquid into a large flask, and swirling very gently. But when Jacob joined him in one experiment, he discovered that his French pipette was too short to reach the liquid in the flask. Pardee did not speak French well at all, but he did understand most of the profanity that Jacob uttered. They switched to using shorter flasks.

With important technical details solved, Pardee set about testing what happened with various combinations of mutations and genes. Before conducting any series of experiments, Monod liked to map out the possible results and how each could be interpreted. Monod's idea was that the *i*– mutation caused an inducer of the enzyme to be produced. If that was true, then he expected that transferring the *z*+ gene into a *z*– *i*– cell carrying both mutations would result in enzyme being synthesized, because the inducer present in the *i*– recipient cell would act on the transferred *z*+ gene.

Pardee tried the experiment for the first time on December 3, 1957. It worked beautifully. Monod and Pardee were delighted to see that enzyme was produced within minutes of the transfer of the gene.

As they pursued further experiments, however, they did not get the results Monod expected. When Pardee transferred genes in the other direction, moving the *i*– mutation into *i*+ *z*+ recipient cells that contained normal copies of both genes, Monod's model predicted that enzyme synthesis would also take place; the inducer elicited by the *i*– mutation would cause the recipient cells to produce enzyme constitutively. This did not happen: no enzyme was produced.

Moreover, as they repeated the original experiment of transferring the *i*+ and *z*+ genes into *i*– *z*– cells, they noticed that while enzyme synthesis always began promptly, it stopped within about two hours. Pardee and Monod were baffled.

There were two possibilities: either the experimental design was flawed, or their ideas and therefore their expectations were wrong. The experimental controls indicated that the recipient cells were perfectly capable of making the enzyme when a sugar inducer was added to the culture. This result showed that the transferred *z*+ gene was intact and capable of functioning normally.

Since the experiments were not deficient, then perhaps Monod's ideas were.

It would take a special visitor to help Monod see the error in his thinking. In the course of Pardee, Jacob, and Monod's experiments, Leó Szilárd visited the Pasteur. The Hungarian-born physicist had turned to biology after the war and had been very busy as usual. Prior to arriving in Paris, he had just visited Crick's group in Cambridge to discuss the genetic code.

Those at the Pasteur looked forward to Szilárd's visit and braced themselves for his tenacious but provocative questioning. His style was to pepper each scientist with questions and to write down the responses. Once finished with his interrogation, he would date the entry, hand the scientist his notebook, and say, "Sign there!" On future visits he would resume questioning by stating, "On such and such a date, you said this. Is that still true?"

Szilárd was given an office in Monod's laboratory for the duration of his stay, and he gave the customary seminar to the Pasteur group. Szilárd, too, had once worked on enzyme synthesis and stayed in touch with other scientists working in the field. In his seminar, Szilárd stressed an alternative logic controlling enzyme synthesis: instead of inducers working directly to activate enzyme synthesis, Szilárd argued that enzyme synthesis was normally inhibited or *repressed,* and that inducers acted by blocking that repression.

Monod did not like the idea at all. He found it "repulsive" at first. The discussions with Szilárd in Monod's office grew a bit tense at times. But as the experiments with Pardee unfolded, Monod realized that Szilárd was on the right track, and that he was not.

The unexpected results forced Monod, Jacob, and Pardee to reconsider. After Szilárd had left Paris, Monod painted a new picture in a letter to Mel Cohn. He noted the "stupefaction" that he, Pardee, and Jacob had at seeing enzyme synthesis halt at two hours. He also told Cohn how when they inserted the *i*– gene into an *i*+ recipient, "one gets no synthesis of the enzyme at any time!!" Monod then offered a crucial deduction: "It is the gene *i*+ that is active and the gene *i*– (constitutive) that is inactive."

From these observations, Monod stated there were two possible interpretations. He led with what he felt was the "superior" hypothesis: the *i* gene determines the synthesis, not of an inducer, but of a *repressor* that blocks enzyme synthesis. Induction would then be due to the inhibition of the repressor by the inducer. The new model could then explain why enzyme synthesis shut down two hours after inserting the *i*+ gene—because of the accumulation of the repressor over time.

Monod's original logic had been inverted: enzyme induction occurred by the inhibition of repression, not simple activation. Repression did make logical sense: enzymes were not produced when not needed; they were repressed until an inducer de-repressed them.

The flip in logic was a huge step. The repressor idea would be the cornerstone of a new logic of genetic programs. Pardee, Jacob, and Monod wrote a paper for the *Comptes Rendus* of the French Academy of Sciences. The experiments were named for the first two letters of each author's last name, Pa-Ja-Mo; then the preferred moniker became the more memorable PaJaMa. Breaking open the logic of gene regulation was just one of two major dividends of the PaJaMa experiment; the very rapid synthesis of the enzyme in recipient cells was also a clue to Crick's black box between DNA and protein. That dividend would take two more years to cash in.

A Matter of Human Dignity

Pardee and Szilárd were just two of many scientists who were drawn to Monod's clever band in Paris.

While Agnes Ullmann had missed out on rubbing shoulders with Crick, Jacob, and others in England, she still wanted to use her passport to make contact with Western scientists. She decided that she would try to visit a place where she believed that there was interesting work going on. One of the few Western scientists she knew of was Jacques Monod—from both his scientific publications and his article on Lysenko in *Combat* that Györgi Adám had shown her many years earlier. She was determined to find some way to get to Paris to meet Monod.

Getting from Budapest to Paris, let alone getting an appointment to see Monod, was complicated. She would need visas for each country through which she passed. Since she knew the French consul and scientific attaché, she had been able to get a transit visa to travel through France on her way to England. After that trip fell through, but with that paperwork in hand, she figured that she could visit France by taking a train through Austria and Switzerland. So she applied for and received Austrian and Swiss transit visas.

Then there was the matter of paying for the ticket. She could not afford it. A number of colleagues from the time of the revolution paid her way.

Tamás, now out of prison, was amazed that his wife even had a passport. Tamás predicted, "They won't let you leave the country. They will stop you at the border."

Ullmann went nevertheless. She was allowed to carry just five dollars, so she smuggled another twenty dollars in a tube of toothpaste. Just in case she did make it out of Hungary, Tamás had alerted a cousin of his, Eva Ekvall, who lived in Paris.

In early January 1958, Ullmann arrived at the Gare de l'Est railway station in Paris, and found Eva waiting for her with her name written on a piece of paper. Eva took Ullmann to the home she shared with her husband, an American military attaché, and gave her some more money. The couple was about to leave for the United States, so that very same day there was a big farewell party. Ullmann was very happy to enjoy the excellent buffet, but only for a while. When the Hungarian military attaché arrived at the reception, Eva told Ullmann to slip away because it would have been dangerous for her to meet him.

Ullmann had not tried to write to Monod beforehand, so she still had to find a way to get an appointment at the Pasteur. Once in Paris, she tracked down Tamás's uncle, who was also living in the capital. Fortunately for Ullmann, the uncle knew one of Monod's cousins who had been in the Resistance with the scientist. The cousin agreed to phone Monod and to set up an appointment.

The day Ullmann went to the Pasteur, she was very nervous. She waited in the office of Monod's secretary, Madeleine Brunerie. She then heard a man come down the corridor whistling, perhaps a trombone theme from a Brahms concerto. He stopped in, and to her surprise, introduced himself, "I am Jacques Monod." Ullmann introduced herself and explained that she had come to Paris hoping to work for perhaps six weeks in his lab.

Pardee was also waiting for Monod, to show him the latest results in the PaJaMa experiments. Monod was clearly anxious to talk to Pardee, so he suggested to Ullmann that she give a seminar the next day about her doctoral work in Budapest.

She came back the next day and talked to Monod's group about her work on amylase synthesis in the pigeon pancreas. After the seminar, Monod asked Ullmann, "What are you doing in Paris and how long will you stay?" Monod had clearly forgotten about their conversation the day before.

Ullmann, gathering all the courage she could muster, told Monod, "If you would allow me to work in your lab, I can arrange to stay six weeks. But if not, I will have to return to Budapest very soon."

Monod took Ullmann into his office and explained all of the work going on in the lab. He then asked her, "What would you like to do?"

Ullmann said she would like to work with François Gros on the effect of antibiotics on protein synthesis in bacteria. She added cautiously, "If Monsieur Gros accepts me."

Monod burst out laughing, "François? He is the nicest man on earth; he never says no." She started working with Gros the next morning.

ULLMANN WAS THRILLED to be working with Gros and to be part of Monod's group. As days went by, she began to divulge to Gros the troubles she had in Hungary—the crushing of the revolution, the crackdown on intellectuals, the arrests and trials of her colleagues, the secret police and Tamás's arrest. The situation was unbearable and hopeless. She confessed to François, "I want to leave Hungary for good." She added, "I want to come to France, and go on working forever in the lab."

Gros did not know quite what to say. He encouraged Ullmann to discuss her situation with Monod. Ullmann said that she would not dare do so. But Gros insisted, telling her that Monod "was the most understanding and nicest person on earth" and that if she was too scared to talk to Monod herself, he would do so on her behalf. Gros must have mentioned something to Monod, for the next day he invited Ullmann to dinner at his home.

Ullmann went to Monod's apartment on the avenue de la Bourdonnais. Monod's brother, Philo, joined Odette and Jacques for the dinner. During the '56 revolution, Philo had been working at the Ministry of Foreign Affairs and had helped Hungarian refugees who were stranded in Austria. The dinner conversation was dominated by talk of Hungary. Ullmann recounted all of her difficulties.

After dinner, Monod, Philo, and Ullmann retired to a small salon for a smoke. Ullmann knew from Gros and from others in the lab that Monod had been in the Resistance. Yet she was still stunned when Monod told her that he would do everything he could to organize her and Tamás's escape from Hungary. Monod needed to know what the consequences would be if she was caught. He asked, "What will happen to you if they pick you up?"

"Twenty years in prison," Ullmann replied.

"So, do you accept this risk?" Monod asked.

"Yes," Ullmann said.

Monod explained that it would take some effort to organize the plans. Ullmann asked her host, "Why would you help me?"

"It is a question of human dignity," Monod replied.

At the end of the evening, Ullmann thanked Odette for the dinner and left with Philo, who took her back to the lab at the Pasteur, where she still had some late-night work to do. As they were riding in his car, Philo turned to her and said, "My poor child, you have never lived in a democracy, you don't even know what it is."

After that momentous evening and hearing Ullmann's eyewitness account of events in Hungary, Monod wanted her to have access to works that were banned in Hungary, especially those that were damning of Communism. Monod gave her Camus's books; Trotsky's autobiography, *My Life*; and Arthur Koestler's *Darkness at Noon*. He also asked Gros to take Ullmann to see Camus's play *Caligula*.

Between working in the lab, and her crash "education" with all of those books to read, "I didn't sleep for six weeks," Ullmann recalled.

Agnes Ullmann in Paris, winter 1958, headed to a party outside the city, driver is unidentified. (Courtesy of Agnes Ullmann)

ULLMANN WORKED IN the lab right up until the last minute of her stay. A Hungarian-born colleague took her back to the Gare de l'Est. The prospect of returning to Budapest made her miserable. But she could not defect at that moment, for Tamás would surely be punished if she did.

When she arrived home, she told Tamás about the plan hatched in Paris. Tamás was cautious. He said, "We shall see."

CHAPTER 29

MAKING CONNECTIONS

Creating a new theory is not like destroying an old barn and erecting a skyscraper in its place. It is rather like climbing a mountain, gaining new and wider views, discovering unexpected connections between our starting point and its rich environment.

—Albert Einstein, *The Evolution of Physics*

The PaJaMa paper that proposed that enzyme synthesis was controlled by a repressor was just a short note in French—barely three pages long and with only one graph. A much longer paper with more meaty details, and in English, was still a year away. The note appeared at the end of May 1958.

And with its publication came the end of the French Fourth Republic.

Five prime ministers had attempted to deal with the Algerian crisis in the four years since the war erupted. Concerned over a perceived weakening resolve of the central government to maintain Algeria as part of France, a coalition of generals and the Army seized power in Algiers on May 13. Other members of the group plotted a coup d'état in which they would seize Paris, remove the French government, and demand the return of Charles de Gaulle as head of state.

Jacob was summoned to an emergency meeting of the Compagnons de la Libération. It was to be a gathering of the faithful to call for the return of the great leader of the Free French, the only person who, many believed, could save France again from the quagmire. Jacob joined about two hundred fellow Compagnons to listen to fiery speeches and emotional debates. Finally, a vote was taken to confirm the group's support for their general. De Gaulle received all but one vote cast. Jacob, however, did not get caught up in the fever: he abstained.

On May 29, de Gaulle agreed to serve.

—ʊʊ—

SUMMER ARRIVED, AND most of the Pasteur group dispersed for their ritual vacations. By late July, Monod was sailing in the Mediterranean, and Wollman was in the United States. Jacob, however, remained in Paris, overloaded with preparations for a series of talks to which he had committed—an upcoming genetics congress in Montreal, a symposium in Copenhagen, and a set of Harvey Lectures in New York City. The latter was a great honor, a mark of having become one of the leading figures in experimental biology through his work on bacterial viruses and genetics.

One Sunday afternoon, with the children away enjoying their vacation, Jacob was home with his wife, Lise, working on the lectures. While Lise played the piano, Jacob was pacing in his study, trying to formulate his talks on the "Genetic Control of Viral Functions." He was getting nowhere; this would be one of those rare days "with no taste for work." He gave up late in the afternoon and took Lise to the movies.

Jacob was no more interested in the film than he had been in his work. Slumped in his seat, he started to daydream. His mind, however, returned to the subject of the lectures that he had been trying to escape. He shut out the picture on the screen to let the insistent images take shape. Jacob then felt "invaded by a sudden excitement mingled with a vague pleasure . . . And suddenly a flash. The astonishment of the obvious. How could I not have thought of it sooner? Both experiments—that of conjugation done with Elie on the phage . . . and that done with Pardee and Monod on the lactose system, the PA JA MA—are the same! Same result. Same conclusion."

For years, no one had seen any meaningful connection between what had begun and lived at the opposite ends of Pasteur's attic. They were two separate research programs studying two separate phenomena, until that moment in the movie theater.

Provoked by the new idea of a repressor controlling enzyme synthesis, Jacob connected that idea to the control of the prophage—the phenomenon that first prompted Lwoff to admit him into the attic. Perhaps the virus's genes were controlled in the same fashion: "In both cases, a gene governs the formation . . . of a repressor blocking the expression of other genes and so preventing either the synthesis of the galactosidase or the multiplication of the virus. In both cases, one induces by inactivating the repressor, either by lactose or by ultraviolet rays. The very mechanism that must be the basis of the regulation."

Jacob then drew mental pictures of the repressors, one sitting on a bacterial chromosome, the other on a viral chromosome. He thought: "With the phage it is not simply two or three proteins whose synthesis is blocked, but at least fifty . . . Where can the repressor act to stop everything all at once? The only simple answer, the only one that does not involve a cascade of complicated hypotheses, is: on the DNA itself! In one way or another, the repressor must act on the DNA of the prophage to neutralize it, to prevent the activity of all its genes. And by way of symmetry, the repressor of the lactose system must act on the DNA containing its genes."

Jacob now had a new theory with which to work. He was exhilarated, as if he "had climbed a mountain, attained a summit from which [he] saw in the distance a vast panorama." His brainstorm made a cascade of predictions that needed to be tested, on colleagues as well as in the laboratory.

Lise saw him stir in his seat. "You've had enough?" she asked. "You want to leave?"

Out on the broad boulevard Montparnasse, Jacob told her, "I think I've just thought up something important."

But Monod was away and would not be back before Jacob had to leave for New York. He would have to wait more than a month to try his new ideas out on the one person he needed most to convince.

—∞—

JACOB ARRIVED BACK from New York in mid-September. Without pausing to sleep after the overnight flight, he had lunch with his family and then hurried to the Pasteur to see Monod. He'd had weeks to refine his ideas, and had grown more convinced that he was on the right path. But he had not yet tested the full story on anyone. He was very excited, but exhausted, punchy, and bleary-eyed from the long trip.

He started to tell Monod about all of the similarities between induced enzyme synthesis and induction of the prophage. Jacob did not get very far. Monod barely seemed to pay attention. He smiled faintly before breaking into booming laughter. He thought the idea that the two phenomena had a common explanation was silly, even "childish," and told Jacob that he could think of at least five arguments against it. Jacob decided not to press further, and to try again the next day. He went home to get some rest.

The following day, refreshed from a long sleep, Jacob launched again into his story. This time Monod was more patient and receptive. Jacob enumerated the parallels between enzyme induction and prophage induction. In both cases, a set of silent genes was triggered and became expressed. In both cases, the silencing was due to a single gene: in the lactose system it was the *i* gene; in the case of phage lambda, it was due to the "clear" gene that he had discovered when he isolated phage that could not form lysogenic bacteria. In both cases, genetic analysis had shown that the normal function of these genes was to produce a repressor that somehow blocked the expression of the lactose enzymes or the viral proteins. To Jacob, the analogy seemed so strong that a similar mechanism seemed inescapable.

If that were true, Jacob continued, then virus induction seemed to place some constraints on the possible ways that repression could work. Unlike the lactose system, which used just 2 or 3 different proteins, the phage made perhaps 50 to 100 different proteins to produce new copies of itself. Yet the synthesis of all these proteins was blocked by the phage repressor. To Jacob, it seemed very unlikely that the production of each protein would be blocked individually. Rather, it was more logical that repression would operate "as if it closed a single lock to bolt in a single action, the activity of the whole viral chromosome."

Monod continued to listen attentively. Jacob then suggested that the one place that a repressor could lock down a whole chromosome was by acting on the DNA. This idea had an additional implication that there was some order of genetic organization higher than the gene. Jacob called them "units of activity" (for the time being)—perhaps comprised of several genes that were expressed together. All of these ideas were driven by Jacob's insight into the logic of the phage. He asked Monod's opinion about whether he thought that the lactose system might have a similar logic in which the regulation of enzyme production was governed "like a switch, by an all-or-none, stop-or-go mechanism." The analogy had occurred to Jacob while he watched his son Pierre playing with an electric train.

Monod was not sold on the switch idea or most of Jacob's proposals. But he admitted, "Actually, there is no direct evidence either for or against the idea of repression at the level of DNA, and we should keep this possibility in mind." As the discussion went on, Monod grew more animated, getting up to draw on the chalkboard, and pacing back and forth. Monod considered the consequences of the models, and the experimental predictions that followed from them. Jacob sensed that he had hooked his partner.

The conversation lasted into the late afternoon. Jacob was drained. Monod broke out a bottle of scotch, and other lab members joined the lively discussion.

From that day forward, results were considered and experiments were planned in the new light of Jacob's proposal that there was symmetry between the two systems such that phenomena observed in the phage had a counterpart in the lactose system, and vice versa. Jacob and Monod talked nearly every day, often for hours, sketching out experiments and diagrams on a dark-green chalkboard. Jacob's lab was upstairs, still in the attic; Monod's was two floors below on the ground floor. Jacob spent so much time going up and down that he figured he should have an office in the stairwell. Conversations always wound up in Monod's office.

The game was to devise experiments that would rigorously test every facet of Jacob's proposal; it was a game at which Monod excelled. One day, Monod pointed out that if the switch acted as an acceptor site on the chromosome for the repressor, then that site should be subject to genetic mutation. Furthermore, the predicted behavior of mutations in the acceptor site would be that they would break the switch and no longer be subject to repression; they would cause continuous "constitutive" expression of the lactose enzymes. The isolation of such a mutation would be technically daunting but a critical test of the model.

In the middle of the discussion, Jacob realized that he and Elie Wollman had isolated the very equivalent mutation in phage lambda.

Normal phage could not infect a bacterium already containing a prophage; their genes were kept silent like those of the prophage. However, in the so-called virulent mutant that Jacob and Wollman had isolated years earlier, the phage grew on prophage-containing bacteria. This was exactly the property expected of phage in which the repressor could no longer bind to its acceptor site on the phage chromosome, because the site was altered by a mutation. Once he realized that he had held crucial evidence without recognizing it, Jacob thought, "How stupid I had been not to have thought of it before!"

But it was a happy embarrassment; the existence of a mutant that fit the prediction gave Monod and Jacob greater confidence in the model, as well as the impetus to identify its counterpart in the lactose system. Jacob jumped on the search for acceptor-site mutations that affected lactose enzyme synthesis.

Jacob likened his exchanges with Monod to an intellectual Ping-Pong match: serve, return, and volley, all at top speed, until a concrete experiment manifested itself. They divided up the experiments according to their interests and expertise. Then the experiments were performed, the results obtained, and the volleying started all over again. The repartee was more than scientific; there was a lot of good-natured ribbing served in both directions about literary or political preferences; each other's training as a zoologist or medical student; or the merits of the respective generals under whom they served, de Gaulle or de Lattre de Tassigny.

During the winter of 1958–59, there were many new experimental results, and all were remarkably consistent with the various proposals Jacob had made—about repressors blocking the expression of sets of genes, and of the parallels between the control of the lactose system and the control of phage lambda.

In March 1959, Monod and Jacob felt secure enough to call attention to the analogies. They completed the writing of the full-length version of the PaJaMa story—a much more extensive account in English of the short note published in French the year before. The authors were far more expansive about the roles of repressors than in the previous paper. In addition to accounting for their observations concerning the induction of galactosidase, they argued that the repressor model "may lead to a generalizable picture of the regulation of protein synthesis," and pointed to the "formal analogy" between the lactose system and the induction of bacteriophage lambda as being "so complete as to suggest that the basic mechanism might be essentially the same."

The technical prowess of the experiments, the explicit reasoning, the bold connections, and Monod's command of English combined to produce what readers recognized almost immediately as a tour-de-force.

The duo was just beginning to hit their stride.

Extracting Biochemists, Plan A: The Boat

In March 1959, Agnes Ullmann was able to return to the Pasteur for a visit of several months. Monod had arranged some lecture invitations that allowed her to leave Hungary. Tamás, however, stayed behind and remained under the watchful eye of the authorities, subject to arrest and imprisonment again at any time. With Ullmann in Paris for a length of time, there was now a window of opportunity during which, if Tamás could somehow be brought out of Hungary, Ullmann could defect and both would then be free. The great difficulty with that general scenario was that the borders were very heavily guarded; at the time, no one was sneaking out of Hungary.

Monod was committed to helping Ullmann and Erdös, he just did not have any specific ideas of how to go about it. But once Ullmann returned to Paris, an old acquaintance from Hungary came to visit who had some promising connections.

Andre Kövesi was a young journalist and writer who had written articles critical of the Hungarian system before the revolution. Kövesi left the country immediately after the first week of the revolution, when it was still possible to get out. It was his second escape from extreme danger. As a child during the Nazi occupation of Hungary, Kövesi and hundreds of other Jewish children were sheltered in a network of thirty-two homes established by Lutheran minister Gábor Sztehlo, and given papers identifying them as Christians. Despite the Nazis' searching, none of those homes were discovered. Kövesi, however, lost his parents in the Holocaust. After the war, Tamás, who was many years his senior, became a sort of second father to Kövesi.

Now living in Vienna, the tall, handsome, and very personable writer came to Paris to see a girlfriend. Ullmann invited him to the lab and introduced him to Monod. Kövesi was charmed by Monod, and when he learned that Monod wanted to help get his friends out of Hungary permanently, Kövesi offered to be the link between Monod and Ullmann in Paris and Tamás in Hungary. Kövesi was very familiar with the Austria-Hungary border and knew people who had business going in and out of Hungary.

There were several great challenges in getting Tamás out of Hungary. The first was to identify some mode for smuggling him—car, truck, boat, or train. The second was engaging whoever might be willing to carry out the operation. The third was coming up with enough money to pay the smugglers for the risks they would take. The fourth was to communicate with Tamás inside Hungary, to make him aware of any plan, and to notify him when he would need to be ready.

Within a few weeks after returning to Austria, Kövesi had sketched a plan. He knew some characters who were affiliated with a smuggling business in Linz, Austria. This group was willing to arrange to have Tamás smuggled out of Hungary on one of the ships that made regular trips up and down the Danube and traveled through Budapest on their way to or from Austria and Yugoslavia.

Kövesi explained to Monod in a letter from Vienna on April 18, 1959, in his broken English, that the operation would be expensive: "Last year prices are—as far as my informations go—about 3,000 US $ for one, though it is said that the Goverment is getting hard on the business . . . which may mean a rise in prices this year. Also there is no routine price for our way of making it."

Kövesi phoned Monod two days later at his Pasteur office to explain why the costs were necessary. Monod stirred quickly into action, soliciting Pasteur colleagues and contacts abroad for donations. Ullmann, too, wrote to some of Tamás's friends who had fled Hungary. Monod reported to Kövesi on April 21, also in English (as were all of their communications): "We have been busy here trying to arrange financial matters, but we have already arranged to have your account in Vienna credited of 3,000 dollars . . . We hope that more may be coming shortly . . . I[n] any case, you may count on a total of at least 5,000. Good luck to you." The sum was more than half a year's pay for a Pasteur scientist.

One week later, Monod reported that according to his "books," Kövesi would have $4,700 directly available to him, and another $2,500 held in reserve, in case it was needed. The donations came from the Pasteur itself, Tamás's cousin, and other concerned parties. Monod told Kövesi, "I have entire confidence in your ability to perform the operation if you find it feasible or to halt it if you eventually find it unadvisable."

With the money in hand, Kövesi then had to convince a ship's captain to consider a deal to smuggle Tamás aboard and out of Hungary. Kövesi learned that was a hard bargain. In mid-May, he approached a captain during a ship's brief stopover in Vienna. Kövesi described the encounter to Monod:

> *My interview with the captain was rather brief and actually not a very bright one.*
>
> *He was not quite happy about the business, and returned always to the point that he shant touch anything which is connected with politics or politicians. We talked about the possibilities all in conditionell [sic] form and parted with a formal agreement without obligations.*

The captain told Kövesi that "it is against the law to smuggle out somebody of the country, but if somebody climbed his ship and smuggled himself in, it is no affair of his—as far as the police is concerned."

What did concern the captain was his fee, which was $1,000 in advance, and another $4,000 when Tamás safely reached Austria. Kövesi explained that he also agreed to pay an "agent," an acquaintance who'd put Kövesi in contact with the captain. His fee was another $500 up front, and $500 more upon success. Kövesi pointed out to Monod, "If we have no luck, just to start with we may lose fifteen hundred bucks."

Upon making the agreement with what he called the "Linz Crowd," which included the agent and a few other participants, Kövesi was given an address in Esztergom, a small town on the Danube about twenty miles northwest of Budapest, where Tamás ("the cargo") would be picked up. Kövesi informed Monod of the text of a message he sent to Tamás via a courier, hidden inside a piece of Swiss chocolate. The message told Tamás the rendezvous spot, and how he would be given the exact departure date. The idea was for the Linz people to telegram Kövesi, who would then call Monod, who would tell Ullmann, and then Ullmann would phone Tamás and tell him that she was going to Marseille to lecture on a certain date—which was the signal for the departure date from Esztergom.

Just to be sure that Tamás got the message that the operation was on, the Linz people would also send him a telegram:

> TOM TO COME TO ESZTERGOM WITHIN 48 HOURS—PRIVATE VACATION IS PAID.

The operation was planned for between June 10 and June 15. Kövesi, Monod, and Ullmann were very hopeful.

The Danube, however, had other plans. The river flooded, and river traffic was interrupted. The operation would have to wait.

The Operator

Fortunately, work in the lab was making much swifter progress.

Throughout the winter and spring, Jacob had worked to develop the genetic strains that would allow him to search for the hypothetical "acceptor site" mutations in the lactose system. He had also been thinking a lot about the expected properties of the acceptor site. He came up with an analogy—of an airplane carrying bombs that was equipped with a transmitter and receiver. The airplane was the chromosome and the bombs were the genes for the enzymes. The transmitter was the repressor, and the receiver was its acceptor site on the chromosome. Jacob pictured the chromosome carrying the lactose gene as a plane flying around full of bombs, with one transmitter and one receiver. As long as a plane received the signal "Do not drop, do not drop," it didn't drop its bombs (the genes were repressed). But if the receiver (the acceptor site) was broken, it would not get the signal, and the plane would drop its bombs.

By summertime Jacob obtained mutant strains that were candidates for the acceptor site (or receiver) mutations. He and Monod set to work characterizing the mutants in detail to see if the mutations had the properties predicted by the model.

They didn't wait for the analysis before unveiling the next iteration of their rapidly developing model. In a short theoretical paper, Jacob and Monod predicted that mutations in the hypothetical acceptor would cause loss of repression—the plane with a broken receiver would drop its bombs. They proposed to distinguish three genetic elements involved in the regulation of protein synthesis: *structural* genes responsible for encoding the structure of a protein; *regulatory* genes that govern the expression of structural genes; and a new term for the receiver or acceptor site, the site of action of the repressor—the *operator.* They underscored how mutations in each of these elements would have different properties.

The brief paper was a double coup. Monod and Jacob were predicting the existence and properties of an entity that no one had ever conceived (but the evidence for which they would soon discover). And their terminology, introduced at the next reunion of the molecular biology elite in Copenhagen late that summer, was quickly adopted (and continues to be used to this day).

Jacob made it to one other notable meeting in the summer of 1959. On June 17, the anniversary of the founding of the Free French in 1940, he and a few hundred other members of the Compagnons de la Libération met for their annual reunion at the elegant Élysée Palace. They were welcomed by de Gaulle, who was now president of the Fifth Republic. After a short speech, "The General" worked the room, moving from one group of soldiers to another. Suddenly, he was in front of Jacob. They shook hands.

"Ah, Jacob. Pleased to see you again," the president said. "What are you doing these days?"

"Scientific research, sir," Jacob answered.

"Ah, very interesting. In what field?" de Gaulle asked. The president had recently implemented recommendations to support French research.

"Biology, sir," Jacob responded.

"Ah, very interesting. What kind?"

"Genetics, sir."

"Ah, very interesting. Where do you work?"

"At the Pasteur Institute, sir."

"Ah, very interesting. You have everything you need?"

Jacob hesitated, "No, sir."

"Au revoir!" The general moved on.

CHAPTER 30

THE POSSIBLE AND THE ACTUAL

Scientific reasoning is a kind of dialogue between the possible and the actual, between what might be and what in fact is the case.

—Peter B. Medawar, Nobel laureate (1960),
"Induction and Intuition in Scientific Thought"

After the flooding of the Danube earlier in the summer of 1959, the river dropped to very low levels due to a prolonged drought. The combination of extreme weather caused one of the longest interruptions in river traffic in many decades. As the weeks went by without any hope of action by the Linz mob, Kövesi became desperate to find a land route for getting Tamás Erdös out of Hungary and to Paris before Agnes Ullmann would have to return to Hungary.

Plan B: The Ford Fairlane

Kövesi got the idea to modify a car so that Tamás could be hidden securely in a secret compartment and driven across the border. To build a large enough compartment, it would have to be a large car, preferably a large American car. With Monod's approval, and using the money Monod had raised, Kövesi purchased a used Ford Fairlane for about $1,600. He spent another few hundred dollars having a hidden compartment built and installed.

But time ran out.

Ullmann had permission to stay only until the end of August 1959. If she stayed any longer without permission, she would put Tamás at risk of being put back in prison. She had to return to Budapest. Conversely, if Tamás came out while Ullmann was back in Hungary, she would be in similar danger. The operation was canceled.

Monod was committed to getting Erdös and Ullmann to freedom. He and Kövesi would now have to figure out a way to get both of them out together, and to communicate with the two of them in Hungary without alerting the suspicions of the authorities.

Before she left, Ullmann had an idea about how she could correspond with Monod about escape plans while avoiding trouble with the secret police. She had worked for years with the amylase enzyme, which breaks down starch. Ullmann knew that one easy way to detect starch was with a solution of iodine, which turned the starch a deep blue. She suggested to Monod that she could send the most sensitive messages by using a starch solution as an invisible ink, and Monod or his secretary Madeleine Brunerie could then reveal the message by simply putting iodine on the paper. The former resistant loved the idea, so Ullmann demonstrated how it worked.

Monod thought that they should also have a code in case any of the letters he sent were intercepted. He came up with code words that would make the letters appear to be innocent exchanges concerning their respective research. In Monod's "Code Agnes":

"EXTRACTION" = EVASION (ESCAPE)
"H_2O" = BATEAU (BOAT)
"ORGANIQUE" = AUTO (CAR)
"LA SALMONELLE" = AGI (SALMONELLA—AGNES ULLMANN)
"LE COLIBACILLE" = TOM (E. COLI—TAMÁS ERDÖS)

Extraction du Colibacille would refer to the plot for Tamás's escape, *extraction de la Salmonelle* to Ullmann's escape. Additional code words were related to police activity. Ullmann would indicate any harassment by referring to *enseignement* (teaching) or *leçons* (lessons), and to the police themselves as *élèves* (students).

On her way back to Budapest, Ullmann stopped in Vienna to see Kövesi. Both were very disappointed about the failure to get Tamás out during the preceding five months. The two of them sent a joint letter to Monod in Paris. Ullmann tried to be philosophical about the situation. She wrote (in French), "This is like a game of cards. There are days when nothing works and in those circumstances one should know to get up from the table before all is lost. Other times, everything can go well. We have failed for five months, now a better time should come."

She reported that she'd had many talks with Kövesi. The latter told her that there was still a possibility of escaping by boat. She made one request of Monod; "If it is possible to let us know the exact date before the critical hour, that would be best." She added, "I hope that I am not going to cause you too many problems or much bother. Do not worry, if things work that's good and if not, then we will try to find a 'modus vivendi' in the given conditions."

Kövesi scrawled a short note below Ullmann's, telling Monod, "If there is anything urgent I'll write or call you in your office . . . I know I tried my best and I don't feel too bad—just low. And I did not give up the hope."

"X"

Before Monod could tackle the new challenge of getting both Ullmann and Erdös out from behind the Iron Curtain, he and Jacob headed to Copenhagen for a small meeting called to address the mystery inside Crick's black box: the relationship between DNA and proteins. Most of the other leaders of molecular biology also went, including Jim Watson and Seymour Benzer, now both in the United States; Francis Crick from Cambridge; and even the Danish Nobel physicist Niels Bohr.

Jacob and Monod had been wrestling with that black box since the first success of the PaJaMa experiments and the formulation of their repressor theory. It was difficult for them to assert how the repression of enzyme synthesis worked when no one understood how protein synthesis worked. It was known that protein synthesis took place in ribosomes and that these structures were very stable, lasting several cell divisions. The prevailing view was that different ribosomes were each devoted to the production of different proteins.

But this idea did not make sense with respect to the very rapid synthesis of the galactosidase enzyme Jacob and Monod had observed upon entry of the gene into cells in the PaJaMa experiments. They doubted that such a structure could be assembled so quickly. Furthermore, they asked, if ribosomes were stable protein factories, how and why did repression start ninety minutes after mating? Did the repressor work on ribosomes? They doubted that, too. In his presentation to that select, immensely talented audience, Jacob shared the results of their recent experiments, and his and Monod's doubts about the current picture of protein synthesis. In order to reconcile their results on enzyme induction and repression, Jacob ventured a suggestion: There was some unstable intermediate that would carry the information between DNA and proteins. The repressor might work then by repressing the production of the intermediate. Jacob, who had a knack for naming things, called the intermediate "X."

There was no reaction from the audience, not a single question about his proposal. Indeed, Watson had spent so much time reading the newspaper during the presentations that when his turn came to speak, everyone in the audience pulled out newspapers and began to read them. The black box was not opened in Copenhagen.

Plan C: The Furniture Truck

When Monod returned to Paris, there was new correspondence from Ullmann and Kövesi waiting for him. Ullmann wrote to tell him that she was sending him a phonograph record via a Frenchwoman who had been in Budapest, and "at the same time I sent you a chromatogram . . . I am curious what is your opinion of the results." This was code for a message written in invisible ink that Monod was to decipher. She also mentioned that she had received a letter from Kövesi, and she expressed her concern that "he is very upset because of his work."

Kövesi had sent Monod a summary of all the money he had spent over the six months. Between advances to the Linz mob, outfitting the Ford Fairlane, and his expenses going to and from Paris, Kövesi had spent more than $8,000. Of all the money Monod had raised and made available to him, there was now little more than $300 remaining. And Ullmann and Erdös were still in Budapest.

Not only was the money nearly exhausted, but so were Kövesi's nerves. He had become entangled in a dispute with the Linz mob over getting the advance money back. His agent informed him that various members of the mob had personal problems that were preventing the return of the money: the captain's wife had committed suicide due to a quarrel over another woman, and another participant in the scheme had been detained by Hungarian Customs. The agent wanted out of the deal as well, but agreed to introduce Kövesi to a few more ship operators who might be able to help.

Kövesi wrote Monod: "I am not able to do any more waiting . . . I just cannot go on long with this stuff. More than seven months of tension and nothing happening—I must have a break, I really do. I am awfully sorry."

Monod sought to reassure Kövesi and Ullmann, and to solve the money problem.

He wrote an encouraging letter to Ullmann using their code:

> *Thank you so kindly for keeping me informed of your work or rather our work since it is understood that we are continuing our collaboration at a distance. I have not had much new to tell you regarding the best methods of extracting microbes until recently. But fortunately, my colleague and friend, professor Andrews [Andre Kövesi], who has done a great deal of work on these problems, as you know, came to see me recently and we spoke at length. Like us, he has found that the extraction of Colibacille [Erdös] and Salmonella [Ullmann] has posed particular problems and there have been many failures. In confronting our results, we arrived however at the conclusion that by perfecting the previous techniques that you know already since we have employed them together, it ought to be possible to be achieved. He is going to make several attempts very soon to modify the pH and the temperature of extraction [escape], but while keeping H_2O [boat] as the preferred solvent. He will send me a report of his results that I will communicate to you if they are really interesting . . .*
>
> *In any case, continue to keep me informed of your work and I will keep you informed of the work in the lab. You know well how much I wish that this collaboration continues and how much I am counting on you for that.*

Monod then tackled the money problem. He asked André Lwoff for a personal loan of $5,000. Lwoff promptly agreed to provide the money at no interest, payable in a year. Monod informed Kövesi by phone and followed up with a letter encouraging his beleaguered friend: "I am very confident, and if only the good [*sic*] damned Danube will consent to carry the right amount of water for a while, I feel sure that you will succeed and bring them out. I fully appreciate how miserable you must have felt during the last month, but your reward when you bring your friends out will be worth all these miseries. In the meantime, and on account of this enterprise, we have become real friends and this is already a little reward."

In his reply to Monod, Kövesi expressed his gratitude and the same sentiments:

> *Thank you for your good faith and for your friendship. I may fail in my doings but I won't disappoint you.*
>
> *Yours faithfully,*
>
> *Kövesi*
>
> *Love, Andre*

Monod received the Béla Bartók record album that Ullmann had sent weeks earlier. In order to secure the sensitive message to Monod, she had opened the album jacket and written a message in starch solution on the inside cover, then resealed the jacket with a bit of glue. Monod developed the message with iodine solution and deciphered the text. In it, Ullmann voiced her concern about the reliability of the Linz mob. She then told Monod:

> *If you believe that this trip cannot take place this year, then it should be perhaps better to think of other possibilities and Eva [Kövesi's girlfriend] might come to discuss the problems.*
>
> *The situation [police activities] has not changed and it is quite uncertain. It is impossible to tell what might happen until spring.*
>
> *However we are ready to do anything you will find proper.*

Bela Bartok record album sent by Agnes Ullmann to Jacques Monod in October 1959. In order to transmit sensitive information in preparation for her escape from Hungary, Ullmann wrote a message in code in an invisible starch solution on the inside pocket of the album. The message was revealed by exposing the writing to an iodine solution. (Archives of the Pasteur Institute)

Madeleine Brunerie sent a short note to Ullmann to let her know the record and message were received. After the typical greetings from the lab, she told Ullmann, "I adore Mozart. But it may amuse you perhaps to know that I am beginning to discover Béla Bartók with much interest."

Monod shared the decrypted message from Ullmann with Kövesi, as well as a new letter from Ullmann in which she stated that she had received a message from the "former co-workers of Professor [A]ndrews" (the Linz crowd). Ullmann wrote: "They are still working in the same field as before, but I am afraid they are crooks and their work is a danger for biochemistry."

By late October, the prospects were becoming dimmer for getting Ullmann and Erdös out safely by boat. Not only was the Linz mob proving unreliable, but cold weather was approaching. The Danube would be impassable in the winter. Kövesi explored a new scheme. He described the plan to Monod:

I've met somebody . . . who makes miscellaneous business with the countries behind the Iron Curtain. He is not a small-time smuggler, he exports-imports officially with the authorities behind the Curtain and unofficially with those in this side. He, so he says, never took anybody yet illegal through the borders but . . . he presently considers the idea.

His possibilities are excellent. He takes a big furniture-transport truck, loads it with what-ever-is-his-stuff under the controls [sic] *of a "friendly" custom-officer. The truck passes the Austrian border sealed, undisturbed for Prague or Budapest. There it has a free pass / our man does not even have a visa / and will* [be] *officially opened in a factory or depot. Then there, or at another place it will be loaded again, sealed by the communist custom-officers, and passes the border again without further disturbance or investigation.*

I beleive [sic], *I am going to let him in, if he is willing to try . . . I beleive* [sic] *I know your answer . . . I just need reassurance . . . for facing present disheartedness and future difficulties . . .*

Good night, Jacques. My love to you.

Andre

Monod approved:

This new plan sounds rather good, and I believe you should go ahead with it. I am relieved to know that this will not depend on the God damned Danube . . .

Concerning warning to our friends, I think it would be better if they did get warned that a new enterprise, completely independent of the Linz one (of which they are scared) was being organized . . . The best way would be for me to warn them using the "experimental" code . . . let me know in time when the business is going to be carried out.

Love and blessings,
Jacques MONOD

Kövesi was able to update Monod just a few days later. He wrote that he had given the import-export businessman a "perfectly harmless letter in German," addressed to Ullmann. But on the back side of the letter, there was a message written in starch that Ullmann would know to develop. It told her about the truck plan and the abandonment of the boat scheme:

> IT IS ABOUT A CAR CONTROLLED AND SEALED BY CUSTOMS IN BUDAPEST. IF THE BEARER TALKS OF ANYTHING ELSE REFUSE CLEVERLY. HE IS SUPPOSED TO SUGGEST A CONCRET [*sic*] PLAN. YOU HAVE THE RIGHT TO DECIDE BUT PLEASE DO NOT HESITATE LONG. IF YOU DON'T WANT TO MAKE IT NOW OR THAT WAY: IT DOES NOT MATTER. AS ACKNOWLEDGMENT GIVE HIM A VISIT CARD: "SORRY WE ARE BUSY." YOUR MOTIVES YOU'LL EXPLAIN LATER. STOP. THE WET EXPERIMENT DOES NOT WORK: HERE WE HAD TO THROW THE WHOLE MATERIAL AWAY, YOU SHOULD NOT TOUCH IT EITHER.

Kövesi explained to Monod that the driver had no idea that he carried the secret message. He would be carrying a Parker pencil that Ullmann had left at Kövesi's home in September, which would identify him as having been sent by Kövesi. If the driver brought Ullmann and Erdös out, he would be paid $5,000. If they refused to come out, he would be paid $750 if he produced a visit card with the "Sorry we are busy" message on it. The driver wanted to do some reconnaissance, so no specific date was set for the operation.

While Monod waited, hoping to hear good news, he had to deal with some new problems. One of the financial supporters of the Erdös smuggling plan was very concerned about the loss of money and the failure to get Erdös out of Hungary. Arpad Csapo was a Hungarian-born scientist, a former colleague, and friend of Tamás's who had emigrated to the United States. In 1959, he was working at the Rockefeller Institute in New York, where Ullmann had contacted him regarding Erdös's plight. Csapo secured $2,500 from the Hebrew Immigrant Aid Society (HIAS) that wound up in Kövesi's account.

In late October, Monod wrote Csapo with a detailed account of the history and status of the operation—the bad luck with navigation on the Danube, Ullmann's having to return to Budapest, the creation of the communication codes, the new plan that did not depend on the river, and the money spent. Monod told Csapo that "the main difficulty is or rather was that almost all the investment which had been put into the enterprise is virtually lost." But since money was required, Monod informed Csapo that he had personally borrowed $5,000 and put it at Kövesi's disposal. Monod told Csapo that he was certain that he would recoup the lost money from institutions once Erdös and Ullmann were free. Monod expressed his determination to succeed regardless of the financial or logistical difficulties: "I refuse to believe that all of this effort might remain fruitless and that our friends would have to continue living miserably, in danger and in anxiety for nobody knows how many years."

Csapo, however, was not reassured. He wrote Monod to explain that the loss of the money had put him in a "most difficult and embarrassing situation," as the money was borrowed from the HIAS, and it had to be repaid. Csapo explained that he believed that the HIAS funds would only be used once Erdös was out of Hungary. He told Monod that, in his view, Monod was responsible for the money and the design of the arrangements in how it was used: "I must therefore ask you for reimbursement, since I'm already greatly embarrassed about the delay in returning borrowed money."

Monod wrote back immediately to Csapo:

Dear Dr. Csapo,

Thank you for your letter of October 28. I am afraid there is a fundamental misunderstanding between us which I would like to clear up.

You generously offered to A. to help toward T.'s escape. I have no part in the asking or in pursuing you to make this offer. As I understand it, in making this offer, you were moved both by your personal friendship with T. and also by the fact that he helped you efficiently under very similar circumstances. However, you felt that you needed to be certain that this contribution would be used towards this and entirely and exclusively. This is the only guarantee that I could and did give you. In doing so, I took towards you the same responsibility that I have taken towards several other people who also made contributions. All these people were aware of and accepted the risk inevitable in such an attempt . . .

If it was at all possible at the present time to reimburse you, it would be done. But this possibility does not exist at present.

The logician then spelled out the remaining possibilities to Csapo in the same way that he and Jacob spelled out potential results from experiments, and weighed what each would mean for their models:

There are therefore only two alternatives to be considered:

a)—T. and A.'s escape does succeed. The first thing they will discover when they arrive in the free world is that their escape has been financed by a number of people, including yourself. Life would be miserable for them if they had to remain in debt towards several people for many years. I therefore propose that when they come out, all their friends . . . would try to collect a sum of money large enough for them to pay their debts and to start feeling really free in the free world . . .

b)—The escape fails once more and any renewed attempt has to be abandoned. In this case, any fund left over should be reimbursed to contributors . . .

In the meantime a number of people, and in particular yourself and myself, will have to remain in debt. I fully realize how unpleasant your situation is in carrying such a burden alone . . .

Let me again state my reason faith that our friends will be free and that our efforts will not have been in vain.

After a week, there had been no news about the furniture-truck operation. Kövesi wrote to Monod to let him know that he was coming to France and would be staying north of Paris in Épinay-sur-Seine. He was anxious to get together.

Monod told Kövesi, "I am fantastically busy these days, but I certainly would like to talk to you."

The Switch and the Operon

Monod was indeed extremely busy. There had been some outstanding developments in the lab. The possible acceptor-site mutations Jacob had isolated in the summer, now called "operator" mutations, were proving to have the characteristics that they had predicted.

One prediction was that operator mutations would cause constitutive expression of galactosidase regardless of whether an inducer was present. This was found to be the case.

A second prediction was that if the operator was the site for interaction with the repressor, and the operator was necessary for repression of adjacent structural genes, then operator mutations would affect the expression of multiple, adjacent genes on the same chromosome. In the PaJaMa paper published a few months earlier, Jacob and Monod had reported that the structural genes for galactosidase (*z*) and a second gene, called permease (*y*), that encoded a protein that allowed lactose into cells, mapped extremely close together on the chromosome, with no known genes in between them. To their delight, Jacob and Monod then discovered that the same operator mutation caused constitutive expression of both galactosidase and permease—both bombs dropped together.

Changing analogies, the effect of the operator mutations on both galactosidase and permease expression was exactly what was expected if the operator acted like a "switch" through which the genes involved in lactose metabolism were turned on and off. The repressor then acted like a "hand" on the switch, keeping the genes off when no inducer was present, but then releasing the genes to be expressed when an inducer was present.

The very close proximity on the chromosome of the galactosidase and permease genes, and of the *i* gene that controlled the synthesis of the repressor, was another striking observation. Three genes all dealing with lactose metabolism were very nearby, apparently right next to one another. This discovery was the first confirmation of Jacob's notion that there was some level of genetic organization higher than the individual gene (his "units of activity"). The coordinated expression of lactose-metabolizing structural genes, controlled by a common operator and repressor, was further evidence of a higher level of organization. Monod's and Jacob's separate studies of another set of enzymes involved in the synthesis of the amino acid tryptophan had also revealed closely linked structural genes under the control of a repressor. All of this evidence suggested that these higher genetic units, groups of genes, were physically organized together to control specific processes.

A new term entered Jacob's and Monod's lexicon. Jacob christened these units, comprised of structural genes controlled by a common operator and repressor, *operons.*

There was still more work to do. More results to check and recheck, and the writing of the paper that would introduce the new concept. But in just a little over a year since he first presented his ideas to Monod, the by-product of a daydream in a movie theater, Jacob had seen how the possible became the actual. The entity of the operon was the reward of very clever experimentation, but more important it was a triumph of the imagination. Jacob realized that, contrary to what he had first believed, "the process of science does not consist in explaining the unknown by the known . . . it aims on the contrary to give an account of what is observed by the properties of what is imagined. To explain the visible by the invisible. And it is . . . through an appeal to new hidden structures, with hypothetical properties, that science proceeds."

No one had seen or purified any of these new entities—the repressor, the operator, or the operon. Those who first imagine such hidden structures infer their properties, then convincingly demonstrate their reality, are very rare in science. Such originality, creativity, and insight may be called genius.

CHAPTER 31

UNFINISHED

Two things rob people of their peace of mind: work unfinished and work not yet begun.

—SOURCE UNKNOWN

EVEN WITH THEIR CONQUEST OF THE OPERON, MONOD AND JACOB had important business that remained. For Jacob, the structures hidden inside Crick's black box, his mysterious "X," still eluded the brightest minds in biology. And for Monod, there was the as-yet-unfulfilled promise of getting Ullmann and Erdös out of Hungary.

Monod's problems were more than logistical. Arpad Csapo was still perturbed by the turn of events, and continued to question Monod. He wrote Monod a long letter in which he detailed their "fundamental misunderstanding." Csapo complained again that he had only been asked to provide a deposit for Ullmann and Erdös's use once the two were free in the West; the money was not intended as operating funds for the escape plan. Csapo claimed that he did not authorize Monod to pay the money before Erdös and Ullmann escaped, and for Monod to do so was against Csapo's judgment. Csapo put two options on the table: either ask the HIAS for an extension of their loan, while informing them in detail of where their money had gone, or repay the $2,500 immediately.

Monod tried to calm Csapo. He first updated him on where the operation stood. Monod explained again that he had secured $5,000 for fresh attempts to aid Ullmann's and Erdös's escape and that there were three potential scenarios for doing so: the first was the current furniture-truck scheme; the second would be the use of the modified Ford Fairlane; and the third was a new possibility. Monod had been invited to lecture at the Hungarian Academy of Sciences in early 1960.

Monod suggested to Csapo, "If the previous attempt has failed by that time, my trip may help to organize a new one."

Monod suggested that Csapo request an extension of the loan and permitted him to use Monod's name to do so. But Csapo asked for further financial assurances that Monod could not provide.

Monod was fed up with Csapo's complaints and demands, and he strained to reply with courtesy:

> *Must I point out that T. is your friend, not mine, I have never even seen him. He is your compatriot, not mine. As far as I am concerned, trying to help him for me is a matter of principle, of human interest, and of simple decency.*
>
> *I do not regret what I have done, and as you know, I am trying to go on with the job as best as I can—that is to say to the extent of helping materially and morally A.K. who, living in Vienna and being a devoted friend of T., is the only person I know who can do something towards organizing their escape. A.K. is a Hungarian refugee like yourself. He has not acquired, but has applied for Austrian citizenship. Your "inquiries" therefore might hirt [sic] him very much, and force him to abandon this enterprise.*
>
> *If you are prepared to take over this job from A.K. and myself, we would both be greatly relieved, and I would place at your disposal the little capital which I have reserved for this purpose. This would mean for you to take a three or six month leave. I believe it might not be impossible.*
>
> *However, if you are not prepared to do this, then for Heaven's sake, let us continue with this work until it is done or given up. Then we shall see where we stand?*

After firing off his letter to Csapo, Monod took a break from both the Hungarian operation and the lab in order to enjoy the New Year's holiday.

Searching for the First Man

Camus was also feeling the struggle between his commitments and his creative work. As he had feared, the Nobel Prize had brought additional and unwelcome expectations to both facets of his life. The laureate was even more in demand on political matters, especially concerning Algeria. He had continued to provide his private support to the accused and condemned, and he intervened in scores of cases. He wrote appeals to President René Coty, and when it appeared that de Gaulle might be brought back to power, he paid a private visit to the general. Camus talked with him about the risks if Algeria was lost and of the resulting fury of French Algerians. The general did not reveal his intentions regarding the conflict.

Camus had also taken up a new campaign: the protection of French conscientious objectors who faced long prison terms for refusing to serve in Algeria. He helped draft legislation to establish special status for objectors. After de Gaulle became president, Camus wrote to ask for his support for the measure, and de Gaulle pledged to look into the matter.

Camus continued to pay attention to Hungary. He made a public plea to spare the lives of Sándor Rácz and Sándor Bali, two leaders of the Greater Budapest Workers' Council who were among the first arrested in Kádár's roundups in December 1956. Camus stated: "To keep quiet would be to give a free hand to . . . Kádár." He also wrote the preface to *The Truth About the Nagy Affair*, a book that dissected the lies propagated by the Kádár regime.

All of these political activities did little for Camus's frustration with his writing. His doubts about his creative powers, which stalked him before the Nobel, had only grown as months and then more than a year passed without progress on his next book.

While struggling with his original work, Camus managed to pen an adaptation of Dostoyevsky's novel *The Possessed* (*Les Possédés* in French; also known as *The Devils* or *Demons* in English) for the stage. Camus considered the novel prophetic in its criticism of the nihilism of left-wing revolutionaries (the devils) willing to commit political murder. The play, also directed by Camus, was a three-and-a-half-hour spectacle that debuted in late January 1959. On Monod's copy of the published play, Camus offered a little humorous twist: he wrote over the title *Les Possédés* (the devils) to make it *Ces Possédés* (*these* devils) and added:

> *qui nous persécutent encore*
> *avec la fidèle amitié*
> *Albert Camus*
> [who still persecute us / with faithful friendship, / Albert Camus]

THE PACE OF life in Paris and the constant demands on Camus's time drained his energy and aggravated his frail health. Camus recognized that he needed to restore both in order to work. He wrote himself a long series of prescriptions in his journal:

> Enough of "you must."
> Completely depoliticize the mind in order to humanize it . . .
> Remain close to the reality of beings and things. Return as often as possible to personal happiness . . .
> Recover energy—as the central force.

He reminded himself, "My job is to make my books and to fight when the freedom of my own and my people is threatened. That's all."

In order to do that job, he decided that he must find a refuge away from Paris, a retreat where he could create in solitude. He found it in the quaint village of Lourmarin in Provence, far from Paris in southeastern France. He and Francine bought a house in the middle of the village with the prize money from the Nobel. It was purchased, coincidentally, from Dr. Olivier Monod, a distant cousin of Jacques's.

In the spring of 1959, after the opening of *The Possessed,* Camus escaped to Lourmarin to once again attempt a fresh start at writing a new book. The only way to conquer his doubts, as well as those of his critics, was to create something worthy of his prominence. Camus had been outlining and making notes for a novel for more years than he wished to count. He had conceived of the title, *The First Man,* and sketched a general plan for the book as early as 1954. It was to be an autobiographical story of his journey from youth to adulthood, and would feature all of the people who had shaped his life—family, friends, teachers, and lovers. He described the book to Jean Grenier as "a 'direct' novel . . . something that is not, like the previous ones, a kind of organized myth. It will be an 'education,' or the equivalent." Camus had in mind a coming-of-age story.

Over the years, Camus's ambition had grown in scope to make the novel more than just his own history, thinly veiled by fictitious character names. Inspired by Tolstoy, one of his major literary heroes, Camus envisioned *The First Man* as an epic novel of Algeria. Just as *War and Peace* told the story of the lives of its characters through the period of the Napoleonic invasion of Russia and the decline of the tsars, Camus intended to set his story against the backdrop of the saga of his people, the Pieds-Noirs who settled in Algeria and made it their homeland.

The comparison with Tolstoy's masterpiece was conscious and explicit. He had told a close Algerian friend that he wanted to write a "fresco of the contemporary world," like *War and Peace*. Camus was perpetually mindful of the productivity and careers of the great writers he admired—Dostoyevsky, Nietzsche, and Tolstoy—and often noted what they had accomplished when at a similar age as he was. When he was thirty-six, Camus jotted Tolstoy's particulars in his notebook: "He wrote *War and Peace* between 1863 and 1869. Between the ages of 35 and 41."

Despite his frustrations and uncertainty about the still-unwritten book, Camus had been unusually open with the press about the project. As early as February 1957, he surprised an interviewer for the *New York Times* by divulging its title. Later that year, on the day before the Nobel ceremony in Stockholm, he was also asked about what he was writing. He replied that it was a traditional novel—the first time he applied that term to any of his books. He added, "It's also the novel of my maturity, if you like. In consequence, I attach more sentimental value to it than to other books." Talking about the book, of course, only raised expectations.

Camus's plan was for *The First Man* to be part of his third cycle—

on love—after the previous cycles on the absurd (*The Stranger, The Myth of Sisyphus, Caligula*) and rebellion (*The Plague, The Rebel, The Just*). This third cycle was also to include a play and an essay.

Camus both relished and agonized over the solitude of Lourmarin. He knew that isolation was the only way he could create. He wrote Jean Grenier: "I am finding a little peace and inner silence again after a season full of work and problems. I needed solitude like bread. Or rather solitude was my last chance to find the way back to my personal work, all the others having failed." He likened Lourmarin to a monastery, even signing letters to friends "Frère Albert, O.D. [Order of the Dominicans]."

He desperately missed the company of family and friends. Francine was teaching in Paris, and the twins were enrolled in school. They could be together in Lourmarin only during holidays. Camus combated his loneliness both by writing to and about his closest friends. Since the book was all about them, they were never out of his mind, nor were the ghosts of his literary heroes. He told Jean Grenier, "Nietzsche is here . . . in this book, and you too." Lourmarin held special significance for both the master and his pupil. Grenier had also been inspired by the place and published the essay "The Wisdom of Lourmarin" in 1936. He was even married in Lourmarin City Hall. Camus told Grenier, "I put my footsteps in yours."

Camus read for inspiration. From newly crowned Nobel laureate Boris Pasternak, whom he admired for his creativity in the oppressive climate of the USSR, he uncovered a pearl on autobiography: "The greatest works in the entire world, while speaking of the most diverse things, in fact tell us of their own birth."

He was refreshed by walks among the "marvelous roses," blooming rosemary, and violet irises of his garden, and by jaunts into the village, where he enjoyed mingling with the townsfolk. He made some progress before returning to Paris at the end of May, but he had not conquered his doubts. He told a friend, "I think it's all over. It's not coming anymore."

—~—

Nevertheless, on and off during the summer and fall, in between engagements in Paris, Camus returned to Lourmarin to attempt to write. By November, sequestered in his "monastery," he was finally gaining momentum. He had developed the main character, Jacques Cormery, who, at age twenty-nine (the same age at which Camus published *The Stranger*), Camus described as "ailing, tense, stubborn, sensual, dreamy, cynical, and brave." Cormery also just so happened to have "four women at the same time and thus is leading an *empty* life."

Yet Camus's life was not so empty and was far from monastic. He took turns writing to each of his own four women: a new lover, Mi, a young painter; the actresses Maria Casarès and Catherine Sellers, who had recently starred in *The Possessed*; and his wife, Francine. He updated his progress, shared his frustrations, and occasionally reported small triumphs.

He wrote to Mi: "I have never worked with such dense material, and this afternoon I had the fleeting impression that my characters had taken on that density and for the first time in the twenty years that I have been searching and working, I've finally arrived at the truth of art. It was a delicious lightning bolt to the heart, but a fleeting one, followed by blind work and constant doubt."

The book's title carried a double meaning. The First Man was Camus/Cormery, and as Camus told a journalist that summer, every man is the first man in his own story. But the First Man was also any man who left France to colonize Algeria and thus started a new life from scratch.

One of the crucial incidents that inspired the novel was Camus's visit to his father's grave in Saint-Brieuc in 1947. His father had gone off to fight for France without ever having seen the country, and died from injuries at the Battle of the Marne. Camus's mother had begged her son to visit the grave. In chapter 2 of his nascent novel, Camus had Jacques Cormery recount the shattering experience.

While the search for his father would be a major element of *The First Man,* Camus's love for his mother, whose heart was "the best in the world" was always brimming. On the first page of the manuscript, he scribbled his dedication to his illiterate mother: "To you who will never be able to read this book."

As THE NEW Year approached, Camus paused to enjoy the holiday with his family and close friends. The Gallimards, among others, were due to come to Lourmarin. He had written 144 pages. He still had a long way to go. He told a friend from the theater that it would be a while before he could return to the stage: "I must finish the first draft of my enormous story, and I am far from completing it . . . I have eight months to finish this before getting back to the theater, only eight months . . . I will hang on as long as possible."

But the progress had lifted his mood. That and the prospect of seeing his lovers had him looking forward to his next visit to Paris just after New Year's.

On Tuesday, December 29, he wrote to Mi to tell her that he would see her in Paris shortly: "By the time you read this letter, we will only be separated by two or three days."

The next day he wrote to Maria Casarès to tell her the same: "Alright, this is a last letter just to say that I'm arriving Tuesday by car, leaving with the Gallimards on Monday. They are passing this way on Friday, and I'll phone you when I arrive, but maybe we can already set a dinner date for Tuesday. Let's say Tuesday in principle, taking into account surprises on the road, and I'll confirm dinner by phone . . . I kiss you and hug you tightly until Tuesday, when I can start all over again."

On December 31, he wrote to Catherine Sellers: "This is my last letter, my tender one, to wish you a heart-fulfilling year . . . see you Tuesday, my dear." He added, "As long as this monstrous book is not done, I'll have no peace."

The Gallimards—Michel, Janine, and their daughter Anne—arrived on New Year's Day. The two families celebrated Anne's upcoming eighteenth birthday. Camus gave Anne, whom he fondly called "Anushka," a book on contemporary theater.

The next day, they all lunched together in the village before dropping Francine and the twins at the station in Avignon so they could catch the train to Paris. Camus, too, had a ticket, but the Gallimards had convinced him to make the trip back with them by car. They wanted to wrap up their holiday with a more leisurely two-day journey, and to stop to enjoy several restaurants en route.

Before departing Lourmarin on January 3, the Gallimards and Camus gassed up Michel's sporty Facel Vega HK500 at the local garage. The owner had been holding on to a copy of *The Stranger* for Camus to sign the next time he stopped in. Camus obliged, telling him, "You shouldn't have bought it, I'd have given you as many as you want."

The plan was to break the 470-mile journey into several legs. The first took them to the village of Thoissey, where they had reservations to stay overnight and to dine at the Chapon Fin. The second leg the next day, January 4, took them 180 miles to Sens, where they lunched at the Hôtel de Paris et de la Poste, a restaurant Camus knew.

The four travelers then started the last, sixty-five-mile leg to Paris.

Shortly before two p.m., they were fifteen miles out of Sens on National Highway 5 when the Facel Vega suddenly swerved left across the road. It slammed into one of the trees lining the route, then wrapped around another tree forty feet away. The violent crash broke the car into pieces and scattered parts for as far away as five hundred feet.

Michel, Janine, and Anne were each hurled from the wreck.

Camus, sitting in the front seat, was thrown into the rear window.

He died instantly.

—~~—

WHEN THE POLICE arrived on the accident scene, they found Michel on the ground, bleeding and badly injured (he would die in a Paris hospital six days later). Janine and Anne were also taken to a nearby hospital; they would survive.

The police gathered up the luggage and other personal items that were strewn about the scene. On the road, near one of the trees, was Camus's mud-caked briefcase. It and Camus's body were taken to the nearby village of Villeblevin. In his coat pocket, examiners found his unused return train ticket to Paris. And inside his black leather briefcase there were some letters, Camus's passport, some personal photographs, a copy of Nietzsche's *The Gay Science,* and the unfinished 144-page manuscript of *The First Man.*

Camus was placed on a cot in the Town Hall and covered with a sheet, and a bouquet of flowers was placed on his body. Members of the town council watched over Camus while his family and friends, France, Algeria, and the rest of the world learned of his death.

THE SHOCK AT the sudden loss of a national figure reverberated through the government and the capital. André Malraux, de Gaulle's minister of cultural affairs, quickly issued a statement: "For over twenty years the work of Albert Camus was inseparable from the obsession with justice. We salute one of those through whom France remains present in the hearts of men." Word of the accident reached Paris in the evening, around theater time. The director of the Théâtre de France promptly closed his doors and refunded his patrons' money; other theaters observed a minute of silence before their performances.

"Absurde" was the single-word headline in *Paris-Presse.* The large, bold headline across the special edition of *Combat* declared ALBERT CAMUS IS DEAD. Although his association with the newspaper had ended thirteen years earlier, the front page was devoted to Camus, with the two lead stories, entitled "A Conscience Against Chaos" and "The Best of Us." The *New York Times,* in addition to reporting Camus's death on its front page, offered a short editorial that summarized the writer's philosophy as "a creed which calls on men for the most heroic kind of affirmation of life," and concluded that Camus's memory was assured "such immortality as mere men can give."

As the stunning news spread, business as usual in French literary and political circles came to a halt—the criticism that had long dogged Camus gave way to grief and eulogies over an extraordinary life that had been cut short. François Mauriac, who had been critical of Camus's silence on Algeria, said that Camus's death "is one of the greatest losses that could have affected French letters at the present time. A whole generation became aware of itself and its problems through Camus . . . And it is all youth that mourns him at this moment."

Even Sartre, who had neither seen nor spoken to Camus in the eight years since their split, joined the public mourning with a tribute to Camus's work and life: "We shall recognize in that work and in the life that is inseparable from it the pure and victorious attempt of one man to snatch every instant of his existence from his future death."

But to Camus's friends and longtime admirers the loss was, as Germaine Brée put it, "an irreparable catastrophe." Said another critic, "We are not just weeping over the premature disappearance of one of the greatest talents of our time; with the death of Camus, the very value of man seems diminished. Thanks to him, we were less hesitant, less uncertain in the confusion that always surrounds us."

Those who knew that he had resumed writing and knew of his great ambitions for his novel—Grenier, Bloch-Michel, and Monod—grieved the loss both of their friend and of the work that was still to come.

Kövesi offered his condolences to Monod in a short note, the first time he had written any words to Monod in French: *"Je suis triste pour toi; pauvre Albert Camus."* (I am sad for you; poor Albert Camus.)

Monod replied, "The death of Camus has been a terrible blow for his friends. Those who knew him well knew that he had not yet produced his greatest work."

CHAPTER 32

MESSENGERS

Happiness depends on being free, and freedom depends on being courageous.

—THUCYDIDES, *The History of the Peloponnesian War*

FOR TWO MONTHS, THERE HAD BEEN NO PROGRESS WITH THE furniture-truck scheme. On February 18, Monod received a type-written postcard from Kövesi in Vienna. The message on the back read:

Dear Jacques,

There are good hopes for our friends on the 29th of February or on the 1st of March. I do not bother you with details—please keep your fingers crossed, in about ten days I can tell you more.

Love, Andre

PLAN D: THE CIRCUS

It was just as well that Kövesi did not give Monod the details beforehand. This time Kövesi had concocted something that, at least to Ullmann, seemed right out of an Ionesco play. The basic idea was to smuggle her and her husband, Tamás Erdös, out of Hungary under the cover of a traveling circus troupe that was headed for Poland and Czechoslovakia. Ullmann met the circus's contact in Budapest to get the details. The contact proposed that she would travel in the lion's cage . . . with the lion! The animal was to be drugged with sleeping pills, and the contact assured Ullmann that the great cat would not move. Ullmann thought at first that the man was joking and asked him, "What would happen if the lion wakes up and is hungry?"

Ullmann told Tamás that she thought it was a crazy idea. The director of the circus ultimately decided that there were too many risks involved and scratched the plan.

After the scheme had been abandoned, Kövesi informed Monod, who tried to be encouraging: "A. and T. . . . now have experimental proof that things are moving and that they have not been abandoned and that you are still busily working for them. It should be a comfort and give them some patience to wait for the next turn."

Monod had been thinking about taking more responsibility upon himself by traveling to Hungary to set up an operation. On March 7, he let Kövesi know that he had received an official invitation from the Hungarian Academy of Science to lecture in Budapest. He asked Kövesi, "Would you let me know as soon as you can whether there would be any organization or reason to prefer one or another week in May or June?"

"X" Is a Message

There was much for Monod to do in addition to preparing for the trip to Hungary and hatching a plan with Kövesi. His and Jacob's continuing transatlantic collaboration with Arthur Pardee had borne new and important rewards.

The very first results of the PaJaMa gene-transfer experiments in late 1957 had demonstrated that enzyme synthesis began very promptly after the transfer of the galactosidase gene into cells—within minutes. Ever since those observations, Monod, Jacob, and Pardee had doubted the prevailing idea that individual ribosomes were devoted to the production of particular proteins. It did not seem plausible that ribosomes could be assembled so quickly and then dedicated to the making of a new protein. Their doubt had led Jacob to propose the existence of the unstable intermediate "X" that carried information between DNA and protein.

The challenge for Monod, Jacob, and Pardee was to come up with some way to determine whether genes were responsible for the production of intermediates, whether stable or unstable, that were necessary for protein synthesis. The basic experimental idea was to find out whether enzyme synthesis would continue in a cell after a transferred gene was somehow removed. If it stopped, then that would suggest that ribosomes could not keep churning out protein without the gene being present, and that would rule out the role of a stable intermediate. It would also support the idea that the intermediate between DNA and protein was unstable. The problem with testing these ideas was that there was no way to physically remove a gene once it was transferred.

After he returned to the University of California–Berkeley, Pardee and his graduate student Monica Riley solved the problem by executing an ingenious but technically difficult experiment. The trick they employed in their experiment was to use radioactivity to destroy the transferred galactosidase (*z*+) gene after mating and then to see whether enzyme production continued.

Riley grew the donor cells in a medium in which the only source of phosphorus was the radioactive form phosphorus-32 (denoted ^{32}P). The donor cells' DNA thus became highly radioactive. The cells were then mated with *z*– *i*– recipient cells. After allowing time for mating, the transfer of the radioactive *z*+ gene, and the start of enzyme synthesis, Riley interrupted the mating and froze the bacteria in a preservative. While metabolism stopped in the bacteria, radioactive decay of the ^{32}P in their DNA continued. The emission of beta particles by the ^{32}P shattered the bonds that held together the DNA of the transferred *z*+ gene. Over many days in the freezer, the transferred gene disintegrated.

Riley then thawed out the mated cells and compared enzyme synthesis in them with control cells that had received a nonradioactive gene but had been otherwise treated similarly. She found that enzyme synthesis was in fact compromised in the bacteria in which the transferred gene had been inactivated. The results meant that the continued presence of the gene was required for protein synthesis; there was no stable intermediate.

Two possibilities still remained: either there was no intermediate at all and DNA directly instructed the formation of proteins, which seemed unlikely since it was known that in cells with nuclei, such as animal cells, the DNA is located in the nucleus while protein synthesis takes place outside of it in the cytoplasm; or, more likely, the intermediate was unstable. Identifying that intermediate would open Crick's black box.

In mid-April, Jacob traveled to London for a microbiology meeting and then took a side trip up to Cambridge to talk with Francis Crick and Sydney Brenner, a South African geneticist who had moved to Cambridge, about the latest developments. On Good Friday afternoon, he and a handful of other visitors gathered in Brenner's apartment at King's College. To Jacob, it felt more like an examination than a discussion, with Crick leading the pack of inquisitors with rapid-fire questions and comments.

Jacob knew, however, that he was in a position of strength: he had the data. He went over the operon model, which had only recently been published in French. Then he described the new experiments and results from Riley and Pardee. Jacob explained all of the care and controls built into the ^{32}P gene-destruction experiment. Brenner and Crick knew that Pardee and the Pasteur group were superb experimentalists, but they were still impressed. When Jacob reported the results and the crucial interpretation—that there was no stable intermediate—Crick and Brenner immediately realized the significance of the experiment. "That's when the penny dropped," Crick later recalled.

Brenner and Crick leaped to their feet and began to argue, back and forth or at the same time, and at such a speed that Jacob could not follow all of the exchanges. Several minutes passed before the gist of the battle became clear. Brenner, Crick, and most other molecular biologists had long been hung up on the idea of the stability of ribosomes and of ribosomal RNA. Ribosomes were stable structures, true, but the new experiments proved that information coming from the gene was not stable. Crick realized that ribosomes were merely machines for reading information coming from the gene; their RNA did not carry the actual genetic information, as many had believed.

What was that information, then? In their frenzied reaction to Jacob's data, Brenner and Crick recalled experiments by two American researchers at the Oak Ridge National Laboratory in Tennessee, Elliot Volkin and Lazarus Astrachan. It was work whose significance was not at all clear at the time it was published in 1956, but now it was a vital clue. Volkin and Astrachan had found that when a bacterial virus called the T2 bacteriophage infected *E. coli* cells, RNA was made whose composition matched that of the viral DNA and was very different from the composition of ribosomal RNA. Crick and Brenner realized that newly made viral RNA must contain the information coming from the genes of the viral DNA.

Jacob's "X" had to be an unstable RNA intermediate between DNA and protein.

Jacob was a bit embarrassed. Why hadn't he thought of the Astrachan and Volkin viral RNA when he came up with the idea of an unstable intermediate? It was a big clue that he, Monod, and Pardee had somehow overlooked. But that was the nature of the game of science—of having to make connections between seemingly unrelated observations. And it was not as if these connections were that obvious. After all, Crick and Brenner were in Copenhagen when Jacob proposed "X," and nothing had clicked then. Now, seven months later, the puzzle pieces fit together.

Jacob took satisfaction in how, until that afternoon, something that had been a mere abstraction was now materializing, how "once again, a creature of pure reason came to life."

"X" was materializing, but it was still a long way from being proven.

That same evening, there was a party at Crick's home. Brenner and Jacob were too revved up to socialize with the crowd, so they grabbed some beer and sandwiches and huddled in a corner to figure out the next steps that were needed to look for the unstable RNA intermediate. They soon discovered that they had each been invited separately to visit the California Institute of Technology that summer: Jacob had been invited by Max Delbrück, and Brenner by Matthew Meselson; the latter was a master at radioactively labeling and separating large biomolecules. Brenner and Jacob realized that at Caltech they would have exactly the tools and expertise available to help them chase down the evidence they sought: that newly synthesized RNA carried genetic information and then associated with preexisting ribosomes.

They decided to team up at Caltech to do the experiments. Amid the noise and chaos of the party, Brenner wrote out reams of notes—

page after page of diagrams and calculations to design the experiments. Jacob felt spoiled, for he now had a second brilliant scientific partner.

The next step was to ask their hosts in Caltech for their help. Back in Paris, Jacob wrote to Meselson and spelled out the idea he proposed to test: "A possible model would be that no genetic information is contained in the RNA particles [the ribosomes], but that the gene sends to these particles RNA unstable molecules which act as *message* [emphasis added] to the particles, brings the genetic information to them for the synthesis of a particular protein . . . This hypothesis can be probably checked easily in bacteria infected with phage T2. I think that Sydney will write to you in detail about this."

Brenner also wrote to Meselson, and he asked his host to provide all of the radioactive labels and equipment they would need: "You know exactly what we will have to use."

Plan E: The Camping Trailer

As Jacob was planning his trip to California, Monod was finalizing the details of his mission to Hungary in mid-May. Ullmann wrote to him to say that they were excitedly awaiting his visit and that everyone in Budapest wanted to meet him. She told Monod that the police were still keeping a close eye on her and Erdös. She added, "I believe that your visit will not be long enough for us to do any experiments." This was code that there would not be time to execute an escape.

Nevertheless, after almost giving up due to all of the previous failures and setbacks, Kövesi was crafting an entirely new plan, one that Monod could help set up during his visit. He described the new scheme to Monod, as well as his somewhat fragile state of mind, in a letter at the end of April.

Kövesi explained that one of the jobs he had done was to repair a used twenty-two-foot-long British camping trailer. He realized that if he bought it using the funds Monod had provided, it could be used for smuggling Ullmann and Erdös. All he needed was a driver willing to take on the risks. After being cheated again and again by criminals in the previous attempts, he thought that he should find an honest worker who needed some money. He wrote:

> *I talked to one of my new friends and he did not hesitate very long. His name is Helm, is about thirty-six, an automobil-mechanic-monteur . . . The operation is simple and there is not one misterious factor . . . The trailer has two si[n]gle beds and a double one, two rooms, a kitchen and a bath./ I havnt decided yet, wether I should build a hide place or two, or just use the room under the single beds./ Helm takes a one week vacation with his wife, two children, and mother in law to Yugoslavia and visits the Budapest Fair. He passes the border each time at early night when the children are already sleeping upon the two single beds. Of course the family knows nothing of the real zeal of the voyage . . .*
>
> *On the last evening, he sends his family sight-seeing gets our friends into their places, puts the children into bed, and is in four-five hours on Austrian soil.*
>
> *That is about it.*
>
> *We are in no hurry what so ever, we can wait quietly until the Helms get their visas etc. But it would be good if you could get down there as soon as possible—you see why.*
>
> *Please, let me know, if you have any ideas, or remarks and the exact date of your visit, and your possible arrival to Vienna.*

Monod immediately dashed off a note to Ullmann, telling her that he looked forward to visiting her and Erdös "so that we can discuss anew the experiments we are preparing."

In order to understand the operation in detail, and so that he could explain it to Ullmann and Erdös, Monod scheduled to fly first to Vienna to see Kövesi. He also planned to see Kövesi on his way home from Hungary in order to share any ideas or instructions that Ullmann and Erdös had for Kövesi.

Before Monod left for Hungary, Madeleine Brunerie took the precaution of making a copy of his address book—without Kövesi's and Ullmann's entries—in case Monod was searched by the Hungarian authorities. He told Madeleine that if he had to extend his trip in order to work on arranging the escape, he would call her and ask her to inform Kövesi.

There was one more important task to complete before he left—to wrap up their paper with Riley and Pardee that described the gene-destruction experiment and proposed an unstable intermediate between DNA and protein—the proposal that Jacob would soon dash off to California to test. The collaboration entailed many drafts of the manuscript being sent back and forth and much correspondence between Berkeley and Paris. Finally, after seven months of writing and rewriting, the paper was sent to the *Journal of Molecular Biology*.

Late on the afternoon of Friday, May 13, with all of the arrangements made, Monod gathered his packed suitcase from his Pasteur office. In the hallway, all of the Pasteur staff who knew about the mission, including Lwoff, lined up to see him off.

Sightseeing in Budapest

After seeing Kövesi in Vienna and going over the various parts of the operation, Monod flew to Budapest on May 14. He was to visit various institutes and the University of Szeged, to meet with scientists and students, and to give a lecture before the Hungarian Academy of Sciences.

Ullmann and Erdös were to serve as Monod's hosts for the entire visit, which would give them the opportunity to talk privately, to discuss every detail of the operation, and to scout possible locations around Budapest where Kövesi's driver and the two would-be escapees might rendezvous.

But talking without their conversations being overheard and surveying locations without arousing suspicion would not be easy. The first day in Budapest, Monod went to Ullmann and Erdös's apartment. Once Monod started to talk candidly, Ullmann quickly put a pillow over the telephone, assuming that it was bugged.

Monod understood her signal. He started to laugh and said, "Oh, let's go and have a walk because there are such nice surroundings here." It was like being in the Resistance again. The three conspirators went for a walk in a nearby park.

Monod and Erdös had never met prior to Monod's visit. The

two men had an immediate rapport. Erdös was very impressed with Monod's obvious intelligence and physical fitness. Monod found Erdös to be very calm, particularly under the circumstances of constantly facing the threat of imprisonment. Monod sounded Erdös out regarding his determination to escape. Erdös was emphatic that he wanted to get out of Hungary, even knowing that he faced life imprisonment if caught. At the time, Hungary's borders were tightly sealed; no one was slipping out of the country.

They next had to work out various details of the operation. One of the most important matters was to find a place where Ullmann and Erdös could meet Helm and get into their hiding places in the camping trailer without being seen. Monod could not rent a car, for there were none available in Hungary, so Ullmann arranged for a car and chauffeur from the Hungarian Academy. Monod told the chauffeur that he wanted to see the Danube. Under the guise of a sightseeing trip, Monod, Ullmann, and Erdös looked for good spots along the river for a clandestine rendezvous.

The chauffeur could not understand Monod's interest in the river. Monod kept asking to stop, saying, "Oh, I should like to have a look at this place."

The chauffeur kept telling Monod, "There is nothing here."

Monod replied, "I like rivers."

Along the shores of the Danube, there were many long stretches where there were no settlements, just trees and small clearings. It was important to find a well-concealed place that was close to a bus stop so that Ullmann and Erdös could get to it on their own. Monod selected one spot and pointed it out to Ullmann and Erdös. They got out of the car and memorized the kilometer marker on the stone.

The chauffeur later remarked that he was impressed by the French professor's love of nature.

Erdös thought of more items for the operation checklist. One great concern was whether the border guards were likely to search the trailer. He suggested to Monod that Helm first make a test run to see how the guards reacted. Monod agreed and later passed the idea on to Kövesi, with the instruction that Helm should then go see Ullmann and Erdös to let them know how the test went.

The official part of Monod's visit to Hungary was a smashing success. In his lecture given in English before the Academy, it was the first time since 1948, in a country still under the influence of Lysenkoism, that the word "gene" was pronounced. Lysenko had continued to deny the role of DNA in heredity, even years after the discoveries of Watson and Crick. And very few scientists in Hungary were familiar with bacterial genetics, so Monod brought an entirely new perspective. Those in attendance said that they had never before heard a lecturer like Monod, who held the audience spellbound.

Monod had several other official obligations. He had lunch with the French ambassador to Hungary, who cautioned him to be very careful when he spoke with Ullmann because she was certainly a spy of the Hungarian secret police! Monod found this hilarious and could not stop laughing when he later told the story to Ullmann.

On his visit to the University of Szeged, Monod enjoyed a dinner with some colleagues, serenaded by an excellent Gypsy violinist. When Jacques learned that the musician could not read music, he asked whether he could play by just listening to whistling. The Gypsy said yes. Sure enough, after Monod whistled Mozart and Bach tunes, the violinist played them perfectly.

AFTER STOPPING IN Vienna to convey the messages from Ullmann and Erdös to Kövesi, Monod returned to Paris on May 20.

Monod wrote to Ullmann and Erdös in code to let them know that the operation with Kövesi and Helm was in motion: "Our extractions, in particular, appear to be progressing methodically and favorably. In this regard, Tom's advice, drawn from his own experiments with mycobacteria, will be of great help to us."

Monod wrote to Kövesi that he was transferring 50,000 Austrian schillings to his account to pay for the camping trailer and Helm's fee. He also asked Kövesi to stay in close contact:

> *I hardly need to tell you that I am constantly thinking about our plans and how they will work out. I will be grateful to you for any information you will find time to send me regarding the dates,*

the organization and especially the testing of the hiding places, and so on.

The more I think about it, the more I feel that the two tests the trailer will go through at the two boarders [sic] is of the utmost importance. Helm should be instructed to let A. know precisely how it went so that they can make the decision by themselves whether to go ahead or not.

The former Resistance officer was also concerned about maintaining secrecy:

Also, may I mention again that leaks about some enterprise browing [sic] around Laubichler Garage [where Kövesi and Helm worked] may be most unlikely, but they represent a potentially very great danger. Helm must be made to understand that clearly.

My love to you and Eva. I hope to God I will see all four [of you] pretty soon.

To Austria Under a Bathtub

In early June, Helm made a preliminary trip to Budapest and visited Ullmann's institute in order to introduce himself. When he arrived, Ullmann was summoned to meet her German-speaking guest. Afraid that Helm would arouse suspicion among her colleagues, Ullmann had a handy alibi: she was teaching in German to first-year medical students from East Germany, so she explained that Helm's visit was related to her teaching. Nevertheless, she wanted to get rid of Helm as quickly as she could.

They briefly discussed the details. On the day of the operation, Ullmann and Erdös would have only a few hours' notice to get ready. Helm was to come to Ullmann and say that he was in Budapest and would be at the rendezvous spot on the road between Budapest and Esztergom at a given time. Ullmann and Erdös would take a bus to the rendezvous. They would leave only with the clothes on their backs.

—∿∿—

The operation was slated for Saturday, June 18.

Helm and his wife checked into a hotel in Budapest on the seventeenth. Helm then went to Ullmann to let her know that he had arrived and would see them at the rendezvous point. Ullmann and Erdös knew that their disappearance would arouse suspicions. In order to throw the secret police off their trail, they told some friends that they were going for a tour around a lake.

They took the bus out of Budapest toward Esztergom and got off at the rendezvous point. Helm was not there. Ullmann and Erdös knew that timing was critical. It was a long drive to the Austrian border, and they had to reach it before dark.

Finally, with Ullmann and Erdös waiting anxiously, Helm and his wife arrived in their car, pulling the camping trailer behind. Ullmann and Erdös got into the car. Ullmann took a tranquilizer, both to calm her nerves and to suppress any coughing once she was hidden in the camping trailer.

Two hours before they reached the border, Helm pulled off the road. His wife took the two children for a walk and snapped their pictures along the Danube. Helm opened the camping trailer for Ullmann and Erdös.

The inside of the trailer was large. It had a kitchen, a sitting room, a bathroom, and a bedroom with two beds. Helm had planned that Erdös and Ullmann would hide in compartments under each of the beds. When Ullmann saw the arrangement, however, she vetoed that idea right away, telling Helm, "I will not get in there. That's the first place they will look."

There was not much time to arrange any alternative. Ullmann noticed a large box in the bathroom. She asked Helm, "What's in this box?"

"Just some things," he replied.

Ullmann asked, "Isn't there enough space for two people?"

Helm hesitated. "It is difficult." Then, he decided, "Yes, it might be secure."

The box was a sort of closet. It was full of old clothes and towels, with a bathtub perched on top of it. It looked to Ullmann like the bathtub had not been used. Helm quickly threw the clothes and towels out of the trailer to make a space for Ullmann and Erdös. There was just enough room for the two stowaways to cram themselves in. They each had a small canteen of water and a straw through which to sip it. Ullmann covered her head with some white linen. Erdös said to Ullmann, "Don't move, and don't say anything until the last moment."

Helm closed the box, fixed the bathtub into place, returned to the car, and started for the border.

AT THE BORDER, two customs officers stopped the vehicle. They told Helm that they wanted to see inside the camping trailer. Helm obliged while his wife stayed in the car.

One officer had an electric light that he shone into each part of the camping trailer. He asked Helm about each of the rooms, "What's this?"

"The sitting room," Helm replied.

"And what is this?" the policeman continued.

"The kitchen," Helm answered.

Ullmann and Erdös could hear the entire exchange. Ullmann's heart was racing.

"And what is this?" the officer asked.

"This is the bathroom."

"You have a bathroom?" The policeman was surprised.

"Yeah, we have a bathroom." Helm added, "I'll show you."

The bathroom was too small for both men to enter; Helm occupied almost the entire room and could barely move around.

The officer pointed to the compartment under the bathtub. "What's this in the bathroom, under the bathtub?"

"Nothing," Helm said. "Just a place where we store dirty things." Helm knocked on the wood.

"Now open it," the policeman demanded.

Helm pried it open half an inch. Ullmann realized that the beam from the officer's light brushed her hand. She tried to pull it back gently. The officer did not see her but could see the white linens that Ullmann had used to cover her head. The officer wanted to know why the linens were in the space below the bathtub and asked Helm to dismantle the box.

"I am not going to open it," Helm protested, claiming that he had already opened it once while crossing into Yugoslavia. "I will cut into it for you," he offered.

Helm grabbed a saw and then made a big show of cutting into the compartment.

"OK, enough." The officer was satisfied.

The officers next went into the bedroom and made Helm dismantle the beds—exactly as Ullmann and Erdös had predicted.

Satisfied by their search, the officers left the trailer.

Inside their compartment, Ullmann and Erdös felt the trailer start to move very slowly. But then it stopped again.

Ullmann could hear the officer say some paperwork had to be fixed.

After nearly an hour at the border, the trailer started moving again. Then, after a short while, Ullmann smelled gasoline. She knew there were no gas stations on the Hungarian side—they were in Austria!

Once they were far enough from the border, Ullmann and Erdös, their limbs numb from being confined for hours, climbed out of their cramped hiding space and stepped into the fresh Austrian air, and freedom.

It was night before the caravan reached Helm's house, so the exhausted stowaways decided to sleep there. They called Monod and Kövesi to let them know they had made it out.

MONOD SENT TELEGRAMS to Pardee in Berkeley, Jacob in Pasadena, Csapo in New York, and other supporters of the mission to spread the good news.

When Madeleine arrived at the Pasteur on Monday morning, Monod was already there. He shouted out to her, "Agnes and Tom are in Vienna!"

Madeleine was so overjoyed she could not help but run up to Monod and give him a kiss.

A few days later, Monod received a letter from Ullmann in Vienna:

You have organized something that appeared absolutely impossible. But you have not only organized it, you have always thought about all of the details. How many painful hours have you spent? I had remorse over stealing your time and energy. You are an extraordinary man. If I did not know that you really existed, I would imagine that you were a fiction who exists only in dreams. There are some things for which one can never thank someone enough and never forget. I cannot yet fully appreciate what it means that I am free . . . free thanks to you.

CHAPTER 33

SYNTHESIS

Our observation of nature must be diligent, our reflection profound, and our experiments exact. We rarely see these three means combined, and for this reason, creative geniuses are not common.

—DENIS DIDEROT, "On the Interpretation of Nature"

JACOB WAS NOT IN PARIS TO CELEBRATE THE GREAT NEWS OF Ullmann and Erdös's liberation. He was still in Pasadena with Brenner, and feeling discouraged. After three weeks of experimenting, they had accomplished nothing. It was probably unrealistic to think that two people who had not worked together before could fly to a laboratory six thousand miles away and make a technically complex experiment work in a short period. For Jacob, it was the first time he had ever worked outside of the Pasteur.

Moreover, few believed in his and Brenner's idea anyway. Before arriving in Pasadena, Jacob gave a series of seminars along the West Coast. While his work on gene regulation with Monod was well received—including their concepts about repressors, operators, and operons—the idea of an intermediate carrying information from the gene to the ribosome was not. Max Delbrück, who had invited Jacob to Caltech, threw up his hands and said, "I don't believe it."

Neither Brenner nor Jacob was dissuaded, however, from trying their experiment. The idea was to mark or "label" newly synthesized phage RNA with radioactivity, and then to see whether it became associated with already-existing ribosomes. The plan was to exploit Meselson's unique expertise in labeling molecules in growing cells, and in separating large molecules and complexes in high-speed centrifuges according to their density.

Nothing went smoothly. They had difficulties getting the bacteria to grow on heavy isotopes of carbon (^{13}C) and nitrogen (^{15}N) that were to mark the ribosomes. The ribosomes also fell apart in the high concentration of cesium chloride in which they were centrifuged. Brenner and Jacob kept modifying the experimental conditions in order to set up for the crucial test, but time was against them. Moreover, Jacob's war-injured legs were aching in the California heat. It also did not help when Delbrück poked his head into the lab every now and then and asked sarcastically for any news about "X."

Their skepticism aside, Brenner and Jacob's Caltech hosts had made sure that they felt welcome and that they were well fed; they invited them to many parties, lunches, and dinners. But with nothing to show for their efforts, Brenner and Jacob discussed whether they should stay and slog on or return home. They decided they would leave in just a few more days, at the end of June.

With time running out and little hope left for success, they decided to take a day at the beach. Stretched out on the sand, Jacob and Brenner continued to mull over all of the modifications they had done. Suddenly, Brenner yelled out, "It's the magnesium! It's the magnesium!" Brenner realized that magnesium held the ribosomes together and that the very high concentration of cesium chloride, about 8,000 times that of magnesium in the experiment, was displacing the magnesium.

He and Jacob hurried back to the lab to repeat the experiment with higher levels of magnesium. This would be their last chance. Brenner asked Jacob how much magnesium they should add. He suggested ten times more. Brenner said, "All right," but he suggested they also try one tube with one hundred times more: "Put in a lot; can do no harm."

They were rushing to get the experiment started and to get the very long centrifuge run going. In his haste, Jacob missed a tube with his pipette and spilled a large amount of radioactive phosphorus into a water bath. They tried to rinse the bath out, but it was still radioactive, so they crept down into the building's basement and stashed the contaminated equipment behind a Coca-Cola machine.

Then yet another problem arose. In the middle of its long run, the centrifuge broke down. In order to salvage the experiment, they had to move the heavy centrifuge rotor that held the tubes with their experimental samples to another centrifuge. The risk in doing so was that the process for separating the ribosomes and RNA in the centrifuge had to be done in a cold environment. If the rotor and samples warmed up, the experiment, their last chance, would be ruined. Brenner managed to swiftly and carefully carry the rotor down to another cold room and place it into another centrifuge.

At the end of the run, he had to reverse his steps and bring the rotor back again to the lab. The sample had been separated into layers like a liquid parfait. If Brenner shook or bumped the tubes, he could wreck the separation. He carried the rotor through the halls like it was a holy relic.

Nervously but skillfully, Brenner then punctured the bottom of the tubes so that each drop from the bottom to the top of the tube could be collected and measured for radioactivity in a Geiger counter. If the experiment worked *and* Brenner and Jacob were right, the radioactive RNA would be associated with the heavier, preexisting ribosomes and not the lighter ribosomes. They watched the samples get counted one by one. In the first tube with extra magnesium, they saw the radioactive counts rise in the drops containing the heavy ribosomes, and then fall with the light ribosomes. They both shouted and then danced a double jig.

They had their first evidence that an RNA intermediate carried information from the genes in DNA to already existing ribosomes.

Their California gamble had paid off, and just in time. Crick's black box was now open.

Jacob flew back to Paris two days later.

Masterpiece

Monod was not in Paris to congratulate Jacob on his triumph. He was in Cannes for a month's vacation, but only after having taken care of arrangements for Ullmann and Erdös. Monod saw to it that the two escapees, who had arrived without immigration papers, money, or a place to live, would have what they needed.

His task of obtaining French visas for them was made considerably easier by his brother Philo's position within the Ministry of Foreign Affairs. As soon as Ullmann and Erdös safely reached Austria, Monod asked Philo to arrange for their visas, which he did. Philo was about to take up a post as the French ambassador to Australia, so he generously offered his Paris apartment to the couple for a year.

As for money to live on, some of the funds Monod had raised had been set aside for the two escapees once they made it to freedom. It was enough to support them in Austria while their immigration papers were pending and for a few months after they settled in Paris.

After waiting for their visas, the two stowaways finally arrived at the Gare de l'Est in Paris on Wednesday, August 10, almost two months after their escape. Madeleine Brunerie met them at the station and took them to breakfast at a nearby café. Erdös savored the taste of what he would thereafter refer to as his "croissant of liberty." The next day, Ullmann and Erdös went to the Pasteur for a celebratory lunch with Monod, now returned from vacation.

Of course, the two scientists would eventually need jobs. Monod once again sprang into action, writing colleagues both near and far. Monod soon obtained a fellowship from the Rockefeller Foundation for Ullmann to work at the Pasteur. Erdös secured a position at a research institute just outside of Paris.

With the Hungarian operation settled, and Jacob back from his California adventure with Sydney Brenner, Monod and Jacob turned again to their collaboration.

In three years, they had made astonishing progress. Jacob's inkling of a connection between the induction of bacteriophage and the induction of enzyme synthesis had led to concrete experimental evidence for repressors, operators, operons, and a general model for the logic of gene regulation. And now, thanks to Brenner's ingenuity, there was evidence for a new entity—Jacob's "X" RNA—that carried the information for making proteins from the gene to the ribosome.

But the progress had been in some ways perhaps too swift, as the scientific community had not yet assimilated what Monod and Jacob had discovered and proposed. Since the first PaJaMa paper with Pardee, a succession of papers had appeared every few months that tested a recent prediction, coined a new entity, and made new predictions that were already being tested in the lab before the ink was dry. Given that several key papers were in French, it was understandable that even some privileged insiders had not yet been convinced of or seen the full implications of their work, let alone biologists in general. Indeed, the new experiments with Brenner were not yet written up for publication and would not be until the following year.

Monod and Jacob decided that they should attempt to write one comprehensive article, in English, that would distill all of their data, integrate all of their ideas, place them into historical context, and explore their broader implications.

It meant that a lot of time would be spent at Monod's blackboard, hashing over what they would say and who would write which parts. The challenge was to forge one coherent story from all of the parts of their work, with the dual goal of persuading others of their ideas and catalyzing new research.

Monod had much more practice at such persuasion. He had succeeded earlier in recasting enzymatic adaptation as a matter of the induction of enzyme synthesis. He had triumphed in the Lysenko debate. Thanks to his mastery of the language, Monod was confident, even stylish, when writing in English. Jacob thought his own English was stiff and awkward. He felt overmatched. The challenge of writing in English and the vast scope of the article meant it would take months to complete.

That was all right, for both men sensed that the stakes were high, and perhaps growing higher. Madeleine Brunerie, who would type the whole manuscript, knew that as well. A year earlier, in October 1959, Mel Cohn had written to her in secret, asking her for Monod's curriculum vitae and for reprints of his papers. At first he told her the documents were to support his nomination of Monod for a scientific prize to be shared with André Lwoff. Madeleine was suspicious and pressed Cohn as to whether the documents concerned the Nobel Prize. Knowing their shared admiration for Monod, Cohn confirmed her hunch and pledged her to secrecy.

Madeleine told no one about their pact and waited hopefully for the announcement of the Prizes. A year later, on the day the Prizes were to be announced, she was in Monod's office taking dictation when the phone rang. It was a journalist asking about two immunologists who had won the 1960 Prize.

The next morning, Monod asked Madeleine to draft letters of congratulations to the two scientists. Then he asked her, "Do you know, Madeleine, who were two other candidates for the Nobel?" In a low voice, he added, "They were Monsieur Lwoff and me."

"I knew that." Madeleine smiled.

Monod seemed startled.

"I have known that for ten months," she said.

"Who told you?" Monod asked.

For once she had the upper hand on "Le Patron." Madeleine was enjoying the game. She replied, "Excuse me, monsieur, but I don't believe that I am allowed to tell you."

Then Madeleine turned the tables and asked Monod who told him.

"Cohn!" Monod said.

Her boss then mused about the Prize. "It would have given me pleasure, sure enough: I could have bought a boat . . . It would have been good for public opinion. But there are also disadvantages to those who are awarded it."

Monod explained that the Prize generated rivalries among scientists. He also feared, like his late friend Camus, that the Prize would mark the end of his creative career. Monod told Madeleine that he did not think he was finished, nor did he think that he would ever win the Nobel. Madeleine told him she was sure that he would.

Every October thereafter, a Nobel vigil would begin at the Pasteur in anticipation of each year's Prize announcement.

Monod could not be too disappointed about not winning. After all, Watson and Crick had still not won the Prize, and their enormous discovery was now seven years old. Monod's work, including all that he had accomplished with Jacob, was much younger. Indeed, some of it was still in gestation and needed to be delivered.

THE WRITING OF their opus entitled "Genetic Regulatory Mechanisms in the Synthesis of Proteins" consumed the fall of 1960. It was a comprehensive review of the state of knowledge concerning the crucial question of how protein synthesis is controlled. The article was also much more—it was an extraordinary display of the hypothetico-deductive scientific method.

Monod had mastered the style in previous papers. The presentation took the form "If X is true, then Y should be observed. But Z is what we observe, so X must not be true, and an alternative A must be considered." And then the ramifications of A would be examined in turn. Throughout their article, as Monod and Jacob addressed each scientific question, they explicitly stated each alternative hypothesis, enumerated the predictions made under each hypothesis, and reviewed in detail the experimental evidence that then either corroborated or falsified each hypothesis. After drawing each major conclusion, the scientists moved on to the next issue. Their new picture of gene regulation was thus built up piece by piece, thoroughly reasoned conclusion by thoroughly reasoned conclusion, illustrating the roles of all of the entities they had imagined, discovered, and named in the previous three years (repressor, operator, operon, structural gene, regulatory gene), plus one more. Although Jacob's work with Brenner on "X" RNA was not yet even written for publication, he and Monod incorporated that crucial new evidence into their synthesis. The Frenchmen introduced a new term for the unstable RNA intermediate between DNA and protein: "messenger RNA." (For further details and an overview of the science in the book, see Appendix.)

After the rigorous, exhaustive development of their model for gene regulation in bacteria, the two men were keen to make sure that the wider implications of their work for other great mysteries were appreciated. They drew attention to cell differentiation and embryonic development in particular. In animals, for example, liver cells make different proteins than do kidney cells, skin cells, or other cells, but each cell type generally contains the same DNA, the same collection of genes, or "genome." They drew the connection to the fundamental problem of embryonic development and once again framed how the central issue is "to understand why tissue cells do not express, all the time, all of the potentialities inherent in their genome." Monod and Jacob suggested that their discoveries in bacteria applied to this general problem because just as bacterial genes are repressed when they are not needed, the genes that distinguish each cell type are thus somehow repressed in other cell types. They would later summarize that assertion with the quip "anything found to be true of *E. coli* must also be true of Elephants."

The duo pressed to complete and submit the manuscript before breaking for the holidays. Finally, on Christmas Eve, they finished writing and sent the article to the editor of the *Journal of Molecular Biology,* the leading chronicle of the still-young discipline.

While most articles in the journal occupied fewer than a dozen pages, Jacob and Monod's treatise would command thirty-nine pages, by far the most of any paper in the journal's history or for the next many years.

It would be a watershed in modern biology.

The novelty and breadth of their ideas, the elegant logic, and the general explanatory power of their model made an immediate and powerful impact. What Watson and Crick's discovery of the DNA double helix had done to reveal the structure of genes and the mechanisms of inheritance, Jacob and Monod's synthesis did for illuminating how genes were regulated. They were virtually assured that Nobel Prizes were in their future as well.

Their masterful synthesis also marked personal milestones for each scientist. For Monod, it was exactly twenty years since his fateful discussion with Lwoff in December 1940, when he first learned the term "enzymatic adaptation," and after which he resolved to devote himself to understanding the phenomenon. Now he had done just that.

For Jacob, the achievement coincided with the anniversaries of two pivotal moments in his life. It had been twenty years since the young medical student fled France in the agony and chaos of her defeat, perhaps never to return. And it had been ten years since Lwoff admitted the naïve, inexperienced, and wounded would-be physician into his laboratory.

LATE IN THE afternoon on that cold, gray Christmas Eve, Jacob stepped out of the laboratory onto the snow-covered sidewalk of rue Docteur

Roux. He turned left in front of the original institute building erected by Pasteur himself, the snow muffling the sounds of his footsteps on the empty street. As he wound his way home and passed by other landmarks of the past decade of his life in his beloved city, he reflected on the improbable journey that had brought him to that moment.

He turned right onto boulevard Pasteur and left onto the rue de Vaugirard. The streets were now crowded with Parisians finishing their Christmas shopping. Jacob then strode past the Gare Montparnasse, where General von Choltitz had glumly surrendered Paris to "his general" Leclerc on August 25, 1944—a historic moment that Jacob had missed as he lay in a body cast in Cherbourg. He then turned right onto the boulevard Montparnasse and headed toward the movie theater in which he'd had his revelation that summer day in 1958. After he passed La Coupole but before he reached La Closerie des Lilas or the Val-de-Grâce military hospital, where he spent Christmas Eve 1944, Jacob turned left toward the Luxembourg Gardens.

The side streets were again deserted. He entered the gardens, and the snow began to fall again, adding to the pure white blanket that covered the grounds. As he approached rue Guynemer, where his apartment overlooked the gardens, darkness fell. The streetlamps and other lights again cast their magic spell across Paris, his home, and still the most beautiful city in the world.

Epilogue
French Lessons

THE WORLD IS NO SECRET FOR THE WISE MAN. WHY DOES HE NEED TO STRAY INTO ETERNITY?

—ALBERT CAMUS,
NOTEBOOK III

CHAPTER 34

CAMUS IN A LAB COAT

I know men and recognize them by their behavior, by the totality of their deed, by the consequences caused in life by their presence.

—ALBERT CAMUS, *The Myth of Sisyphus*

THE ANNUAL PASTEUR OCTOBER NOBEL VIGIL ENDED AT ONE P.M. on Thursday, October 14, 1965, when Monod, Jacob, and Lwoff each received telegrams from the Karolinska Institute in Stockholm informing them that they were to be awarded the Nobel Prize in Physiology or Medicine "for their discoveries concerning the genetic control of enzyme and viral synthesis."

They were the first French Nobel laureates in science in thirty years, since Frédéric Joliot and Irène Joliot-Curie, the daughter of Pierre and Marie Curie, shared the 1935 Chemistry Prize. In breaking the long drought, the three molecular biologists instantly became celebrities. *Paris-Presse*'s full-page headline blared:

MAGNIFIQUES, NOS 3 "NOBEL"
[Our amazing Nobel Three]

The newspaper's lead article described "The beautiful and hard adventure of three Frenchmen who are a team in life as in their work," and the front page carried a picture of the new laureates beaming for the camera.

The trio had not simply revived French scientific pride; the press also seized immediately on their activities during the war. *France-Soir* trumpeted the "marvelous history of the three French friends" on its front page, and highlighted Monod and Lwoff's activities in the Resistance and Jacob's membership in the Compagnons de la Libération. It showed pictures of Monod, rope in hand and cigarette in his mouth, sailing on his boat with his son Philippe; Jacob with his wife, Lise, and children; and Lwoff with his wife, Marguerite, holding up one of Lwoff's paintings.

Brilliant scientists, brave patriots, family men, rugged sailors, and talented artists to boot—the press could not have imagined better copy. Unlike Camus, who had long been famous when he won the Prize and thus endured a heavy dose of backbiting, the scientists were up till that moment largely unknown outside of academia and received only praise and admiration (except from the Communists, of course). The press's images and stories of the three gallant Pastorians transformed them into national figures.

Requests for interviews flooded in to the Pasteur. Each man was soon called upon to opine upon all sorts of scientific, political, and even philosophical matters. Of the three, no one was more inclined or prepared to hurl himself into the fray than Monod. Indeed, just after the Nobel announcement, Monod sat down for a long interview with *Le Nouvel Observateur,* a new weekly magazine cofounded by Jean Daniel. Monod stated at the outset his conviction that receipt of the Nobel Prize carried with it a responsibility—one that he immediately began to fulfill in a startlingly frank and wide-ranging interview that became the cover story in the very next issue.

Monod's coronation as Nobelist marked not only the culmination of his long scientific quest but also the beginning of a new chapter in his life as a public figure, a period during which the deep influence of his friend Camus on his worldview was acknowledged repeatedly, and articulated in his internationally acclaimed, bestselling book-length essay, *Chance and Necessity*. Monod said, for example, "Camus's existentialism, in the widest sense, is exactly that which I share." Through his public commitments to human rights, individual freedoms, and indeed even the necessity of rebellion, Monod emerged as a new incarnation of his friend Camus, albeit in a lab coat.

Moreover, Monod went so far as to explicitly adopt and to expand upon Camus's philosophical positions. When Camus died, the *New York Times* had described the central theme of his work as "the proper response of the thinking man to the plight that is posed by the gift of life." Monod wholeheartedly embraced Camus's response and added to that foundation a scientific perspective, one gained from the new molecular biology that had revealed the deepest secrets of life.

Commitments

Before Monod and his fellow laureates had even reached Stockholm to collect their honors, the trio made their first major commitment—to accept the joint presidency of the Honorary Committee of the French Movement for Family Planning (Le Mouvement Français pour le Planning Familial, or MFPF). While the contraceptive pill had been available in the United States since 1961, it remained illegal in France. In fact, the prescription, the sale, and the advertisement of all contraceptives were banned by a 1920 law that sought to boost France's birthrate after her devastating losses in World War I.

The ban had been maintained with the support of the Roman Catholic Church, of course, but also the French Communist Party as well as the medical establishment. PCF leader Maurice Thorez objected to measures that limited the potential growth of the working class: "Communists condemn the reactionary notions of those who seek to limit births and seek thereby to turn the working class away from its struggle for bread and socialism." The medical community took the view that their role was to treat sick patients, not healthy ones, so birth control was not perceived as a medical matter. The ban on information about birth control was so strict that French gynecologists completed their training without even learning of the advances in contraceptives.

As a consequence of the unavailability of simple contraception, abortion, although also illegal, was common, with an estimated 400,000 to 800,000 procedures occurring each year—a figure equal to the number of live births. Horrified at the medical risks of abortion and the myriad consequences of unwanted pregnancies, Dr. Marie-Andrée Lagroua Weill-Hallé founded the organization Happy Motherhood (Maternité Heureuse) in 1956 to arouse national awareness. The initiative gathered momentum with the growing support of physicians and morphed into the MFPF.

Shortly after the announcement of their Nobel Prize, Weill-Hallé asked Monod, Jacob, and Lwoff for their support. In accepting her invitation, the trio explained the reasoning behind their endorsement. Opposition to birth control was based on religious and political ideologies. The scientists could not abide any laws that placed ideology over science and personal freedom. This rationale was one to which Monod would return often, and with much force, in the coming years. They wrote:

> Because of scientific and technical developments, the laws which govern relations among men can no longer be founded on an ethic dating back more than twenty centuries. One of the fundamental values of a modern, advanced society is the liberty of the individual under the law. Such a society cannot allow that women live as slaves to outdated principles.
>
> When the movement that you lead reaches its objectives, many women and men will know a more harmonious and balanced existence, many tragedies will be avoided, in particular thousands of secret abortions, even the existence of which is a condemnation of a society.
>
> Those who oppose you and ignore the hard reality, the tragedies, the mutilations and deaths, carry a heavy responsibility. No one should have the right to sacrifice the happiness, the health, or the life of another human being to their own personal principles, however sincere and noble they may be.

Monod was soon asked to contribute a preface to a book on the biological and psychological aspects of contraception. He seized the opportunity to applaud the authors, to register his contempt for the 1920 law, and to point out the gulf between the state of science and the outdated ethics that rejected its advances: "Many of our fellow citizens who are still poorly informed will be, I do not doubt, enlightened by reading your book. They understand that a law, which could be interpreted as repressing the dissemination of scientific information or as prohibiting doctors from acting according to their knowledge and their conscience, would be contrary to the ethics of a modern society and even to the principles of our law."

Thanks in large measure to the campaign by the MFPF, the ban on contraceptives was lifted by passage of the Neuwirth Act in late 1967.

MONOD'S VISIBILITY GREW quickly. Perhaps the most revealing reflection of his new stature involved the Reverend Dr. Martin Luther King Jr.'s visit to Paris at the end of March 1966. Dr. King, who had received the 1964 Nobel Peace Prize in recognition of his leadership of the civil rights movement in the United States and for his commitment to nonviolence, was making a tour of Stockholm, Lyon, and Paris to raise funds for his efforts.

Had Camus been alive, there would have been no more likely candidate to first introduce Dr. King to a Paris audience. The honor went instead to Monod. He rose to the occasion in front of a star-studded audience of nearly five thousand at the Palais des Sports on March 28, 1966:

> "The Country of the rights of man."
>
> "The land of liberty."
>
> These are the beautiful names that their history, and the aspirations of their culture have earned France and America.
>
> Yet, who would dare say that human rights have always been respected, in France or by France, and if they are even today fully assured?
>
> And again, who could say that the equality of all citizens, guaranteed by the constitution of the United States, is the case in fact, in mores, or even at times, in the application of the law?
>
> If however these two Nations, France and America, could acquire . . . these true titles of glory—it is because they have always found among their citizens men ready to devote their talent, their energy, their entire life, even at sometimes heartbreaking cost, to the defense of liberty and what is right.
>
> The man that I have the distinguished honor of introducing this evening is such a man. He is the Reverend Martin Luther King.
>
> Reverend King has devoted his life to the cause of his brothers,

> the blacks of America. He is not the first. He is not the only one. In the United States, thousands of men, black and white, are engaged in this immense effort.
>
> Among them however, Reverend King enjoys an incomparable standing, a unique authority that is derived from his high moral position and politics. Defender and leader of his brothers, he was able to guide them not only in what circumstances required, but also while respecting the freedom, dignity, and the rights of all people. This is what the most beautiful of all Nobel Prizes, the Peace Prize, recognized in 1964 in this disciple of Gandhi.

After Dr. King's assassination on April 4, 1968, Monod was called upon again, this time to eulogize King at a memorial service organized by the Movement Against Racism and Anti-Semitism, and for Peace (Mouvement Contre le Racisme, l'Antisémitisme, et Pour La Paix) at

Jacques Monod shaking hands with Coretta Scott King at the Palais des Sports in Paris, March 29, 1966. Monod introduced King to the audience. Two years later, he eulogized King at a memorial service after King's assassination. To Monod's left are actress Simone Signoret, singer Harry Belafonte, and actor Yves Montand. (AFP/ Getty Images)

the Cirque d'Hiver. Monod resurrected his comparison to Gandhi and revealed that King had confided his expectation to be assassinated:

> Twenty years ago, on January 30, 1948, Gandhi, liberator of India, was assassinated by a Hinduist fanatic. The bloody skin of this scrawny old man sheltered one of the greatest souls that enlightened our terrible century. Such souls are immortal; we knew that when the powerful figure of Martin Luther King was revealed to us by his preaching, his fight for man, for his freedom, for his dignity. Gandhi's soul had found a body. A body that was vigorous yesterday is today cut down. Twenty years after his master, and like him, Luther King has become the martyr for which he was prepared. I know that: he told me so.

Monod praised King's adherence to nonviolence and reminded his audience that the sort of humiliation and injustice suffered by American blacks had also occurred on French soil. He then challenged his audience:

> Let us listen to the lesson of Martin Luther King, who wrote:
>
> "The movement does not seek to liberate blacks at the price of the humiliation of the whites. It wants to liberate American society and to help all people to liberate themselves."
>
> Let us listen to this lesson; it is addressed to us, as well as to other people: have we fully secured this liberty? Have we broken forever the chains of stupid national pride? Have we fully understood that the respect for oneself is never gained by the contempt for others? And do we believe that everyone who lives on our soil is assured justice, brotherhood, and freedom?
>
> The true, the only homage worthy of the great men whom we honor this evening would be, thanks to them, such a conscience among ourselves.

Camus would have, no doubt, stood to applaud his scientist friend.

To the Barricades, Again

King's lesson of nonviolent protest would, however, soon be overshadowed. In May 1968, just a few weeks after King's assassination and Monod's eulogy, Paris and the whole of France were convulsed by protests and riots that came very close to a full-scale revolution. Monod was thrust into (or, perhaps more accurately, stepped into) the political spotlight, as well as the line of fire.

The revolt was precipitated by the closure of the University of Paris at Nanterre on May 2. A series of confrontations between students and the administration over the preceding weeks had provoked the occupation of an administration building, the suspension of classes for a week, the arrest of a student leader, and then closure. On Friday, May 3, enraged over the lockout of their campus, Nanterre students ventured into the Latin Quarter and rallied in the courtyard at the Sorbonne.

They were joined by Sorbonne students. Late in the afternoon, the rector of the university called the police, who moved aggressively to clear the courtyard. The confrontation spilled out into the Latin Quarter. The students counterattacked with paving stones and anything else they could pry loose and hurl at the police. More than five hundred students were arrested and more than one hundred police and students were injured.

The next day, it was announced that classes at the Sorbonne were also suspended. The Students' Union (Union Nationale des Étudiants de France, or UNEF) reacted by calling for a march on Monday, May 6, to protest the police intervention at the Sorbonne, the presence of which was a violation of university rules, and to demand the reopening of the university.

More than 20,000 protestors marched to the Sorbonne and were met by a solid wall of police who were determined to hold their ground. Violence erupted again as the police used tear gas and water cannons to try to disperse the marchers. Hundreds more students were arrested and many injured.

The government was taken by surprise by the ferocity of the student rebellion. The uprising reflected much more than youthful zeal. The harsh police tactics unleashed the students' deep resentments over the antiquated French higher education system. The universities were badly overcrowded, understaffed, and managed by a central bureaucracy that had no accountability. Vast numbers of students were flunked out or quit before attaining a degree. Since a university degree was necessary for securing a good job, disillusionment with the entire system was widespread.

Monod and other professors had long campaigned for university reform. In his inaugural interview in *Le Nouvel Observateur,* Monod had made a candid and devastating assessment of the state of French universities. Monod was keen to ensure that the surge of national pride over the Pasteur trio's Nobel Prizes did not extend to giving credit where no credit was due. He made it very clear that no credit belonged to the current government, or to the preceding ones, for that matter. Monod said that French science was "underdeveloped" in comparison with other Western countries. He attributed France's problems in science to the stagnation of the centrally controlled universities, and the policies governing them that dated all the way back to Napoleon. Monod, for instance, could not even create a course without the approval of the Minister of National Education. The accumulated result of such bureaucracy was the stifling of young talent. Monod's candor had earned him no friends among de Gaulle, Prime Minister Georges Pompidou, or the minister of national education, Alain Peyrefitte.

Monod feared that, even with violence having erupted in the streets, the government still did not grasp the growing magnitude of the crisis. On Wednesday, May 8, he headed a delegation of professors from the Faculty of Sciences that went to the National Assembly. Their aim was to deliver an urgent message to Peyrefitte. Monod asked that the minister state publicly that the students arrested during the demonstrations would be given amnesty. The minister replied politely that the government was intending to do so, but that he could say nothing before the matter was deliberated by the Council of Ministers.

None of the members of the National Assembly seemed to sense the seriousness and urgency of the situation either. "These so-called representatives of the people understand nothing," Monod told the press.

Monod figured that since the delegation from the Sorbonne failed to make an impression at the Assembly, then perhaps a public appeal to a higher authority from him and fellow Nobel laureates might be heard. That same evening of May 8, Monod telephoned Alfred Kastler, Nobel laureate in Physics, to ask him if he wished to join Monod, Jacob, Lwoff, and François Mauriac in addressing a telegram to President de Gaulle. Kastler agreed. The telegram read:

General de Gaulle, President of the République Palais de l'Élysée

WE URGE YOU TO MAKE A PERSONAL GESTURE TO APPEASE THE STUDENT REVOLT—STOP—AMNESTY FOR THE CONVICTED STUDENTS—REOPEN THE FACULTIES—STOP—DEEP RESPECT

SIGNED JACOB—KASTLER—LWOFF—MAURIAC—MONOD

Copies of the text were also sent to French press agencies, newspapers, and radio.

Having provoked no response, Monod sought other ways to get the government's attention and to prevent further violence. The students amassed again on the evening of May 10. At eight o'clock Monod tried to convince the rector of the Sorbonne to threaten the minister of education with his resignation should the government not accede to the students demands. The rector was hesitant.

The protestors then started building dozens of barricades all over the streets of the Latin Quarter. Monod pushed the rector again. All of Paris, indeed all of France, was following events by radio. Correspondents were all over the Latin Quarter. Monod suggested that the rector deliver an ultimatum to the minister over the radio. The rector could not bring himself to do so.

Students injured in skirmishes with the police sought refuge at the university. Monod realized that the rescue services could not get through the blocked streets. He called Agnes Ullmann and asked her to go to the hospital at the Pasteur, get first-aid supplies, and bring them to him at the Faculty of Science (at the Halle aux Vins) at the university.

Ullmann told Monod, "OK, but how can I get over there?" She

knew there was fighting on the boulevard Saint-Michel. "I don't know where my car can pass."

Monod got angry, the first time in her eight years in Paris when she had ever heard him angry. He told her, "Listen. If you did it in '56, you can do it now."

Ullmann was not very happy with her assignment, but she went to the hospital, got the supplies, and found a way through the streets. She got out of her car and made it past the students and to Monod. She stayed at the university through the night.

At a quarter past two in the morning of the eleventh, the police launched coordinated assaults on the barricades. Tear gas filled the streets. Cars were flipped over and set ablaze. Molotov cocktails were flying through the air. It was a full-scale riot.

Monod and Jacob telephoned Minister Peyrefitte. It was an exasperating conversation. Lacking any sense of urgency while the battle raged, the minister took his time to explain his position. Peyrefitte did not speak of the students, but of restoring law and order. He was clearly much more afraid of de Gaulle's reaction than of the riot itself. Monod cut him off at one point and told him, "Mr. Minister, your political career counts little next to the responsibilities you are assuming."

Monod joined the rebels on the barricades. He wanted to take a public stand on the side of the students. He phoned Alfred Kastler in the middle of the riot: "Are you in agreement with our declaring now that we are entirely on the side of the students?" Kastler agreed: "Entirely."

Monod went to see Peyrefitte to give him one last warning that he, Kastler, Jacob, and others would declare their loyalty to the protestors. The minister was unmoved. A short while later, among the barricades and injured students, Monod's declaration went out by radio: "In the present circumstances, it is not possible for men such as Kastler, Jacob, and I not to be near the students. The situation is tragic. The exits are blocked. It is necessary to leave a way out for the students and their injured."

Monod was horrified by the violence and desperate to see to it that the injured received treatment. When he heard that a young woman had been injured by tear gas in the area of the rue Gay-Lussac, Monod made his way to the street, found the student with bandages wrapped around her eyes, and took her by the hand. Early Saturday morning, newsmen caught the image of the Nobelist, still in his shirt and tie, escorting the student to safety.

Monod escorting a wounded student from the Sorbonne during the student unrest of May 1968. Monod played a leading role as an intermediary between the students and the government. (Archives of the Pasteur Institute)

FOUR HOURS OF battle left hundreds injured and the streets of the Latin Quarter strewn with burnt-out automobiles, broken glass, and smoldering piles of debris from the barricades.

Later that day, on the afternoon of the eleventh, Monod presided over a meeting of the Faculty of Sciences. He proposed a brief motion for consideration: "The undersigned professors declare that as of this afternoon, Mr. Alain Peyrefitte, minister of national education, no longer has their confidence." The motion passed with all but seven votes. It was the first such no-confidence vote in the history of the university.

That evening, Prime Minister Pompidou announced on national television that police actions would cease and that the Sorbonne would be reopened.

The concessions were a victory for the students, but the government's response was too little, too late. Unrest spread outside the university. In sympathy for the students, labor unions called a one-day strike and demonstration for the following Monday, May 13. On that afternoon, several hundred thousand workers and students marched from the Place de la République to Place Denfert-Rochereau. In the provinces, students demonstrated at France's nearly two dozen universities, and there were clashes with police in Nantes, Clermont Ferrand, and Le Mans. Revolt was in the air, and spreading quickly.

When the Sorbonne reopened, thousands of students occupied it and began a continuous stream of debates on the reform of the university. Monod was the only chair-holding professor present during the all-night session that began on May 13. One of the students proposed that Monod address the student body. It was a risky situation. The tenor had shifted among the students as more and more of them began to perceive the professoriate as part of the establishment against which they were rebelling, rather than as allies. Monod, with all of his prestige and titles, was suspect.

At three in the morning, Monod told the students that what they had made happen was "prodigious and exceeded the imagination." Then, with an uncharacteristic trembling in his voice, he said that it was now incumbent upon the students that they succeed. He added that he preferred that the first revolution occur within the university rather than there being an immediate general revolution apart from the university.

The students gave him a huge ovation.

Events outside the university, however, escalated. By midweek, strikes had spread to many industrial plants across the country. Then more workers walked off their jobs such that by the weekend the trains and subways were crippled, Air France canceled all flights, and mail delivery ceased. On Monday, May 20, up to ten million French workers, nearly two-thirds of the country's workforce, were on strike demanding higher wages, shorter workweeks, and better benefits. The economy ground to a halt. The student crisis had become a social crisis, and would soon be a political crisis that threatened to bring down de Gaulle's government and the Fifth Republic.

Throughout the demonstrations and riots, de Gaulle had remained absent and silent. He finally returned from a trip to Romania and tried to quell the unrest. But the general was himself perhaps the greatest symbol of the Old Guard—of the rigid, paternalistic French establishment. On May 24, he addressed the nation on television and radio. He admitted that the recent events had signaled a need for change in society, but he warned that change must be orderly: "Otherwise we will tumble through civil war to the most ruinous adventure and usurpations." He pledged to hold a referendum in June on some as-yet-undefined measures to reform the universities and to address worker issues.

De Gaulle's vague intentions not only failed to stem the unrest but also fueled another night of rioting across the city. The stock market was set ablaze, barricades were built at the Bastille, and the police responded again with tear gas and nightsticks. De Gaulle's failure raised serious doubt as to whether his government could survive the revolt.

Realizing its misstep, the government immediately agreed to large wage increases to appease the workers. Then de Gaulle made a much bolder move: he dissolved the National Assembly and called for elections in June.

As the days went by before the elections, the general public grew weary of the everyday disruptions and the violence. The rebellious students were losing the public's sympathy and patience. After a month of unrest, Monod and Jacob published an appeal in *Le Monde* in which they exhorted students to focus on the constructive challenges that lay before them. Monod penned the Camusian statement that served as the declaration's headline: "Si les barricades peuvent être parfois nécessaires, elles ne peuvent être permanentes" (If Barricades May Sometimes Be Necessary, They Cannot Be Permanent). Monod and Jacob urged the students:

> There comes a moment, and it has arrived, where courage is not found in the street, but in declaring opposition to attempts at destruction.
>
> That is why today we solemnly call upon all those teachers and students, faithful to the hopes of May, to express and organize through their continuous presence in faculties and schools, their will to construct together the new university and the future society, which will not emerge from a trash fire.

There would be no more barricades. Two days later, the police retook control of the Sorbonne and cleared out the students who had occupied it since May 13. The general strikes ended, and attention turned instead to the general elections. The French voters, shaken by the upheaval and seeking to restore some order, gave de Gaulle's right-wing party, Union pour la Défense de la République (Union for the Defense of the Republic), its first-ever absolute majority in the Assembly.

MONOD HEEDED HIS own advice to the students and teachers. He spent much of his time and energy in the months following the May uprising securing the reform of the university. Monod and a few colleagues had won the endorsement of the Faculty of Sciences for radical changes, and he continued to press for comprehensive, national action. Sweeping reforms of the universities were passed into law in November.

As for the making of a better future society, that prospect presented a dilemma to Monod. The visibility precipitated by the Nobel Prize, his public commitments and appearances, his activism during the May–June unrest, and his near daily presence in the newspapers during that time had made him a significant figure on the national political landscape. One possible path for pursuing his political commitments was to seek office.

Indeed, some of his longtime political friends, most of whom had also distinguished themselves in the Resistance, tried to persuade Monod that he should run for president in the spring 1969 election. With his scientific eminence, his war experience, his charisma, and his leadership qualities—it was not an impossible idea. But when Monod told Agnes Ullmann about it, she blurted out, "They are crazy!"

"No, no, they put it seriously," Monod replied. "Maybe I will think about it."

He did not run. But he had been thinking a great deal about what science could offer for the betterment of society. And *that* he was willing to campaign for.

CHAPTER 35

CHANCE AND NECESSITY: SISYPHUS RETURNS

A good writer possesses not only his own spirit but also the spirit of his friends.

—FRIEDRICH NIETZSCHE, *Human, All Too Human*

LONG BEFORE HE RECEIVED HIS NOBEL PRIZE, JACQUES MONOD had advocated that scientists bore special responsibilities. He once wrote in the *Bulletin of the Atomic Scientists*: "Whenever objectivity, truth, and justice are at stake, a scientist has the duty to form an opinion, and defend it." And for more than two decades Monod had acted on his convictions. In the Lysenko episode, no one was better positioned than scientists to expose Lysenko as a fraud and to alert society at large to the dangers of blind obedience to Soviet ideology. All that was required was a bit of courage—to face the ire of Communist colleagues—and objective scientific reasoning.

On the other hand, Monod rejected the common perception that a scientist had any duty to contribute to human comfort and happiness. Rather, he thought that the most important contributions of science were what often made humans *uncomfortable,* in what challenged their perceptions of themselves. He told an interviewer for the BBC: "There's always the tendency of the layman . . . of trying to strike from a fundamental scientist some statement about the applications of his work. This stems, I think, from a basic misconception as to the role of fundamental science, which exists in modern societies in particular: that the object of science is to be applied and create technology, when in fact technology and applications are by-products. I feel the most important results of science have been to change the relationship of man to the universe, or the way he sees himself in the universe."

Monod gave astronomy as an example: "The most unnecessary science of all, the one that off-hand you might think had absolutely no possible applications . . . is probably in one sense the most important of all sciences, in that it has taught more to men and modified their outlook more than any other science." Monod asserted, "If we still thought that we lived on a flat disc, created by some being living under the ground or up on a mountain, and that this was all the universe contained, we could *not* be what we are and we could *not* have evolved the societies which exist."

But, the interviewer acknowledged, there was public resistance to the most basic ideas in science. He asked Monod whether that was changing.

"I don't think so, and I think it's a great danger and it's a tragedy," Monod replied. "Science has molded our whole society, both by technology, of course, but even more by the creation of new ideas and new outlooks at the universe. The fact that this is not fully understood and recognized by the general public and the governments and the church and the universities and the philosophers is one of the causes of what we might call the neurosis of modern societies."

The interviewer challenged Monod, "Aren't you, as an eminent scientist with the Nobel Prize, in a strong position to try and change this, along with your colleagues? I mean, shouldn't scientists be doing more to change this?"

"I fully agree," Monod answered.

And indeed, that is exactly what Monod would attempt to do at every opportunity. And thanks to the Nobel Prize, many opportunities were forthcoming. In a series of prominent public lectures, articles, and finally, his book-length essay *Chance and Necessity* (1970), Monod sought to establish the new biology's place at the philosopher's table, as well as in the minds, if not the hearts, of thinking people.

Monod's inquiry into questions of human existence reflected the influence of and his debt to Camus at many turns. Monod's starting point was where Camus left off in *The Myth of Sisyphus*. In the latter,

Camus had characterized the human longing for understanding: "The mind's deepest desire, even in its most elaborate operations, parallels man's unconscious feeling in the face of his universe: it is an insistence upon familiarity, an appetite for clarity."

Twenty-seven years later, Monod echoed his late friend: "The urge, the anguish to understand the meaning of our own existence, the demand to rationalize and justify it within some coherent framework has been, and still is, one of the most powerful motivations of the human mind."

Monod's acknowledgment of Camus was even more explicit and literal. Both his article "On Values in the Age of Science" and *Chance and Necessity* were preceded by the same epigraph—the last two paragraphs from *The Myth of Sisyphus:*

> At that subtle moment when man glances backward over his life, Sisyphus returning toward his rock, in that slight pivoting he contemplates that series of unrelated actions which become his fate, created by him, combined under his memory's eye and soon sealed by his death. Thus, convinced of the wholly human origin of all that is human, a blind man eager to see who knows that the night has no end, he is still on the go. The rock is still rolling.
>
> I leave Sisyphus at the foot of the mountain! One always finds one's burden again. But Sisyphus teaches the higher fidelity that negates the gods and raises rocks. He too concludes that all is well. This universe henceforth without a master seems to him neither sterile nor futile. Each atom of that stone, each mineral flake of that night filled mountain, in itself forms a world. The struggle itself toward the heights is enough to fill a man's heart. One must imagine Sisyphus happy.

This is not to imply whatsoever that Monod had nothing original to offer beyond what Camus had already said in so many ways and works. Camus relied solely on philosophical reasoning, and developed his arguments in the context of previous generations of philosophers and writers—Jaspers, Heidegger, Kierkegaard, Dostoyevsky, Nietzsche, and Kafka, to name a few. Empirical science played no role in shaping the development of his thoughts.

Monod, on the other hand, began with new empirical scientific facts. Always the logician, he then explored just how far that logic could take him into the philosophical realm. Everything he presented about the biology of DNA was completely unknown to the world when Camus wrote *The Myth of Sisyphus* (whose appearance also preceded publication of Schrödinger's *What Is Life?*). Monod drew from that new science the most profound, logical, and unavoidable conclusion. It was the implications of that conclusion that led Monod both into territory well plowed by Camus—how we should respond to the gift of life—and into ground Camus had not touched but was of pressing concern in the wake of the rapid advancement of science since World War II, namely the role of science in shaping modern societies' values.

There were four essential points of which Monod set out to convince his audiences:

1. Biology has revealed that the emergence of humans is the result of chance, and therefore not a matter of any preordained plan.
2. All belief systems that are established on the latter notion are no longer tenable.
3. All ethics and value systems based on such traditional beliefs have no foundation, and create intolerable contradictions within modern societies.
4. Humans must decide how we should live and how we should act. A society that valued knowledge, creativity, and freedom above all would best serve human potential.

The Emergence of Humans Is the Result of Chance

It was Charles Darwin who first changed our concept of humans' place in the living world. Monod acknowledged that the theory of evolution had profoundly affected every domain of human thought—"philosophical, religious, and political." Yet, while the phenomenon of evolution had largely been accepted by the end of the nineteenth century, at least by the scientific community, Monod suggested that it "remained as if suspended, awaiting the elaboration of a *physical* theory of heredity." At the time Monod began his studies of enzyme adaptation, that theory seemed unattainable. "Thirty years ago, the hope that one would soon be forthcoming appeared almost illusory." But after the ensuing revolutionary advances—demonstrating that DNA was the hereditary material, solving the structure of DNA, revealing the logic of gene regulation, and cracking the genetic code—biology had a precise understanding of heredity in the form of the molecular theory of the genetic code.

"The 'secret of life' . . . has been laid bare," Monod declared. "This, a considerable event, ought certainly to make itself strongly felt in contemporary thinking, once the general significance and consequences of the theory are understood and appreciated beyond the narrow circle of specialists."

The general significance of the newly revealed secret of life for Monod's purposes was that it in turn revealed the precise, fundamental basis of evolution—and the interplay between what Monod called "chance and necessity." The phrase and the book's title came from Democritus's dictum that "everything existing in the universe is the fruit of chance and of necessity." It was now understood how variation in DNA, the source of biological change and diversity, arose by mutations in the sequences of bases in DNA. These mutations occur through "accidents"—unpredictable, random errors that occur in the copying of a single molecule of DNA. The errors thus arise by blind *chance,* without any relation to what their effects might be on an organism's function.

Only after the errors are copied through the DNA replication machinery and passed on to offspring is their relative *necessity* weighed at the level of organisms, by the non-random, competitive process of natural selection. A mutation may have a positive beneficial effect, a detrimental negative effect, or no effect at all on organisms' survival and reproduction. In Monod's poetic description, "Drawn out of the realm of pure chance, the accident enters into that of necessity, of the most implacable certainties."

There was no need, and indeed no evidence, in this process for divine intervention or special creation. The power of chance and necessity, of mutation and natural selection, was sufficient to generate and to explain all of the species, and all of the genetic diversity, on the planet, including the existence of humans.

It was now certain that humans depended not only on the same DNA chemistry as all other species but that molecular biology had just revealed that humans and all other species used the very same genetic code in utilizing the information in DNA. The differences in anatomy, physiology, and behavior between humans and other species were due to thousands to millions of changes in DNA accumulated over time.

The most profound consequence from this new understanding of DNA, mutation, and the genetic code was the inescapable, logical certainty that "man was the product of an incalculable number of fortuitous events," that humans had thus emerged through a process dependent upon chance. "The emergence of Man can only be conceived as the result of a huge Monte-Carlo game, where our number eventually did come out, when it might not well have appeared," Monod wrote. "And, in any case, the unfathomable cosmos around us could not have cared less."

Monod admitted that, of course, "this fundamental scientific result is also the most unacceptable" to most people, as it overturns all previous, long-cherished notions of humans' special significance in the universe.

Molecular biology had brought Monod full circle to Camus's territory of the absurd condition—that contradiction between the human longing for meaning and the universe's silence. Monod asked what man should do in the face of his fortuitous existence: "Should he despair? Or reject the science that imposes such conceptions upon us? The despair of the man convinced of being absurd and refusing to be that: the theme that has nourished many of the greatest contemporary works."

He was referring to Camus, of course, who wrote in *The Myth of Sisyphus*: "We must despair of ever reconstructing the familiar, calm surface which would give us peace of heart."

Monod offered his scientific slant on the absurd with a quote from a Scottish philosopher named McGregor: "Each conquest of Science is a victory of the absurd."

It was a pithy remark, but there was in fact no such philosopher. Monod made use of quotes from many luminaries—Heidegger, Pascal, Comte, Nietzsche, Kant, as well as Camus and Democritus—in the course of his writing. But when he did not have an apt quote, he simply made one up and attributed it to "McGregor." It was his Scottish-American mother's maiden name.

All Belief Systems That Are Founded on a Special Place or Purpose of Man in Nature Are No Longer Tenable

For Monod, the philosophical consequences of molecular biology followed from the role of chance in the emergence of humans, and the challenge that presented to all traditional belief systems. Monod explained, "In virtually all the mythic, religious or philosophic systems, Man's existence receives its meaning from being supposed part of some general purpose which accounts for the whole of nature and creation. The 'purpose' may be naively ascribed to a mythic founder-hero, or more grandiosely (albeit less poetically) to some abstract divine intervention; or it may be assumed that the 'laws of nature' are such that the universe in its evolution, could not fail to produce Man and history."

The common flaw in all of these systems, Monod underscored, is that they assume "between Man and the Universe, between Cosmology and History an unbroken continuity, a profound immanent alliance." However, Monod argued, "the scientific approach reveals to Man that he is an accident, almost a stranger in the universe, and reduces the 'old alliance' between him and the rest of creation to a tenuous and fragile thread."

Moreover, Monod asserted that molecular biology had snapped the last thread: "It remained for modern Biology . . . blossoming into Molecular Biology, to discover the ultimate source of stability and evolution in the Biosphere [DNA and mutation], and thus blow to shreds the myth of the old alliance."

As a result, Monod asserted, "none of the gracious or frightening myths that [man] had dreamed, none of the hopes that he had tenaciously entertained, none of the certainties that had formed the structure of his moral and social life for thousands of years, can stand anymore."

All Ethics and Value Systems Based on Such Traditional Beliefs Have No Foundation, and Create Intolerable Contradictions Within Modern Societies

Pressing to follow his logic as far as it could take him, Monod asked what the implications were of this loss of certainty, of the traditional systems that had guided human societies for millennia.

All traditional systems teach values, duties, rights, and prohibitions based on various claimed sources—historical, divine, or natural. Monod recognized that the important psychological function of these teachings was to satisfy individuals' longing for meaning, while their social function was to provide stability. Take away those sources, as Monod argued modern science had done, and both individuals' and society's foundations were undermined.

Monod believed that the greatest threat to society was not science itself, with all of its technological powers; rather, it was the continuing embrace of traditional systems alongside the practice of modern science. In his debut interview in *Le Nouvel Observateur,* he said, "Society by definition does not like things being put into question, it is not for nothing that the societies who wanted the best seats, in their faith in themselves have always had difficulties with science." He cited as evidence: "The Church with Galileo, Stalin with Lysenko, Hitler with 'Jewish science' as they referred to relativity," and the condemnation of Darwin by biblical literalists in the United States.

Monod declared, "Modern societies had accepted the treasures and the power that science laid in their laps. But they have not accepted—they have scarcely heard—its profounder message: the defining of a new and unique source of truth." That source is the objective knowledge provided by the scientific method. Rather than abandon their traditional sources of knowledge and values, Monod lamented, "our societies are still trying to live by and to teach systems of values already blasted at the root by science itself." Monod accused the Western, liberal-capitalist countries of still teaching (and preaching) "a nauseating mixture of Judeo-Christian religiosity, 'Natural' Human Rights, pedestrian utilitarianism and 19th Century progressism" while "the Marxist countries still throw up a stupefying smokescreen of nonsensical Historicism and Dialectical materialism."

"They all lie and they know it," Monod wrote. "No intelligent and cultivated person, in any of these societies can really believe in the validity of these dogma."

And yet, he acknowledged, "no society can survive without a moral code based on values understood, accepted, respected by the majority of its members." The outstanding question, then, was: After having banished all traditional sources, from where could or should those values come?

Humans Must Decide How We Should Live and How We Should Act

"Man must wake out of his millenary dream . . . wake to his solitude, his fundamental isolation," Monod urged. "Now does he at last realize that, like a gypsy, he lives on the boundary of an alien universe. A universe that is deaf to his music, just as indifferent to his hopes as it is to his suffering or his crimes."

Through the facts of molecular biology and his Cartesian logic, Monod had arrived at the same junction Camus had reached through his philosophical journey three decades earlier, when the latter wrote: "A world that can be explained even with bad reasons is a familiar world. But, on the other hand, in a universe suddenly devoid of illusion and lights, man feels an alien, a stranger."

And Monod therefore probed the same question as Camus—of how to live in the face of this knowledge. Camus's reply was given by Meursault in *The Stranger,* who laid his "heart open to the benign indifference of the universe," and by Sisyphus, to whom "this universe without a master seems . . . neither sterile nor futile." Monod argued that, in a scientifically enlightened world, man must realize that there is no external source of meaning or values, "that he alone creates, defines, and shapes them."

And which values, then, should humans choose?

Monod proposed a "supreme value" that combined Camus's view of the role of art with his own vision of the role of science—the dual pursuit of creation and knowledge.

Camus had written, "Of all the schools of patience and lucidity, creation is the most effective. It is also the staggering evidence of man's sole dignity: the dogged revolt against his condition, perseverance in an effort considered sterile." Moreover, he claimed that "the absurd joy par excellence is creation" and that "authentic creation is a gift to the future." Monod embraced Camus's maxims, and held that they applied equally well to scientific creation and the pursuit of objective knowledge. Monod wrote: "And what other ultimate values to choose then, than those creations, born from Men, yet transcending those creators, as existing in the Kingdom of ideas, richer and wider in content than any single man or all men at any one time can perceive? I mean of course the great, ever unfinished, monument of creation and knowledge, that is of Art and Science?"

Monod continued: "A society that would accept these transcendent values as the ultimate standard of all, more immediate human values, and designed itself deliberately to serve them, would have to defend intellectual, political, and economic freedom; to foster education . . . as its primary task" in order to progress toward more freedom, creativity, and knowledge.

"A utopia. Perhaps," Monod admitted in closing his book. "But it is not an incoherent dream. It is an idea that owes its force to its logical coherence alone. It is the conclusion to which the search for authenticity leads. The ancient covenant is in pieces; man knows at last that he is alone in the universe's unfeeling immensity, out of which he emerged only by chance. His destiny is nowhere spelled out, nor is his duty. The Kingdom above or the darkness below: it is for him to choose."

A Surprise Bestseller

Le Hasard et La Necessité appeared in France in October 1970 with the public reception of his essay very much in doubt. Who would read a book whose middle five chapters were laden with technical description, one that needed an appendix with chemical diagrams to boot? And who was eager to hear his message that all religious systems were imaginary constructions? What's more, Monod had taken a considerable risk to his reputation by straying onto the turf that some philosophers and most theologians would no doubt feel was reserved for themselves.

The book was a sensation.

It struck chords, and nerves, across France and quickly became a bestseller. Almost 200,000 copies were sold in the first year, and for many weeks *Chance and Necessity* trailed only the French translation of Erich Segal's *Love Story* at the top of book sales. Foreign translations quickly followed, and the book was a bestseller in Germany and Japan as well.

The book garnered scores of reviews in France alone, as well as reviews of its English translations in *The Economist, Atlantic Monthly, LIFE,* and *Newsweek.* Monod received two reviews in the *New York Times,* which also published two interviews with him.

The critics ran the gamut from condemnation to high praise. Few were neutral. Not surprisingly, various clergy and church representatives protested, as did the Communists in *L'Humanité.* Most writers were ill-equipped to understand, let alone to debate, Monod's core scientific argument about the role of chance in humans' biological origins. Most criticism therefore focused on denying the implications for religious thought, and on questioning Monod's authority and reasoning on ethical matters.

Less partisan writers, however, lauded the book. Said one reviewer in *Le Figaro,* "I read Jacques Monod with the same passionate and anxious attention as I would a book brought from a faraway star and from which the author was, for millions of years, the clear and careful witness of the birth of life and of humanity on our Earth." The same reviewer, though he did not agree with all of Monod's conclusions, praised "the depth of his sensibility, the lucidity of his experience, and the force of his courage." Another reviewer wrote, "Even the most cautious reviewer, however nervous of debasing the currency of adjectives, must sometimes use the word great. This is a great book, sinewy, lucid, and intelligible."

The success of the book and the extensive press coverage created great demand for radio, television, and newspaper interviews, which gave Monod the opportunity to clarify his positions, to extend his arguments, and to acknowledge again Camus's influence on the larger message of his essay. One common lament was the perceived pessimistic message of the book, that Monod's vision was too cold, too austere. One interviewer asked Monod: "One could ask oneself effectively why live, why have children only in order to condemn them to death, why this thirst for knowledge?"

"On that, I do not have an original response to this question," Monod answered. "I will answer you with the citation that I put at the beginning of my book and that I will ask you to read because I meet Camus exactly on this text, which is admirable besides."

The interviewer read Monod's epigraph from *The Myth of Sisyphus*. He then said, "This text is admirable but it does not provide what I would call the pedagogical means to arrive at serenity."

Monod countered, "Myself, I do believe that it does, I do believe that it does; I believe that it provides the only means." He continued, "I believe that it provides that so long as one demonstrates the grandeur of this idea. I believe that man has a need above all, he has a need for transcendence. And there is, in the existential attitude of the kind that Camus has offered to him, a deliberate transcendence . . . that I find so beautiful that it ought to be acceptable."

Last Words

The success of *Chance and Necessity* surprised no one more than Monod himself. However, he was not able to enjoy it fully. A few months after publication, Odette was diagnosed with terminal cancer. At the time of her illness, Monod had been contemplating taking a sabbatical abroad, or even going on a long sailing voyage. Instead, he took on a new challenge that would keep him near home and able to support Odette—he became the eighth general director of the Pasteur Institute. Monod's parents had once speculated whether their talented, promising son would become the next Beethoven or the next Pasteur. He had now joined a short, direct line that reached back directly to the great scientist.

When the Institute celebrated the founder's 150th birthday in 1973, the honor fell to Director Monod to offer a few words about Pasteur. The celebrated scientist, author, statesman, and philosopher could have been speaking of himself when he said:

> Where does genius come from? Often we are contented with attributing it to a unique, exceptional, and mysterious resource of mind. On the contrary, in the case of Pasteur, we see clearly that the power of his genius comes from multiple sources, very much in opposition to his intelligence, character, and temperament. He was an artist and a dreamer. He would allow himself to be fascinated by mirages of an imagination that always tended to go beyond the horizons of knowledge. He was ambitious and dominating and would be satisfied only with real and complete victories. He was rigorous and demanding towards himself. At the same time he would spare no efforts to be severe and disciplined.

The years of his directorship, however, were difficult. Odette died just eleven months after her diagnosis, and the following year Monod became severely ill with viral hepatitis. He recovered, but his own inclination to be demanding and disciplined led to considerable contention. Monod had accepted the leadership of the Pasteur out of a great sense of duty and debt to the institution, which had allowed him to pursue a style of science that was not possible elsewhere in France. The Institute was facing severe fiscal challenges that threatened its survival. Monod hurled himself into the task of saving the Pasteur, which led him to instigate major operational and structural changes that displeased several formerly close colleagues.

The burden of administration forced Monod away from his own science, and he missed the close rapport he had with his scientific team. A couple of years before the end of his six-year term, Monod decided that he would not seek reappointment as director. He wanted to return to creative pursuits. He told Agnes Ullmann, still at the Pasteur fifteen years after her escape from Hungary, that he was thinking of writing a new book entitled *Man and Time* (*L'Homme et Le Temps*).

In late 1975, however, he developed aplastic anemia, a condition in which the body fails to replenish red blood cells. His prognosis was dire. Although easily fatigued, he continued his duties as director, sustained by frequent blood transfusions.

Despite all of the competing demands on his time, and his illness, Monod found moments to reply to correspondence that came from many different quarters. In January 1976, a handwritten letter arrived from Grenoble:

> *Sir,*
>
> *I am a thirteen year-old boy who is very interested in research. I know that you are one of the greatest researchers in the world (our professor of science told us so).*
>
> *Excuse me for bothering you, but I would like to know what maxim guides your life. Perhaps I could apply that when I grow up.*
>
> *Could you also send me a signed photo of yourself so that I can display it in my bedroom . . .*
>
> *Good-bye M. Monod and Happy New Year 1976.*
>
> *Bruno*

Monod replied:

> *My Dear Bruno,*
>
> *Thank you very much for your letter that has interested me a great deal. However, you have posed a difficult question because I do not think that one can find one maxim that, alone, allows one to conduct an entire life and to govern the sometimes painful choices with which one can be confronted.*
>
> *All I can tell you are the qualities that appear most important to me. If one were to pose this question to me, I would reply*

without a doubt that they are: courage, as much moral as physical, as well as the love of truth, or rather, the hatred of lies.

I prefer to speak of the hatred of lies rather than the love of truth, since one is never sure of holding the truth, whereas with lies, one is almost always able to detect them, to discover them, and to denounce them.

As you requested, I am sending you a signed photo.

Thank you for the New Year's wishes. Happy New Year and good wishes to you.

Jacques Monod

The disease symptoms waxed and waned. Feeling better in May 1976, Monod went to Cannes to spend a weekend relaxing at the Monod family home with friends. On Thursday, May 27, he went to a reception at the famed film festival with author Jerzy Kosiński, whose book *Being There* was published at the same time as *Chance and Necessity*, and whose lead character was named Chance. Monod, in black tie and light-colored jacket, was his usual elegant and charming self. He enjoyed drinking, smoking, and joking with the other party guests.

The next day, Friday, Monod chatted on the phone with Ullmann in Paris about his upcoming travel plans to a meeting in the States and a research paper they were working on together. They would continue their conversation when Monod returned to the Pasteur after the weekend. "See you Monday in the lab," he said.

Ullmann was entertaining dinner guests at home Sunday night when the phone rang. It was Philo calling from Cannes; he told her that Jacques had been hospitalized and that his condition was grave. She would need to come quickly.

Ullmann, Erdös, and Monod's sons, Olivier and Philippe, raced to Cannes. Olivier and Philippe took the overnight train; Ullmann flew early the next morning.

AT THE HOSPITAL in Cannes, Monod's cardiologist and Philo were keeping a close watch as Monod battled for his life.

Philo heard his brother say very faintly between breaths, "Odette . . . Pasteur . . ."

Then, after a pause, "*Je cherche à comprendre*" (I am trying to understand).

He never regained consciousness.

After his funeral, a few rough pages of *Man and Time* were found on his desk.

APPENDIX: **THE SCIENCE**

Monod's and Jacob's critical observations, discoveries, and ideas are spread over many pages in this book. Since most readers are not biologists, I thought it might be helpful to provide a short summary of the relevant science that could be consulted to refresh certain points. This information is divided into (1) what was known about genes, DNA, and proteins as they began their work; (2) Monod's and Jacob's discoveries that are covered in the book; and (3) important subsequent, related discoveries.

GENES, DNA, AND PROTEINS

Chromosomes and Genes

From the pioneering work by Thomas Hunt Morgan (Nobel Prize 1933), it was known that chromosomes are the physical entities responsible for heredity, and that specific genes reside at specific positions along chromosomes. Oswald Avery's studies (1944–1946) demonstrated that deoxyribonucleic acid (DNA) is the chemical component of the chromosome responsible for inheritance. Each chromosome contains a long molecule of DNA.

DNA

James D. Watson and Francis Crick deciphered the structure of DNA in 1953. DNA molecules in cells are made of two strands of four distinct *bases*. These chemical building blocks are denoted by the single letters A, C, G, and T. The strands of DNA are held together by strong chemical bonds between pairs of bases that lie on opposite strands—A always pairs with T, C always pairs with G—as shown below:

Solving the structure of DNA immediately revealed how the two fundamental processes of inheritance and mutation worked at a molecular level. That is, a DNA sequence could be faithfully copied and passed on because the base present at each position on one strand determined its complement on the other strand. Mutations result from errors in this copying process, where the wrong base or an extra base(s) gets inserted, or a base(s) may be deleted, generating a change in the DNA sequence.

Proteins

Proteins are the molecules that do all of the work in cells, breaking down nutrients, assembling cellular components, copying DNA, and so on. Proteins are made up of chains of building blocks called *amino acids.* There are twenty different amino acids. The chemical properties of these amino acids, when assembled into long chains averaging about 400 amino acids in length, determine the unique activity of each protein. Enzymes are proteins that catalyze specific chemical reactions.

At the time of Monod's and Jacob's seminal work, the relationship between the sequence of bases in DNA and the sequence of amino acids in proteins was not understood. Crick asserted that the main function of DNA was to encode proteins, but how that information was decoded and the nature of the "genetic code" were unknown (Crick's black box) until the early 1960s.

MONOD'S AND JACOB'S DISCOVERIES

Diauxy

The observations that started Monod on the path to fundamental insights and the Nobel Prize concerned the growth of bacteria in the presence of simple sugars. When bacteria are grown in the presence of a single sugar, such as glucose, they grow exponentially until the sugar is exhausted. But Monod noticed that when bacteria were grown in the presence of certain combinations of two sugars—glucose and lactose, for example—they grew exponentially, then paused briefly before resuming exponential growth. He called this phenomenon "double growth" or "diauxy." By shifting the relative ratios of the sugars, he found that he could shift the relative length of each part of the double growth curve. From that observation, he deduced that the bacteria were using up one sugar before utilizing the second, less-preferred sugar.

Enzyme adaptation and enzyme induction

The time lag in bacterial growth on a less-preferred sugar was interpreted as an example of "enzyme adaptation," in which growth is delayed briefly until the enzyme required to break down a nutrient appears.

Monod specifically focused on the control of lactose metabolism. Lactose is a disaccharide made up of the two monosaccharides glucose and galactose. Lactose itself cannot be used as an energy source in *E. coli* bacteria; it must be broken down into glucose and galactose, which is the function of the enzyme ß-galactosidase. Importantly, the enzyme is normally not produced in the absence of lactose, but appears when lactose is the sole energy source provided to the bacteria (a case of enzyme adaptation). Since the appearance of the enzyme occurred in the presence of the sugar it broke down, Monod and others renamed the phenomenon "enzyme induction," and substances that were able to elicit the enzyme were dubbed "inducers."

The lactose operon

How a simple bacterium "knew" which sugar to use, and when to produce ß-galactosidase—were the crux of the mystery Monod set out to solve.

Crucial to his progress was his decision to tackle the problem using genetics. During the war, Monod and Alice Audereau found that strains of *E. coli* bacteria that were unable to utilize lactose occasionally gave rise to colonies that *could* grow on lactose, and that these colonies were due to genetic mutations. This was crucial evidence that the ability to metabolize lactose was genetically determined. Therefore, components of the enzyme-induction process could be identified through mutations.

In order to figure out where these mutations occurred on the *E. coli* chromosome, and to sort out how they affected enzyme induction,

Monod teamed up with Jacob in 1957. Jacob had been studying a bacterial virus, or bacteriophage, called *lambda*. Jacob was a pioneer in developing methods for mapping bacterial genes. He was particularly interested in the phenomenon of *lysogeny*, whereby a virus hides out in a bacterial host but can be induced to emerge by certain treatments. Jacob proposed that enzyme induction and virus induction were analogous—that in each case a repressor kept genes off unless an inducer was present.

From 1957 to 1960, working with key collaborators such as Arthur Pardee, Monica Riley, and Sydney Brenner, Monod and Jacob identified several components involved in the control of lactose metabolism in *E. coli*, and coined several general terms that have remained in use to this day:

Structural gene, which encodes the structure of a protein, such as an enzyme
Regulatory gene, which governs the expression of a structural gene
Repressor, which turns off enzyme production, such as the protein encoded by the *i* gene
Operator, the acceptor site for the repressor on DNA
Operon, a set of structural genes controlled by a common operator and repressor, and usually involved in the same biochemical pathway
Messenger RNA, an intermediate that carries information from genes in DNA to the ribosome for the synthesis of specific proteins.

Monod and Jacob figured out the logic of the genetic switch by deciphering what happened in certain mutants. For example, mutations in the repressor gene *i* disabled repression, but not if a second, normal, copy of the gene was present in the cell. Mutations in the operator (o) also disabled repression, regardless of whether another operator was present. Mutations in specific structural genes blocked the production of specific enzymes.

From these observations, they were then able to construct a general picture of the process of gene regulation that looked like the following:

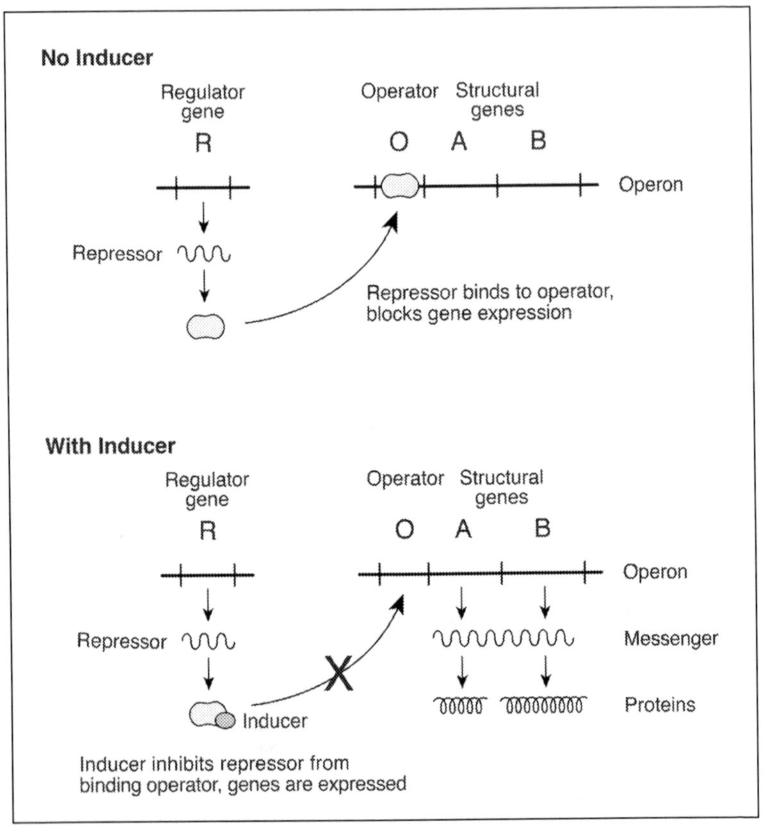

Monod-Jacob General Model of Gene Regulation. The key to the genetic switch is the interaction of the inducer and repressor. Top, when no inducer is present, the repressor binds to and occupies the operator that lies adjacent to the structural genes A and B, keeping them switched off. Bottom, when an inducer is present, the repressor is inhibited from binding to the operator, allowing genes A and B to be switched on, leading to the production of messenger RNA that carries information for the synthesis of proteins A and B. The figure is modified from Jacob and Monod's synthesis published in 1961. (Drawing by Leanne Olds)

The diagram, a version of which appeared in their landmark paper, depicts how the operon is regulated in the absence or in the presence of an inducer, such as the sugar lactose. The key to the logic of this genetic on/off switch is the interaction of the inducer with the repressor. When no inducer is present, the repressor binds to the operator and keeps the genes of the operon turned off. When the inducer is present, it inhibits binding of the repressor to the operator, and the genes of the operon are switched on. In this way, for example, certain enzymes are made only when a certain nutrient is available.

SUBSEQUENT DISCOVERIES

The repressor protein, a critical component of the lactose regulatory system, was not isolated until 1966. Walter Gilbert and Benno Müller-Hill of Harvard University achieved that coup, and demonstrated that the repressor bound specifically to the operator sequence of the lactose operon. Moreover, the binding of the repressor to DNA was inhibited by inducers. The latter was an example of *allostery,* in which the binding of a molecule alters the shape and activity of a protein.

Allostery was yet another powerful idea conceived and coined by Monod. When he realized that the activities of some proteins, such as repressors, might be regulated by the binding of substances, he declared, "I have discovered the second secret of life!" Indeed, allostery underlies much biological regulation, such as how hormones regulate physiology.

The repressor for bacteriophage *lambda* was also isolated in 1966, by Mark Ptashne—also at Harvard. Ptashne demonstrated that *lambda* repressor bound specifically to the operator region of the bacteriophage DNA. The similarities between the two systems of regulation were in fact as great as Jacob had imagined years earlier.

The elucidation of the lactose and *lambda* phage regulatory systems provided critical tools for the early days of molecular genetics, in both basic and applied research. Many of the first practical advances in genetic engineering relied on knowledge developed from study of these two systems.

The general principle revealed by study of the lactose operon and bacteriophage was that genes were switched on and off by the action of proteins that bound specifically to DNA sequences near genes (operators). While the details are a bit different in more complex organisms, the principle is the same from *E. coli* to elephants to humans. As Monod anticipated in 1947, and as Jacob and Monod explicitly argued in their masterful synthesis published in 1961, the development of complex creatures and the differentiation of the many kinds of cell types in animal bodies are orchestrated by the turning on and off of different sets of genes. Most of that action is controlled through the binding of regulatory proteins to specific DNA sequences around genes.

Acknowledgments

How does a middle-aged American biologist wind up writing a story about world-turning events and remarkable people in France more than seventy years ago?

That is a bit of a long story, although I promise not nearly as long as the book!

In explaining how he selected topics for books, the great biographer David McCullough admitted, "I sometimes get the feeling that the subject picks me." I recognize that feeling here. My preparation for the book was a matter of chance. Writing it eventually became one of necessity.

The seeds were planted a very long time ago. In 1977, Simon Silver, my freshman adviser at Washington University in St. Louis, slid a book on the operon in front of me. I understood later that he was testing me, to see whether I was curious enough to do some extracurricular reading. I was. Simon later offered Zhores Medvedev's book on Lysenko to me, and told me about the treatment of dissident scientists in the Soviet Union.

At this same time, I had enrolled in a French class. The university required that, in order to receive Advanced Placement credit in a given subject, students had to pass the next, higher-level course in that subject. I signed up for one course, with no intention of taking any more French—ever. Professor James Jones's talent and charisma, however, changed my plans. I pursued (and almost completed) a second major in French that introduced me to many great thinkers from Diderot and Voltaire to, of course, Sartre and Camus. That endeavor also gave me some language skills that, even though badly rusted, later turned out to be very handy.

To complete the trifecta of my accidental preparation for this book, it also happens that I am a lifelong World War II history buff. I had read scores of books on the conflict and visited many key battlefields and museums without the slimmest notion of ever writing about the subject.

These ingredients simmered in the background for more than thirty years as I pursued my career as a biologist. I have admired Jacob and Monod's scientific work since my graduate school days, and my own research on the genetic control of animal development and evolution has in some ways followed in their footsteps. But when I read brief mentions that Monod was "in the Resistance" and that he was "friends with Albert Camus," or that Jacob was nearly killed in Normandy, I became curious to know what was left untold. How did their war experiences shape their future work and worldview? What did being in the Resistance entail? What was daily life like for Monod during the Occupation of France? Over many years, the questions expanded: What drew Monod and Camus together? How close was their friendship, and what did it mean to each man? What common ground did the writer-philosopher and the scientist find?

I had to know more.

Eventually, I was able to track down and meet some of the people who could tell me the answers, and so began one of the most exciting and gratifying adventures of my life. Indeed, I could not have told any of this story without the generous assistance and, most important, the trust of several extraordinary individuals who appear in it. In particular, I am indebted to Agnes Ullmann for sharing the story of her daring escape from Hungary, and for putting me in touch with many people connected to Monod and Jacob; to Geneviève Noufflard for permitting me to relate stories from her unpublished memoir of the war years, and for sharing numerous original documents from the Resistance; and to Olivier and Philippe Monod for giving me access to troves of private family letters, documents, and photos, for helping me understand their significance, and for sharing the story of their remarkable father and family. It has been a great privilege to meet such extraordinary people, and an honor to be welcomed into their homes.

Many others in France have also provided vital assistance. The exceptional dedication and efforts of the late Madeleine Brunerie, Monod's longtime secretary, has preserved decades of invaluable documents at the Archives of the Institute Pasteur. There, I benefited enormously from the generous assistance of Dominique Dupenne and Daniel Demellier of the Service des Archives de L'Institut Pasteur, who went far out of their way to accommodate me during my visits and to help locate large numbers of critical dossiers.

I also thank Marcelle Mahasela of the Fonds Albert Camus, Bibliotheque Méjanes, Aix-en-Provence, for her assistance in obtaining materials, and Catherine Camus for her permission to quote several original letters and numerous passages of her father's writings.

Special thanks to Isabelle Tarisca of the Cabinet du Préfet of the Préfecture de Police in Paris for her assistance in locating police files related to the Nordmann case, as well as Jacques Monod's participation in the Resistance in 1940. I also thank Liora Israel and Julien Blanc for correspondence relating to the existence of these files. Thanks also to Severine Maréchal and the staff of Le Centre de Documentation et de Recherche du Memorial du Maréchal Leclerc–Musée Jean Moulin in Paris for generous access to their collections.

Many individuals generously agreed to be interviewed for the book or responded to written queries. Many thanks to the late Tamás Erdös, Madeleine Brunerie, Françoise Benhamou, Yves-Marc Achenbaum, Melvin Cohn, Georges Cohen, Donald Brown, David Hogness, Stuart Edelstein, and Arthur Pardee for their generosity.

I also had the benefit of interviews conducted long ago. I thank Olivia and Nicholas Judson for graciously granting access to transcripts of interviews produced by their late father, Horace Freeland Judson, in the course of his writing *The Eighth Day of Creation*, the definitive history of the early history of molecular biology. I thank Charles Greifenstein, associate librarian and curator of manuscripts at the American Philosophical Society in Philadelphia, for providing the files.

Closer to home, I am deeply indebted to several individuals who provided crucial assistance to the project. Dr. Héloïse Dufour, a member of my laboratory at the University of Wisconsin, was a guide, translator, interpreter, and researcher throughout the development of the book. Her assistance in navigating Paris and its institutions, conducting or assisting with interviews and meetings, initiating correspondence, unearthing information in various archives, and deciphering key documents has been immeasurable. Her knowledge of French history, culture, and customs helped to open doors to many rewards.

Dr. Benjamin Prud'homme, a former member of my laboratory, catalyzed the launch of this project by locating Geneviève Noufflard and conducted a pivotal initial interview with Mme. Noufflard.

Throughout the preparation of the book, I have been assisted by Megan Marsh-McGlone, who tracked down countless books and articles, shouldered the Herculean task of curating the bibliography and sources in the book, as well as securing permissions for illustrations and the quoting of copyrighted material. And thanks to Leanne Olds for preparing the illustrations.

I have been very fortunate to have the attentive guidance and encouragement of my agent, Russ Galen, throughout the conception, development, and writing of the book. Special thanks to my editor at Crown, Domenica Alioto, for championing the book, and for her thoughtful input throughout the editing process. I am also especially grateful to Héloïse Dufour, Megan Marsh-McGlone, Steve Paddock, and Jim Carroll, who provided detailed feedback on earlier drafts.

This book would not have been undertaken without the unwavering support, encouragement, and understanding of my wife, Jamie Carroll. Jamie not only put up with my ramblings but also read every word of the first draft of the book, identified the thickets and brambles, and made countless great suggestions. No amount of Parisian chocolate can express my love and gratitude.

Notes

Documents obtained from the archives of the Pasteur Institute Paris are documented as follows: item, date, location, Fonds, SAIP.

All archival letters originally in French are translated by either Sean B. Carroll (SBC), Héloïse Dufour, or both.

Interviews were conducted by the author unless otherwise noted.

Prologue: Chance, Necessity, and Genius

1 **On October 16, 1957:** Todd (1997), 371.
1 **"One wonders whether":** Lottman (1979), 601.
2 **"My dear Monod":** Letter, A. Camus to J. Monod, 11/18/1957, courtesy of Olivier Monod.
2 **"I have known only one":** Brunerie (2008), 162.
4 **"Frenchmen, the French Resistance":** Camus (2006), 9; *Combat* 58, July 1944.
5 **"at least share":** Camus (2006), 2; *Combat* 55, March 1944.
6 **"Four years ago":** Camus (2006), 17; *Combat,* August 25, 1944.
6 **"To risk one's life":** Bernard (1967), 173.
7 **"those accidents which":** Aronson (2004), 36.
7 **the talk of Paris:** Ibid., 46.
7 **"judging whether life":** Camus (1991a), 3.
7 **"Being aware of one's life":** Ibid., 62–63.
7 **"The first teaches him":** Ibid., 66.
8 **"the struggle towards":** Ibid., 123.
8 **"One must imagine":** Ibid.
8 **"In the depths of winter":** Ibid., 202.
8 **"between hell and reason":** Camus (2006), 237.
8 **"Camus taught me":** "Roger Grenier: Camus m'a appris des raisons de vivre," NouvelObs.co, January 4, 2010.
8 **"the most elevating form":** Jacob (1988), 274.
8 **"an admirable conjunction":** Aronson (2004), 37.
9 **"made things that were":** Judson (1979), 22.
9 **"taste" and "elegance":** Cohn, as quoted in Ullmann (2003), 93.
11 **"ideological terrorism":** Medvedev (1971), Monod preface.
11 **"make his life's goal":** Cohn, as quoted in Ullmann (2003), x.

11 **"Never lacking in courage":** Crick, as quoted in Ullmann (2003), 23.
12 **"The urge, the anguish":** Monod (1969), 19.

Chapter 1: City of Light

17 **well below freezing temperatures:** *Le Matin*, January 1, 1940; *Le Figaro*, January 1, 1940.
19 **"extremely powerful bombs":** Letter, A. Einstein to F. D. Roosevelt, August 2, 1939, Manhattan Project Heritage Preservation Association, Inc. Available at http://www.mphpa.org/classic/COLLECTIONS/MP-Einstein~Sachs/Pages/Einstein-Sachs-001.htm.
20 **"Par precaution":** Song: "Paris Sera Toujours Paris." Words by Albert Willemetz, music by C. Oberfeld, 1939.
23 **"I return with":** Shirer (1969), 403.
23 **"There is not a woman":** Ibid., 404.
24 **"We can never deal":** May (2000), 182.
24 **"There is nothing more":** Ibid., 187.
24 **"We should not devote":** Ibid., 187.
25 **"War has been imposed":** *L'Intransigeant,* September 2, 1939, cited in Shamir (1976).
25 **"The Nazis have compelled":** *Le Populaire,* September 4, 1939, cited in Shamir (1976).
25 **"the modern Attila":** *La Croix,* September 3, 5, and 6, 1939, cited in Shamir (1976).
25 **"brilliant attack":** *Le Figaro,* September 10, 1939.
27 **"Throughout this night":** *Le Figaro,* January 1, 1940.

Chapter 2: Plans

28 **"Since France, the deadly enemy":** Hitler (1939), 14.
28 **"The year 1939 was so dramatic":** Goebbels (1939).
29 **"On September 2":** Ibid.
29 **"It would be a mistake":** Ibid.
29 **"harangue":** *Le Matin,* January 1, 1940.
31 **"How can anyone believe":** Shirer (1969), 186.
31 **"must go into Belgium":** Ibid., 185.
32 **"It is impenetrable":** Ibid.
33 **"the whole operation":** Ibid., 581.
33 **"The enemy would take":** Ibid., 555; De Gaulle (1964), 29.
34 **"At the end of":** McIntire and Burns (2008), 339.
37 **The *Pourquoi-Pas?* sank:** Debré (1996), 71.
37 **"There will be no war":** Letter, Jacques Monod to his father and mother, August 31, 1939, private archives, Monod family.
38 **"I would like to raise":** Ibid.

38 **He wanted to serve:** Letter, Jacques Monod to Odette Monod, January 15, 1940, private archives, Monod family.

38 **So, rather than waiting:** Letter, Odette Monod to Lucien and Charlotte Monod, January 31, 1940, private archives, Monod family.

38 **If he was accepted:** Letter, Jacques Monod to Odette Monod, April 15, 1940, private archives, Monod family.

38 **Odette approved of the whole idea:** Letter, Odette Monod to Lucien and Charlotte Monod, January 31, 1940, private archives, Monod family.

38 **In February, Monod learned:** Letter, Jacques Monod to Odette Monod, February 7, 1940, private archives, Monod family.

39 **"I demonstrated dizzying panache":** Letter, Jacques Monod to Odette Monod, February 29, 1940, private archives, Monod family.

39 **"The laboratory has been":** Letter, Jacques Monod to Philo Monod, January 9, 1940, private archives, Monod family.

40 **"Should one accept life":** Todd (1997), 21.

42 **"They have all betrayed us":** Camus (1963), 139.

42 **"Never have left-wing":** Todd (1997), 89.

42 **"For my works":** Ibid., 95.

43 **"Why must one love":** Ibid., 97.

43 **The issue on Camus's day:** Todd (1996), 238.

43 **"Now that everything":** Camus (1963), 176–77.

44 **"to arouse, reassemble":** Shirer (1969), 553.

44 **"The ironfields":** May (2000), 338.

45 **"When we embarked":** *The Times*, April 5, 1940.

Chapter 3: Misadventures in Norway

46 **At two a.m. on April 3:** Haarr (2009), 66–69, 81–84.

47 **The mining took place:** Ibid., 90–95.

47 **A Polish submarine:** Ibid., 97–99, 135–36.

48 **"You are wrong":** Shirer (1969), 561.

48 **"Will the lesson":** *Le Figaro*, April 10, 1940.

48 **After the successful troop landings:** Haarr (2009), 342–71.

48 **"The situation is thus better":** *Le Figaro*, April 10, 1940.

48 **"It would be absurd":** Shirer (1969), 556.

49 **"defensive on land":** Ibid.

49 **"It is extremely doubtful":** Ibid.

49 **"Have you been":** Liebling (2008), 587.

49 **"Don't worry about me":** Letter, Jacques Monod to Odette Monod, April 16, 1940, private archives, Monod family.

50 **"Except for rotten luck":** Letter, Jacques Monod to Odette Monod, April 15, 1940, private archives, Monod family.

50 **"Training started seriously":** Letter, Jacques Monod to Odette Monod, April 17, 1940, private archives, Monod family.

50 **Jacques's letters were unfailingly upbeat:** Letters, Jacques Monod to Odette Monod, April 18, 19, 20, and 23, 1940, private archives, Monod family.
50 **Jacques was able to confirm:** Letter, Jacques Monod to Odette Monod, April 17, 1940, private archives, Monod family.
50 **"I hope after the war":** Letter, Odette Monod to Lucien and Charlotte Monod, April 24, 1940, private archives, Monod family.
51 **"I've been feeling very isolated":** Letter, Jacques Monod to Odette Monod, April 17, 1940, private archives, Monod family.
51 **"What is happening":** Letter, Jacques Monod to Odette Monod, April 25, 1940, private archives, Monod family.
51 **"Events are going":** Todd (1997), 105.
51 **"You can't live here":** Ibid., 102.
51 **"I see the form":** Ibid., 88.
52 **"a desperate man":** Ibid., 104.
52 **"A novel is only philosophy":** Ibid., 84.
52 **"What is the meaning":** Lottman (1979), 217.
52 **"tightrope, in passionate":** Todd (1997), 108.
52 **"at certain moments":** Ibid.
53 **"it seemed a failure":** Ibid.
53 **"I am writing to you":** Ibid., 110.
53 **"I don't care":** Ibid., 110–11.
53 **"As for the risks":** Ibid., 110.
53 **"I am writing to you at night":** Ibid., 109.
54 **One force was landed:** Churchill (1948), 605–52.
54 **A second force:** Ibid.
54 **"man for man showed themselves":** *The Times*, London, May 8, 1940.
54 **"I confess that I did not":** Kersaudy (1991), 189.
55 **"We cannot go on":** May (2000), 342; Kersaudy (1991), 189–90.
55 **"nothing which can contribute":** Churchill (1948), 660.
55 **"I have had enough":** May (2000), 379.
55 **"certain to lose":** Shirer (1969), 603.
56 **"As I cannot make my":** May (2000), 379–80.
56 **"I can only think":** Letter, Jacques Monod to Odette Monod, May 9, 1940, private archives, Monod family.

Chapter 4: Springtime for Hitler

57 **"We have assured all":** Murphy et al. (1943), 241.
57 **"Columns marching westward":** Shirer (1969), 603.
58 **"The attack that we had foreseen":** *Le Matin*, May 11, 1940.
58 **"The Boches have business":** Liebling (2008), 74.
59 **"They will see we":** Ibid.
59 **"The real roughhouse":** Ibid., 73.
59 **"It's good that it's starting":** Ibid.
59 **"That's it, Hitler has":** Lottman (1992), 2.

59 **"Three free countries":** *Le Figaro,* May 11, 1940.
59 **"satisfied that they have":** Ibid.
60 **"If I have not heard":** Letter, Jacques Monod to Odette Monod, May 11, 1940, private archives, Monod family.
60 **Jacques briefed them:** Letter, Odette Monod to Lucien and Charlotte Monod, May 14, 1940, private archives, Monod family.
60 **"We have unlimited confidence":** *Le Figaro,* May 11, 1940.
60 **"Let us have confidence":** *Le Matin,* May 11, 1940.
61 **on May 11:** *Le Figaro,* May 12, 1940.
61 **on May 12:** *Le Figaro,* May 13, 1940.
61 **"brilliantly prepared and executed":** *The Times,* May 13, 1940.
61 **The French military communiqué:** *Le Figaro,* May 13, 1940.
61 **Seven panzer divisions:** Battistelli and Hook (2011), 20.
62 **"Victory or defeat":** Shirer (1969), 650.
62 **"Our front has been pushed":** May (2000), 413.
64 **"hurling counterattacks":** *Le Figaro,* May 15, 1940.
64 **"France has many trump cards":** Todd (1996), 251.
64 **"If we are to win":** Lottman (1992), 91.
64 **"We have been defeated":** Churchill (1949), 42.
65 **"Where is the strategic reserve":** Ibid., 46.
65 **Churchill needed to rouse:** Ibid., 49–51.
65 **"where the enemy":** *Le Figaro,* May 16, 1940.
65 **"For the moment":** Ibid.
66 **"It is in the best interests":** *Le Figaro,* May 17, 1940.
66 **Odette had stayed in Paris:** Letters, Odette Monod to Lucien and Charlotte Monod, May 15, 1940; and from Jacques Monod to Odette Monod, May 18, 1940, private archives, Monod family.
66 **"In any case":** Letter, Jacques Monod to Odette Monod, May 18, 1940, private archives, Monod family.
66 **"It seems to me":** Letter, Jacques Monod to Odette Monod, May 20, 1940, private archives, Monod family.
67 **"Do as I do":** Letter, Jacques Monod to Odette Monod, May 21, 1940, private archives, Monod family.
67 **"At the sight":** De Gaulle (1964), 36.
68 **"What have we come":** *Le Figaro,* May 20, 1940.
69 **"As the days go on":** Todd (1997), 112.
69 **"in the middle of an almost":** Ibid.
69 **"This war has not":** Ibid.
69 **"The homeland is in danger":** *Le Figaro,* May 22, 1940.
69 **"I think that I missed my vocation":** Letter, Jacques Monod to Odette Monod, May 20, 1940, private archives, Monod family.
69 **"My dear angel":** Letter, Jacques Monod to Odette Monod, May 21, 1940, private archives, Monod family.
71 **"with vigor":** *Le Figaro,* May 31, 1940; Lottman (1992), 156.
71 **"We must be very careful":** Churchill (1949), 115.

72 **"Since I am convinced":** Shirer (1969), 767.
72 **"The disproportion between":** De Gaulle (1964), 53–54.
72 **"convince the English":** Ibid., 54.
72 **"The Battle of France":** Shirer (1969), 762.
73 **During these desperate days:** Jackson (2003a), 179–80.
73 **"We are at the end":** Shirer (1969), 769.
73 **"The necessity of asking":** Ibid., 770.
73 **"no honorable armistice":** Ibid.
73 **"The safety of the nation":** Lottman (1992), 222.
73 **"a day of agony":** Shirer (1969), p. 771.
74 **"In 24 hours":** Shirer (1969), 771.
74 **"We are in the sixth day":** *Le Figaro*, June 11, 1940.
75 **"Mr. President":** "The President of the French Council of Ministers (Reynaud) to President Roosevelt, 10 June 1940," Mount Holyoke College, https://www.mtholyoke.edu/acad/intrel/WorldWar2/reynaud.htm. Source: U.S. Department of State, Publication 1983, *Peace and War: United States Foreign Policy, 1931–1941* (Washington, D.C.: U.S. Government Printing Office, 1943), 548–49.
76 **"Should the Germans":** Barber (1976), 35.
77 **"somewhere in the north of France":** Saint-Exupéry (1942), 68.
78 **After several days of confusion:** Account of Monod's evacuation in letters from Jacques Monod to his parents, June 14, 1940, and to Odette Monod, July 11, 1940, private archives, Monod family.
79 **"The only things that matter":** Letter, Jacques Monod to his parents, June 14, 1940, private archives, Monod family.
79 **"to abstain from all hostile acts":** Lottman (1992), 315.

Chapter 5: Defeated and Divided

81 **"a river of torment":** Jacob (1988), 98.
81 **"indestructible framework":** Ibid., 73.
81 **"a whole nation disintegrate":** Ibid., 99.
81 **"the country, the Republic":** Ibid.
82 **"Everything I believed in":** Ibid., 98.
82 **"foaming at the mouth":** Ibid., 99.
82 **"governments of bunglers":** Ibid.
83 **"If an armistice":** Shirer (1969), 787.
83 **"The duty of the government":** Ibid., 799.
83 **"For the last three days":** Ibid., 805.
84 **"At this most fateful moment":** Ibid., 825.
85 **"would have placed France":** Ibid., 830.
85 **"You imagine that by capitulating":** Champoux (1975), 287–88.
85 **"the greatest disappointment":** Shirer (1969), 831.
85 **"go and ask Marshal Pétain":** Ibid., 836.
85 **"I'm told he has":** Ibid., 840.

85 **"There is my government":** Ibid., 842.

86 **"Frenchmen! On the appeal":** *Times* (London), June 18, 1940.

86 **"the traitors, the crooks":** Jacob (1988), 101.

87 **"We are not going to shrink":** Ibid., 101–2.

87 **"If you can":** Ibid., 103.

88 **"concentrating all his power":** A. Beevor, "Rallying Call: A Mesmerizing Oratory," *The Guardian*, April 29, 2007, http://www.guardian.co.uk/theguardian/2007/apr/29/greatspeeches.

88 **"The leaders who":** De Gaulle (1964), 83–84.

90 **Odette and the twins:** Letter, Dominique Dreyfus to Patrice Debré, July 9, 1996, copy provided by Olivier Monod.

90 **"For all military inquiries":** Jacob (1988), 103.

90 **"Don't make yourselves":** Ibid.

90 **"Ever hear of de Gaulle?":** Ibid., 105.

91 **"It is the bounden duty":** Charles de Gaulle, June 19, 1940, "The Flame of French Resistance," *Great Speeches of the 20th Century*, http://www.guardian.co.uk/theguardian/2007/apr/29/greatspeeches1.

91 **"You will not leave":** Champoux (1975), 293.

92 **"What is the value":** Shirer (1969), 889.

92 **"The French government, after":** Charles de Gaulle, June 22, 1940, "The Flame of French Resistance," *Great Speeches of the 20th Century*, http://www.guardian.co.uk/theguardian/2007/apr/29/greatspeeches1.

93 **"that such or similar terms":** Shirer (1969), 887–88.

93 **"The French government and people":** *The Times*, June 24, 1940; Shirer (1969), 888.

93 **"M. le Maréchal":** *The Times*, June 27, 1940.

94 **"turn over this dark page":** *The Times*, June 26, 1940.

94 **"beneath the German jackboot":** *The Times*, June 27, 1940.

95 **Joseph Meister, Louis Pasteur's first patient:** Diary of Eugene Wollman, June 24, 1940, Fonds Elie Wollman, SAIP. Note: the often-repeated story of Meister's suicide has him shooting himself with a revolver sometime around June 14–16 after refusing to open Pasteur's tomb to the Germans. Neither that means, the date, nor the catalyst are likely correct. The account here is based on a newly available diary (discovered by my colleague Héloïse Dufour) of Eugene Wollman, a Pasteur scientist, who described the atmosphere in Paris in June 1940. His entry for June 24, 1940, reads, in part: "This morning, Meister was found dead. He committed suicide with gas. He was very depressed these last days and as he was in the lodge (his wife and children are away), his meals were being delivered to him." Wollman and his wife, Elisabeth, were murdered by the Nazis in 1943; their son, Elie, served in the Resistance and worked at the Pasteur Institute after the war, making fundamental advances in genetics with François Jacob.

Chapter 6: Regrouping

99 **In six weeks:** Ousby (2000), 111.
99 **After a twenty-four-hour journey:** Details from letter, Jacques Monod to Odette Monod, July 11, 1940, private archives, Monod family.
100 **"Mon amour, I don't have any hope":** Letter, Jacques Monod to Odette Monod at Dinard, June 26, 1940, private archives, Monod family.
100 **"My Dears, Will this letter":** Letter, Jacques Monod to his parents, June 26, 1940, private archives, Monod family.
100 **A week later, he was comforted:** Letters, Jacques Monod to his parents, July 3 and 6, 1940, private archives, Monod family.
100 **On July 7, a telegram arrived:** Letter, Odette Monod to Lucien and Charlotte Monod, July 4, 1940, private archives, Monod family; letter, Jacques Monod to Odette Monod, July 7, 1940, private archives, Monod family.
101 **"My dear angel":** Letter, Jacques Monod to Odette Monod, July 11, 1940, private archives, Monod family.
101 **"French armed forces":** "Franco-German Armistice: June 25, 1940," from the US Department of State, Publication No. 6312, Documents on Foreign Policy 1918–1945, Series D, IX (Washington, DC: Government Printing Office 1956), 671–76, available at the Avalon Project: Lillian Goldman Law Library, Yale Law School, http://avalon.law.yale.edu/wwii/frgearm.asp.
101 **"Hold my little ones":** Letter, Jacques Monod to Odette Monod, July 7, 1940, private archives, Monod family.
101 **Camus thus had a ringside seat:** Todd (1997), 114.
102 **While Pétain had earned:** Ousby (2000), 79.
102 **"a vase on the mantelpiece":** Pryce-Jones (1981), 16.
102 **"The National Assembly gives":** Shirer (1969), 919.
103 **"the arteries of a man of forty":** Todd (1997), 114.
103 **"What we are going to experience":** Ibid.
103 **"the last French soil":** Ibid., 115.
103 **"Work, Family, and Homeland":** *Le Temps*, July 12, 1940.
103 **"Parliamentary democracy lost":** Shirer (1969), 928.
103 **"We have only one":** Ibid.
104 **"Cowardice and senility":** Todd (1997), 115.
104 **"Pro-German policies":** Ibid.
104 **"Around the Marshal":** Shirer (1969), 932.
105 **There was some good news:** Summarized from letters of Jacques Monod to Odette Monod, July 1940, private archives, Monod family.
105 **Finally, he was officially demobilized on July 29:** Letter, Jacques Monod to Odette Monod, July 29, 1940, private archives, Monod family.

Chapter 7: Ill Winds

106 **The Majestic Hotel had become:** Pryce-Jones (1981), 31, 34.
106 **The Monods could not avoid them:** Debré (1996), 116.

107 **"The French Government will bear":** "Franco-German Armistice: June 25, 1940," from the US Department of State, Publication No. 6312, Documents on Foreign Policy 1918–1945, Series D, IX (Washington, DC: Government Printing Office, 1956), 671–76, available at the Avalon Project: Lillian Goldman Law Library, Yale Law School, http://avalon.law.yale.edu/wwii/frgearm.asp.

107 **The 400 million francs:** Jackson (2003b), 169.

107 **Much more stringent rationing:** Ousby (2000), 116; Pryce-Jones (1981), 94.

108 **On July 12:** Jackson (2003b), 150.

108 **On July 22:** Ibid.

108 **Public servants had to swear:** Ibid., 151.

108 **On August 17:** Ousby (2000), 325.

108 ***Au Pilori:*** Pryce-Jones (1981), 77.

108 **Jews were obliged to register:** S. Klarsfled (1996), "A Chronology of Major Events in the War Against the Jews and the Deportations of Jewish Children From France 1940," *French Children of the Holocaust: A Memorial,* The Holocaust History Project, http://www.holocaust-history.org/klarsfeld/French%20Children/html&graphics/T0009.shtml.

109 **"For some time, the Jews":** *Le Matin,* October 2, 1940.

109 **These actions were preludes:** "Régime de Vichy: textes officiels. 2. Lois antisémites," Encyclopedie, http://www.encyclopedie.bseditions.fr/article.php?pArticleId=160&pChapitreId=24023&pSousChapitreId=24024.

109 **The new regulations upended:** Olivier Monod, interview, Paris, August 17, 2010.

109 **Almost 150,000:** "Le Recensement des Juifs," *Un Livre du Souvenir,* http://www.unlivredusouvenir.fr/recensement.html.

109 **Odette's mother and her sister:** Françoise Benhamou, phone interview with Héloïse Dufour, October 19, 2010.

109 **Jacques even had to register:** Handwritten note, MON. Bio 02, item 6, Phillipe Monod dossier, Fonds Monod, SAIP.

109 **Algerian Jews were stripped:** S. Klarsfled (1996) "A Chronology of Major Events in the War Against the Jews and the Deportations of Jewish Children From France 1940," *French Children of the Holocaust: A Memorial,* The Holocaust History Project, http://www.holocaust-history.org/klarsfeld/French%20Children/html&graphics/T0014.shtml.

110 **The publishers cooperated:** Pryce-Jones (1981), 77; Ousby (2000), 177.

110 **"All the Jews are being thrown":** Todd (1997), 116.

110 **"So I am going to choose":** Ibid., 116.

110 **"I have no joy":** Ibid., 117.

110 **"I don't plan to publish anything":** Ibid.

110 **"All of this is particularly unfair":** Ibid.

111 **Vichy acted swiftly:** "Chronologie Détaillée de la Vie du Général de Gaulle, 1890–1970," La Fondation Charles de Gaulle, http://www.charles-de-gaulle.org/pages/l-homme/accueil/chronologies/chronologie-detaillee-de-la-vie-du-general-de-gaulle.php#1940.

111 **"ardently that the British":** Cornick (2000), 70.

111 **"France has never":** Shirer (1969), 918.
112 **"ex-Frenchman" and the "ex-general traitor":** *Le Matin,* September 24, 1940.
112 **"horrifying impudence":** *Le Matin,* September 25, 1940.
112 **Moreover, as a reprisal:** Ibid.
112 **"France follows your resistance":** Ibid.
113 **"By the will":** *Le Matin,* September 27, 1940, trans. SBC.
113 **In late October:** Delpla (1997); Shirer (1990), 814–15.
114 **"I know that personally":** *Le Matin,* October 27, 1940.
114 **He explained that he had invited:** A detailed account of this meeting was later published by Paul Schmidt, Hitler's interpreter. Available at http://pages.livresdeguerre.net/pages/sujet.php?id=docddp&su=300&np=951.
114 **"Frenchmen, Last Thursday":** "Discours de Pétain apres entrevue de Montoire," Encyclopedie, http://www.encyclopedie.bseditions.fr/article.php?pArticleId=160&pChapitreId=24028&pSousChapitreId=24032, trans. per Ousby (2000), 86, and SBC.
115 **"This policy is mine":** Ibid.
115 **A key exhibit:** *Le Matin,* November 2, 1940.

Chapter 8: An Hour of Hope

116 **The second was de Gaulle's:** "Chronologie Détaillée de la Vie du Général de Gaulle, 1890–1970," La Fondation Charles de Gaulle, http://www.charles-de-gaulle.org/pages/l-homme/accueil/chronologies/chronologie-detaillee-de-la-vie-du-general-de-gaulle.php#1940.
117 **The BBC devoted:** Luneau (2005), 74; "Maurice Schumann," http://www.ordredelaliberation.fr/fr_compagnon/911.html.
117 ***Les Français Parlent aux Français:*** Luneau (2005), 64.
117 **"that did not return to base":** *Le Matin*, August 17, 1940.
117 **59 British and 120 German planes:** Two-day total for August 15–16; Battle of Britain data accessed at http://www.bbc.co.uk/news/uk-11029903.
117 **"The war between Germany and England":** *Le Matin*, August 17, 1940, trans. SBC.
118 **"Thus, de Gaulle, traitor":** *Le Matin*, September 26, 1940.
118 **"We have not only fortified our hearts":** "Premier's Review of the War," *The Guardian,* http://century.guardian.co.uk/1940-1949/Story/0,,128255,00.html.
119 **"Français! C'est moi, Churchill":** "Discours de Churchill: Homage à la France (21 Octobre, 1940)," Jalons, http://www.ina.fr/fresques/jalons/fiche-media/InaEdu00281/discours-de-churchill-hommage-a-la-france-21-octobre-1940.html.
119 **"Frenchmen! It is me, Churchill":** English text in Churchill (1949), 510–11.
119 **"I tell you what you must truly believe":** Ibid.
120 **"antinational broadcasts" in public places:** Luneau (2005), 104.
121 **Léon-Maurice Nordmann:** Blumenson (1977), 19–20.
121 **Nordmann had a lot of very bright friends:** Debré (1996), 117.

121 **Avocats Socialistes:** Military File, 1945, MON. Bio. 02, Fonds Monod, SAIP.

122 **Another leaflet was spread:** Chemins de Mémoire, http://www.cheminsdememoire.gouv.fr/le-11-novembre-1940.

122 **"Public organizations and private enterprises":** *Le Matin*, November 10, 1940.

122 **"On the graves of your martyrs":** "11 Novembre 1940: La Résistance au Grand Jour," *Action Républicaine*, http://action-republicaine.over-blog.com/article-13706200.html, trans. SBC.

122 **"On November 11":** Luneau (2005), 109.

122 **Meanwhile, Nordmann, Weil-Curiel:** Schoenbrun (1980), 89; Chemins de Mémoire, http://www.cheminsdememoire.gouv.fr/le-11-novembre-1940.

123 **Students and teachers formed ranks:** Chemins de Mémoire, http://www.cheminsdememoire.gouv.fr/le-11-novembre-1940.

123 **The Germans were surprised:** Chemins de Mémoire, http://www.cheminsdememoire.gouv.fr/le-11-novembre-1940; Luneau (2005), 110–11.

123 **"incompatible with the dignity":** *Le Matin*, November 16, 1940.

123 **They closed the universities:** Ibid.

123 **required that every student register:** *Le Matin*, November 13, 1940.

123 **dismissed the rector:** Chemins de Mémoire, http://www.cheminsdememoire.gouv.fr/le-11-novembre-1940; Luneau (2005), 110–11.

124 **a teacher and writer, respectively:** Humbert (2008); Blumenson (1977).

124 **Monod's neighbors at 30 rue Monsieur-le-Prince:** Humbert (2008), 37.

124 **Rivet introduced Cassou:** Blumenson (1977), 80.

124 **It was decided that Humbert:** Ibid., 91.

124 **"Many of us will be shot":** Ibid.

124 **Life had already changed:** Humbert (2008), 19; Blumenson (1977), 106.

124 **The two groups decided to collaborate:** Blumenson (1977), 112–13.

124 **Humbert was the typist:** Letter, Monod to Mme. Thieuleux, June 2, 1970, MON. Bio. 02, Fonds Monod, SAIP; Debré (1996), 118; Humbert (2008), 24–25; Blumenson (1977), 118.

124 **"Resist! This is the cry":** Blumenson (1977), 117–18.

125 **"January first will offer to all French":** "Discours de Gaulle," http://www.mediaslibres.com/tribune/?post/2007/12/11/396-discours-de-gaulle-decembre-1940#23decembre1940, trans. SBC.

126 **The second issue of *Résistance*:** Blumenson (1977), 132–33.

126 **"The hour of hope":** Ibid.

126 **The call was repeated:** Luneau (2005), 118–21.

127 **For Nordmann, however:** Sources concerning the role, exposure, arrest, and interrogations of the Aubervillier group include http://pcfaubervilliers.fr/spip.php?article512; L. Israël, *Robes noires, années sombres*, 116–23; J. Blanc, *Au commencement de la Résistance*, 237–40, 390–91, 400; and the Archives de la Préfecture de Police in Paris, Dossier BA 2443 (formerly the dossier "Dissolution du PC No. 60. Propaganda étrangère"). These references differ in the details and sequence of events and outright contradict certain other published accounts of the Nordmann episode, but as they cite or comprise official records

in the Archives de la Préfecture de Police in Paris, they are the most reliable. The author thanks Mme. Isabella Tarisca of the Service de Memoire et des Affaires Culturelles of the Cabinet du Préfet for her assitance in locating and accessing the records concerning the Nordmann episode.

129 **"Jacques MONOD, Laboratoire de Zoologie":** Dossier BA 2443, Archives de la Préfecture de Police in Paris.

129 **"Mr. Monod is involved or":** Ibid. The author is indebted to Ms. Liora Israel for information about the records concerning Monod in the Archives.

129 **The inspectors then proceeded:** Ibid.

129 **warrants were issued for Nordmann:** The pursuit and arrest of Nordmann from Blumenson (1977), 137–40.

130 **Early one morning, Monod:** The account of Monod's interrogation relies on that of Debré (1996), 118. As to how the Gestapo obtained his name, the most likely explanation is that they received it when the French police turned over the Nordmann case files in January 1941. Monod stated in 1970 that his name was found among Nordmann's notebooks (letter to Mme. Thieuliex, June 2, 1970, MON. Bio. 02, Fonds Monod, SAIP), and that may also have been the case. The time of that interrogation is not clear, but based upon the Nordmann case being handed over to the Germans in January 1941, and the events involving Nordmann, it is most likely that it was in January 1941.

131 **On February 8:** Blumenson (1977), 144.

131 **On February 10:** Ibid., 149–50.

131 **On February 11:** Ibid., 151.

131 **Vildé and Dexia eluded:** Ibid., 163, 165, 173.

Chapter 9: Waiting and Working

132 **The United States continued:** "Neutrality Act of November 4, 1939," Mount Holyoke College, http://www.mtholyoke.edu/acad/intrel/WorldWar2/neutrality.htm.

133 ***"Nécessité absolue Trouver":*** Notebook 9, 67, November 28, 1940, MON. Lab. 01, November 28, 1940, Fonds Monod SAIP.

133 **Monod found that:** Monod (1941a).

134 **Specifically, instead of one growth phase:** Monod (1941b).

134 **"What could that mean?":** Ullmann (2003), 4.

134 **"That could have something":** Monod (1965), 188.

134 **It appeared as if the bacteria:** See L. Loison (2012), "Enzymatic Adaptation: Monod, Lwoff, and the Legacy of General Biology," unpublished manuscript.

135 **For example, just by changing the ratio:** Monod (1942), 167.

135 **"What Monod is doing":** Ullmann (2003), 5.

136 **The two exchanged brass wedding rings:** Todd (1997), 118; Lottman (1979), 229.

136 **"There is but one":** Camus (1991a), 3.

136 **"I see many people die":** Ibid., 4.

137 **"Man is mortal":** Ibid., 18.

137 **"a universe suddenly divested":** Ibid., 6.
137 **"is not worth the trouble":** Ibid., 5.
137 **"of which man is the sole":** Ibid., 117.
138 **"The struggle toward the heights":** Ibid., 123.
138 **"Finished *Sisyphus*":** Camus (1963), 189.
138 **"*L'Étranger* is very successful":** Todd (1997), 129.
138 **"Very sincerely":** Ibid., 130.
139 **"*L'Étranger* is obviously":** Ibid., 131.
139 **"I read *L'Étranger*":** Ibid., 135.
139 **"The link between *Sisyphe*":** Ibid., 133.
139 **"What matters is that":** Ibid., 134.
139 **"The problem of paper":** Ibid., 133.
140 **Gallimard told Camus:** Lottman (1979), 247; Todd (1997), 136–37.
140 **"Sisyphus, or Happiness in Hell":** Todd (1997), 150.

Chapter 10: The Terror Begins

142 **Vichy dissolved the PCF:** Ousby (2000), 325.
142 **On August 19:** *Le Matin,* August 21, 1940.
142 **On the morning of August 21, 1941:** Pryce-Jones (1981), 118; Ousby (2000), 223; "1941: L'attentat au Métro Barbès," Les Communistes, http://www.lescommunistes.org/spip.php?article339.
142 **Three men were executed:** Laub (2010), 116, frontispiece end.
142 **"Beginning August 23":** Pryce-Jones (1981), 120; *Le Matin,* August 23, 1940.
143 **Another attack followed soon after:** Laub (2010), 119.
143 **In response to three more:** Ibid., 120.
143 **On September 16:** Pryce-Jones (1981), 120.
143 **After the first two attacks:** Laub (2010), 119.
143 **"a German soldier is worth more":** Ibid.
143 **Hitler thought that a ratio:** Ibid.
143 **Von Stülpnagel was deeply concerned:** Ibid., 128.
143 **On October 20:** Ousby (2000), 225; Laub (2010), 136.
144 **Forty-eight hostages:** Laub (2010), 139.
144 **The consequences for attacking:** http://www.cheminsdememoire.gouv.fr/page/affichecitoyennete.php?idLang=en&idCitoyen=13
144 **"Frenchmen, two shots":** Laub (2010), 142.
144 **A hand grenade was thrown:** Pryce-Jones (1981), 121–22.
144 **Altogether, there were sixty-eight:** Ousby (2000), 225.
145 **"Within the occupied territories":** "Night-and-Fog Decree," http://www.yale.edu/lawweb/avalon/imt/nightfog.htm.
145 **"the adequate punishment for offences":** Ibid.
145 **Although he was not a Communist:** Letter, J. Monod to Mme. Thieuleux, June 2, 1970, MON. Bio. 02, item 7, Fonds Monod, SAIP. Although Monod never disclosed the name of the organization, based on his descriptions and the timeframe, it was likely Université Libre. See Jackson (2003b), 421–22.

145 **In addition, Monod helped:** Letter, J. Monod to Mme. Thieuleux, June 2, 1970, MON. Bio. 02, item 7, Fonds Monod, SAIP.

146 **On February 23, 1942:** Chemins de Mémoire, http://www.cheminsdememoire.gouv.fr/le-11-novembre-1940.

146 **The Monod twins:** Letter, Charlotte Monod to Winnie Eschweiller, April 1942, courtesy of Olivier Monod.

146 **As Easter 1942 approached:** Ibid.

147 **"the same vivid":** Ibid.

147 **Through 1941:** Laub (2010), 158.

147 **"I intend to order only a *limited*":** Ibid., 161.

148 **"the position of the MBF":** Ibid., 164.

148 **On June 1, 1942:** Ibid., 168, 196.

149 **"1. A DISTINCTIVE SIGN FOR THE JEWS":** "Être Juif en France," USC Shoah Foundation, http://dornsife.usc.edu/vhi/french/etoilejaune.

149 **"Never would one have thought":** *Le Matin,* June 8, 1942.

149 **The number of Jews in all of France:** Laub (2010), 220, 213.

150 **They could shop for food only:** Zuccotti (1993), 91; *Le Matin,* July 10, 1942; *Le Matin,* July 18, 1942.

150 **and they were to ride only:** Zuccotti (1993), 94.

150 **Nevertheless, Odette and Jacques decided:** Olivier Monod, interview, Paris, September 8, 2011.

150 **At four in the morning:** Zuccotti (1993), 105.

151 **But neither those arrested:** Ibid., 110–11.

151 **Indeed, just three days:** Ibid., 110.

151 **There had been roundups before:** In early May 1941, more than six thousand immigrant Jewish men had received postcards from the Prefecture of Police asking them to report in person for "an examination" of their "situation." More than half complied and were interned at Pithiviers and Beaune-la-Rolande, where their families were able to visit and to give them packages. Another four thousand foreign Jews were rounded up in Paris in August 1941 and sent to Drancy. Zuccotti (1993), 81–83.

152 **Laval and Secretary-General of Police:** Laub (2010), 230–31.

152 **At the time of the July roundups:** Françoise Benhamou, phone interview with Héloïse Dufour, October 19, 2010.

152 **After the roundups:** Olivier Monod, interview, Paris, September 8, 2011; Odette "Brulle" identity card, private archives, Monod family.

Chapter 11: The Plague

154 **"All were separated":** Camus (1965), 51.

154 **"I thought it was all over":** Lottman (1979), 257.

155 **"*The Plague* has":** Camus (1965), 36.

155 **"The first thing":** Ibid.

155 ***la peste brune:*** See, for example, D. Guérin, *La Peste Brune* (1935).

155 **"I want to express":** Camus (1965), 53–54.
155 **"1342—The Black Death":** Camus (1963), 201.
156 **"let us admit":** Camus (1991b), 3.
156 **"everyone is bored":** Camus (1991b), 6.
156 **"it will be easily understood":** Ibid.
156 **"After that I will return":** Todd (1997), 152.
156 **He asked Pascal Pia:** Lottman (1979), 264.
156 **"In the space of five days":** Crémieux-Brillhac (1975), 11.
157 **"The circumstances being such":** *Le Matin*, November 12, 1942.
157 **"Caught like rats!":** Camus (1965), 38.
157 **"In short, the time of the epidemic":** Henry (2007), 113, and translated from "Les Exiles Dans La Peste," in Qulliot Theatre Recits etc d'Albert Camus, 1962, 1951.
157 **"Make separation the big theme":** Camus (1965), 60.
158 **As the Soviet campaign:** Laub (2010), 248.
158 **Some 275,000 French laborers:** Ibid., 251.
158 **He appointed longtime:** Ibid., 249.
158 **Only 53,000 workers:** Ibid., 255.
158 **The number of workers sent:** Ibid., 257.
159 **And on February 16, 1943:** Ibid., 258–59.
159 **"Workers and bosses":** Crémieux-Brillhac (1975), vol. 3, 83.
159 **"One's sacred duty":** Ibid., 98.
159 **"Frenchmen, do not go there!":** Ibid., 105.
159 **FRENCHMEN! STAND AGAINST SLAVERY:** *Le Franc Tireur*, March 16, 1943.
160 **Many men did go to Germany:** Laub (2010), 260.
160 **While more than 600,000 French workers:** Ibid., 120–37.
160 **The longer Camus stayed:** Henry (2007), 109–12.
161 **The two men met regularly:** Ibid., 111–12; Lottman (1979), 273.
161 **At his boardinghouse:** Henry (2007), 110.
161 **to which Pascal Pia also belonged:** Lottman (1979), 269.
161 **Fayol and Camus listened to the BBC:** Todd (1997), 161.
161 **Camus received his first reply:** Lottman (1979), 278.
161 **"In the chapter on the isolation camps":** Camus (1965), 60.
161 **"concerned with man and freedom":** Lottman (1979), 287.
162 **Three thousand copies:** Ibid.
162 **"They felt the profound sorrow":** "Les Exiles dans La Peste," in Camus (1962), 1956; Camus (1991b), 73.

Chapter 12: Brothers in Arms

163 **"Plague. All fight":** Camus (1965), 82.
163 **"With the victory":** "La 'Colonne du Tchad,' s'empere de Koufra et du Fezzan," France-Libre, http://www.france-libre.net/2e-db/historique/koufra-fezzan.php, trans. SBC.

164 **"Certainly, it is on the youth":** De Gaulle, February 25, 1943, speech, http://www.mediaslibres.com/tribune/?post/2010/02/15/Discours-de-Gaulle-Fevrier-1943#25fevrier1943.

165 **"I have noticed":** Noufflard (unpublished), 84, copy of notice, trans. SBC.

165 **One spring evening:** Debré (1996), 120.

167 **They would not allow non-Communists:** Judson (1979), 359.

167 **Monod had long held:** Debré (1996), 120.

167 **But he wanted to get more:** Judson (1979), 359.

167 **Marchal introduced Monod:** Geneviève Noufflard, interview with Benjamin Prud'homme, Paris, January 2010; Debré (1996), 122.

167 **"Do not forget":** Geneviève Noufflard, interview with Benjamin Prud'homme, Paris, January 2010.

167 **"He is in great form":** Letter, Odette Monod to Charlotte and Lucien Monod, May 20, 1943, private archives, Monod family.

168 **The all-Bach program:** Program of May 21, 1943, private archives, Monod family.

168 **Monod's new conductorship:** Toulmond (2005), 19.

168 **Toward the end of September:** Ibid.; Debré (1996), 121.

168 **In early October:** Françoise Benhamou, interview with Héloïse Dufour, October 19, 2010.

169 **He spent every Wednesday night:** Letter, Odette Monod to Lucien and Charlotte Monod, October 22, 1943, private archives, Monod family.

169 **Marcel Prenant had made repeated appeals:** Prenant (1980), 196.

169 **Philo heard de Gaulle's:** Guillain de Bénouville (1949), 168–69.

169 **It was not until François Morin:** Debré (1996), 117.

169 **Philo told Morin:** "Claude Bourdet," Ordre de la Liberation, http://www.ordredelaliberation.fr/fr_compagnon/131.html; P. Monod, interview with Serge Ravneal, Conception Jean-Louis Dufour, Realisation Jacques Boliot.

170 **Bourdet invited Philo:** P. Monod, interview with Serge Ravneal, Conception Jean-Louis Dufour, Realisation Jacques Boliot.

170 **Philo subsequently replaced Bourdet:** Frenay (1976), 249.

170 **Shoop appeared to take:** Ibid.

170 **"wanted to know":** Ibid., 250.

170 **De Bénouville, who was in charge:** Ibid.; Guillain de Bénouville (1949), 173.

170 **The Frenchmen handed over:** Dulles et al. (1996), 53.

171 **Technically, he would not:** Frenay (1976), 253.

171 **Philo soon secured:** Ibid., 263.

171 **De Bénouville set up:** Guillain de Bénouville (1949), 175.

171 **The MUR provided intelligence reports:** Guillain de Bénouville (1949), 184–86.

172 **In October 1943:** Guillain de Bénouville (1949), 270–71; R. Belot and G. Karpman (2009) 282; Guérin (2010), 288.

172 **Jacques Monod was chosen:** Letter, J. Monod to Mme. Thieuleux, June 2, 1970, MON. Bio. 02, item 7, Fonds Monod, SAIP; Guillain de Bénouville (1949), 273.

172 **"I have something to ask you":** Geneviève Noufflard, interview by Benjamin Prud'homme, Paris, January 20, 2010; Judson (1979), 363.

172 **De Bénouville had set up a system:** For details see Belot and Karpman (2009).

172 **After arriving at the train station:** Guillain de Bénouville (1949), 271; Guérin (2010), 286.

173 **Along with Jacques and de Bénouville:** Guillain de Bénouville (1949), 271; Guérin (2010), 288.

173 **The delegates brought:** Guillain de Bénouville (1949), 271.

173 **To attain their potential effectiveness:** Ibid., 355–66.

173 **Jacques was promised arms:** Letter, J. Monod to Mme. Thieuleux, June 2, 1970, MON. Bio. 02, item 7, Fonds Monod, SAIP.

174 **After paying a heavy fine:** Guillain de Bénouville (1949), 271–72; Belot and Karpman (2009), 282–84.

174 **Leynaud told Camus:** Lottman (1979), 270–72.

174 **Camus joined such notable:** Ibid., 280–81.

175 **"We shall meet soon again":** Camus (1974), 4.

175 **"humiliations and silences":** Ibid., 6.

175 **"I belong to an admirable":** Ibid., 8.

176 **Bernard was an early:** Frenay (1976), 53–54; interview with Serge Ravenel, Conception Jean-Louis Dufour, realisation Jacques Boliot.

176 **Camus's first meeting with Bernard:** Account of meeting compiled from Hardré (1964); Bernard (1967); Lottman (1979), 300–301; Todd (1997), 178–79.

Chapter 13: Double Lives

177 **"When are they landing?":** Noufflard (unpublished), 66.

177 **Just a month after:** "Pierre Arrighi," Ordre de la Liberation, http://www.ordredelaliberation.fr/fr_compagnon/32.html.

177 **Marcel Peck, Pia's Combat chief:** Frenay (1976), 312.

177 **He appointed Joseph Darnand:** *Time*, February 7, 1944.

178 **The penalty for being caught:** Kupferman (2006), 198.

178 **In an interview with *Paris-Soir*:** Germain (2008), 29.

178 **A colleague who knew:** Judson (1979), 363.

178 **Monod shifted his experiments:** Letter, Odette Monod to Lucien and Charlotte Monod, December 4, 1943, private archives, Monod family.

178 **Always happy to see:** Olivier Monod, interview, Paris, August 17, 2010.

178 **Monod did decide:** Letter, Odette Monod to Lucien and Charlotte Monod, October 22, 1943, private archives, Monod family.

178 **"I am living a terribly austere":** Letter, Jacques Monod to his parents, December 13, 1943, private archives, Monod family.

178 **Geneviève Noufflard tracked him:** Noufflard (unpublished), 67. Note: The circumstances of the meeting are not clear: Judson (1979), 363, indicates that it was after a choir rehearsal in January, but a letter from October 1943 indicates

that Monod had quit conducting. Noufflard's memoir does not mention choir practice, and in her interview in 2010 she indicates that she "found" Monod and asked him.

179 **Monod tried to dissuade her:** Judson (1979), 363; Geneviève Noufflard, interview by Benjamin Prud'homme, Paris, January 20, 2010.

179 **Noufflard snuck back and forth:** Noufflard (unpublished), 9, 47; "René Parodi: Ordre de la Liberation," http://www.ordredelaliberation.fr/fr_compagnon/750.html.

179 **After returning to Paris:** Noufflard (unpublished), 57–61.

179 **Among the first American fliers:** Bodson (2005), 149.

180 **On the morning of January 23:** The account of Spence's mission and rescue are from Escape and Evasion Report No. 16, RG 498, Entry UD 134, Box 1 Location: 290/55/20/4, National Archives, College Park, MD; 303rd BG(H) Combat Mission Report No. 11, accessed at www.303rdbg.com/missionreports/011.pdf; "Allied Aviators Passed Through Comet Line via Pyrenees," http://www.cometeline.org/fiche087.html; Noufflard (unpublished).

181 **Only two weeks earlier:** "The Comet Line," http://www.cometeline.org/comethist.htm.

181 **After a long afternoon:** Geneviève Noufflard, interview by Benjamin Prud'homme, Paris, January 20, 2010; "Allied Aviators," http://www.cometeline.org/fiche087.html.

181 **Noufflard led Spence:** Geneviève Noufflard, interview by Benjamin Prud'homme, Paris, January 20, 2010.

182 **Two weeks after arriving:** P. Connart et al., "Bidarray, Larresore, Souraïde," http://www.cometeline.org/PassagesNew.html

182 **After some snags:** "'Franco' Nous a Quitté," http://www.cometeline.org/cometfranco06062008.htm; "Andre de Jongh, Organiser of the Comet Line," *The Independent,* December 6, 2007, http://www.independent.co.uk/news/obituaries/andre-de-jongh-organiser-of-the-comet-line-763264.html.

182 **The Noufflards would continue:** "Conrad Blaylock of the 381st Bomb Group, shot down Febraury 6, 1944," in Noufflard (unpublished), 63–65.

182 **but Geneviève was determined:** Noufflard (unpublished), 67.

182 **She told Monod:** Geneviève Noufflard, interview by Benjamin Prud'homme, Paris, January 20, 2010.

183 **"Okay, all right":** Judson (1979), 363.

183 **To conceal whatever she was carrying:** Noufflard (unpublished), 68–70.

183 **Monod showed Noufflard:** Geneviève Noufflard, interview by Benjamin Prud'homme, Paris, January 20, 2010; Judson (1979), 364.

183 **That was done by giving some prearranged sign:** Noufflard (unpublished), 68.

184 **Despite the emphasis on security:** The account of Prenant's arrest is from Prenant (1980), 201–11.

185 **The interrogation began:** The account of Prenant's interrogation and torture is from Prenant (1980), 201–11; quotes are translations by SBC.

187 **The FTP promoted Georges Teissier:** Toulmond (2005), 19.

187 **At the same time Monod:** Letter, J. Monod to Mme. Thieuleux, June 2, 1970, MON. Bio. 02, item 7, Fonds Monod, SAIP; Monod FFI record "Etat des Services dans bsF.F.I" dated November 26, 1944, MON. Bio. 02, Fonds Monod, SAIP.

187 **Monod was made head:** Letter, J. Monod to Mme. Thieuleux, June 2, 1970, MON. Bio. 02, item 7, Fonds Monod, SAIP; Monod FFI record "Etat des Services dans bsF.F.I" dated November 26, 1944, MON. Bio. 02, Fonds Monod, SAIP.

187 **These would be passed to Geneviève Noufflard:** Geneviève Noufflard, interview by Benjamin Prud'homme, Paris, January 20, 2010.

187 **Monod discovered that:** Schwartz (1997), 134; Judson (1979), 361.

187 **Late in the afternoon of February 14:** N. Chevassus-Au-Louis (2004), 187, 193. Croland was sent to Buchenwald and died in April 1945.

188 **The creation of the FFI:** Letter, J. Monod to Mme. Thieuleux, June 2, 1970, MON. Bio. 02, item 7, Fonds Monod, SAIP.

188 **De Bénouville was very impressed:** Guillain de Bénouville (1949), 294–95.

189 **In fact, Audureau isolated several:** Judson (1979), 362–63; Müller-Hill (1996), 11.

189 **She and Monod found:** Noufflard (unpublished), 75–76.

190 **"has delicate health":** Todd (1997), 170–72.

190 **Waiting outside on the street:** Lottman (1979), 307.

190 **As he settled:** Ibid., 295.

191 **Camus accepted, and they began:** Ibid., 293, 296.

191 **Camus was put in charge:** The date of the reading was March 19, 1944. Todd (1997), 175; Lottman (1979), 297–98.

192 **In appreciation for their efforts:** Brassai (1999), 201.

192 **She met Camus:** Todd (1997), 183.

192 **Though just fourteen:** "Maria Casarès," *The Independent*, December 7, 1996, http://www.independent.co.uk/news/people/obituarymaria-casares-1313344.html#; Lottman (1979), 316.

192 **After the Germans invaded:** Lottman (1979), 316–17.

193 **The network had provided:** Todd (1997), 178.

193 **Jacqueline Bernard was stunned:** Bernard (1967).

193 **Yvette Bauman was subsequently deported:** Guillain de Bénouville (1949), 283–84.

193 **It was actually the second time:** "André Bollier," Ordre de la Liberation, http://www.ordredelaliberation.fr/fr_compagnon/116.html.

193 **He also set up a phony:** Lottman (1979), 300.

194 **After being tortured:** "André Bollier."

194 **In late March:** Guillain de Bénouville (1949), 278–79.

194 **The Germans caught:** Ibid., 302.

194 **She managed to alert:** Ibid., 296.

194 **Claude Bourdet, who had become:** Bourdet (1975), 323.

194 **The Gestapo almost nabbed:** Guillain de Bénouville (1949), 299–300.
194 **Camus authored his first:** The editorial of issue 55, though anonymous, is generally attributed to Camus: Lévi-Valensi in Camus (2006), 1–3.

Chapter 14: Preparations

197 **"If they attack in the west":** Hitler and Domanus (1990), 2,850.
197 **In the spring of 1944:** Buell, et al (1978), 276.
197 **whereas of the roughly 40,000 resistants:** Île-de-France figures from Rol-Tanguy et al. (1994), 76.
197 **From January to March 1944:** Ambrose (1994), 103.
198 **Altogether, from June 1943:** Rottman and Dennis (2010), 36.
198 **In October and November 1943:** Harrison (1951), 204.
198 **For example, the Paris–Brest:** C. Bougeard in Ponty (1996), 292.
198 **Instead, the FFI:** Harrison (1951), 205.
199 **"Here they come!":** Noufflard (unpublished), 72.
199 **The primary military effects:** Harrison (1951), 206.
199 **All German units:** Burleigh (2011), 284.
200 **"At around 11 that night":** *Combat* 57, May 1944; Lévi-Valensi in Camus (2006), 5.
200 **"is increasing his efforts":** *Combat* 57, May 1944; Lévi-Valensi in Camus (2006), 6. Eight soldiers were convicted in 1949 of participating in the mass slaughter and condemned to death, but were later pardoned as part of a reconciliation process. See Gildea et al. (2006), 193.
201 **Eisenhower and his planners:** Dallas (2005), 86.
201 **heavy night raids:** Gilbert (2004), 79.
201 **After bombs fell:** Dallas (2005), 86.
201 **"Montmartre and the northern suburb":** Ibid.
201 **Despite all that had transpired:** http://www.youtube.com/watch?v=508EWoNE4fM.
201 **"Our country is experiencing":** *Le Matin,* April 29, 1944, trans. SBC.
202 **"the threat from the East":** Harrison (1951), 464.
202 **"Only an all-out effort":** Ibid.
203 **"throw the enemy back":** Ibid., 465.
203 **He ordered the flooding:** Rommel defense tactics described in Ryan (1959) 15–30; Ambrose (1994), 112–13.
204 **"If, in spite of":** Gilbert (2004), 79.
204 **The original target date:** Eisenhower (1982), 416.
204 **It offered thirty kilometers:** Ambrose (1994), 72–73.
205 **In fact, Eisenhower was:** Ibid., 82.
205 **It became clear:** Eisenhower (1982), 454.
205 **On May 2:** "Pierre Dejussieu-Pontcarral," Ordre de la Liberation, http://www.ordredelaliberation.fr/fr_compagnon/263.html.
205 **The chief of the Paris region:** "Pierre Pène," Ordre de la Liberation, http://

www.ordredelaliberation.fr/fr_compagnon/758.html; Rol-Tanguy, Liberation de Paris, 62.

205 **The arrests necessitated:** "Henri Rol-Tanguy," Ordre de la Liberation, http://www.ordredelaliberation.fr/fr_compagnon/947.html.

205 **and Monod ("Malivert") was promoted:** Monod FFI record from document "Etat des Service dans les F.F.I." dated November 26, 1944, MON. Bio. 02, Fonds Monod, SAIP.

206 **After more street meetings:** Noufflard (unpublished), 81.

206 **If someone came unexpectedly:** Ibid., 82.

206 **Fewer Parisians walked:** Letter, Odette Monod to Lucien and Charlotte Monod, May 19, 1944, private archives, Monod family.

206 **The Germans had installed:** Guillain de Bénouville (1949), 301.

207 **He put his new look:** Noufflard (unpublished), 86.

207 **The short trip:** Letter, Odette Monod to Lucien and Charlotte Monod, May 19, 1944, private archives, Monod family.

207 **There were German troops:** Letter, Odette Monod to Lucien and Charlotte Monod, April 13, 1944, private archives, Monod family.

207 **air-raid alerts were very frequent:** Letter, Odette Monod to Lucien and Charlotte Monod, May 19, 1944, private archives, Monod family; Philippe Monod, interview, Paris, December 3, 2011.

207 **Odette confessed to Jacques's parents:** Letters, Odette Monod to Lucien and Charlotte Monod, April 20, 1944 and May 19, 1944, private archives, Monod family.

207 **Plan Vert had identified 571 rail targets:** Harrison (1951), 205

207 **On May 10, listeners:** Crémieux-Brillhac (1975), 5:2.

207 **On May 12, commentators:** Ibid., 5:3.

207 **On May 20, listeners received:** Ibid., 5:16.

207 **On May 27, they:** Ibid., 5:24–25.

208 **"Ouvrez l'oeil et le bon":** Ibid., 5:32.

208 **Noufflard heard the news:** Noufflard (unpublished), 90.

208 **On Saturday, June 3:** "Pierre Lefaucheux," Ordre de la Liberation, http://www.ordredelaliberation.fr/fr_compagnon/574.html.

208 **"Isn't it terrible":** Noufflard (unpublished), 88–91.

209 **the Gestapo had in fact:** The genesis of the meeting and the arrests are described by Pierre Bourlier, alias "Guillaume," who attended and was arrested at the meeting. See P. Bourlier, "Ma Resistance," http://chezpeps.free.fr/henri/html/buchenwald_matricule_76888.html.

209 **As Monod approached:** Olivier Monod, interview, Paris August 17, 2010.

209 **"Il est sévère mais juste":** Crémieux-Brillhac (1975), 5:41.

210 **Each sentence was:** It has been widely reported that *the* invasion signals to the FFI were lines from Paul Verlaine's poem "Chanson d'Automne," the first line of which was "*Les sanglots longs des violons d'automne*" (The long sobs of the violins of autumn) and the second was "*bercent mon couer d'une langueur monotone* (fills my heart with a langorous sorrow). See, for example, A. Hall

and T. Hall, *D-Day:Operation Overlord Day by Day,* 100. This is not accurate. It was the case that every branch of the Resistance had a distinct message, and the Verlaine lines were directed to the Ventriloquist action group (Brown [1975], 560). In fact, the Germans had learned about this message and were thus tipped off about the timing of the invasion [ibid.]. According to three different sources who were in the Paris region FFI, including Noufflard's memoir that was written in 1945, they were awaiting the line *"Il est sévère mais juste"* and the other lines stated in the text. Also see accounts of G. Gilbert, "Rapport de Mon Activité Clandestine," http://aacvr.free.fr/h_rapport_gg.htm, and P. Boulier, "Ma Resistance."

Chapter 15: Normandy

211 **"The history of warfare":** Churchill (1964), 8.
211 **While one of the two:** Brown (1975), 640–41.
211 **Rommel did not think an invasion:** Ambrose (1994), 88.
211 **On June 3, he had:** Ryan (1959), 37; Gilbert (2004), 110–11.
212 **The Germans had failed entirely:** Harrison (1951), 275.
212 **The weather was:** Ambrose (1994), 183.
212 **The invasion fleet:** Ibid., 257.
212 **"Under the command":** Crémieux-Brillhac (1975), 5:45, trans. SBC.
212 **"Now comes the time":** Noufflard (unpublished), 93–94.
213 **Camus also heard:** Todd (1997), 185; Lottman (1979), 318.
213 **"The whole country is":** "Frankin Roosevelt's Press Conference on D-Day," Our Documents: D-Day, http://docs.fdrlibrary.marist.edu/odddaypc.html.
213 **"So far the Commanders":** "D-Day," Winston Churchill Leadership, http://www.winston-churchill-leadership.com/speech-d-day.html.
214 **"The German and Anglo-Saxon":** *Le Matin*, June 7, 1940, trans. SBC.
214 **Laval compounded Pétain's warning:** *Le Matin*, June 7, 1940.
215 **"The supreme battle":** Crémieux-Brillhac (1975), 5:47–48, trans. SBC.
215 **Eisenhower had composed:** "Message Drafted by General Eisenhower in Case the D-Day Invasion Failed," National Archives, http://www.archives.gov/education/lessons/d-day-message.
215 **at least 4,400 killed and 5,000 wounded:** These numbers vary widely among sources. See D-Day Museum Online, http://www.ddaymuseum.co.uk/faq.htm#casualities.
215 **By the time offloading paused:** "Message Drafted by General Eisenhower."
215 **By the day after D-Day:** Harrison (1951), 206.
215 **Designed to prevent the movement:** Beavan (2006), 144; Asprey (1975), 318.
216 **"SUBJECT: SABOTAGE OF RAIL LINES":** Document courtesy of Geneviève Noufflard, trans. SBC and H. Dufour.
217 **Monod set up an intelligence channel:** Noufflard (unpublished), 98–101.
218 **Monod used his previous:** Noufflard (unpublished), 103–4.
218 **For security, the transmitter-room door:** Olivier Monod, interview, Paris, August 17, 2010.

218 **These compromising papers:** Noufflard (unpublished), 95–97.

219 **The Germans and the French were arresting:** Mitchell (2008), 104–9.

219 **"irreparable loss":** Camus, *Combat,* October 27, 1944; Camus (2006), 92.

219 **a "dreadful death":** Brée (1961), 42–43.

219 **Four days later in Lyon:** "André Bollier," Ordre de la Liberation, http://www.ordredelaliberation.fr/fr_compagnon/116.html.

219 **But before doing so:** Ajchenbaum (1994), 87; "André Bollier"; http://www.lajauneetlarouge.com/article/andre-bollier-38-dit-%E2%80%9C-velin-%E2%80%9D.

220 **In early July:** Lottman (1979), 323; Todd (1997), 185. The exact date of the meeting is uncertain, but it preceded July 11 by some short time.

220 **She saw Camus with his hands:** Todd (1997), 187. The date of the incident is not certain, but it appears to have preceded July 11 because of subsequent actions that Camus is reported to have taken first in response to his close call, and then after Jacqueline Bernard's arrest on July 11; in any case, the incident must have preceded Camus's hasty departure from Paris after Bernard's arrest.

220 **Camus and Marcel Gimont:** Ajchenbaum (1994), 88.

220 **"The time is fast approaching":** Camus, *Combat* 58, July 1944; Camus (2006), 7–9.

221 **Earlier that day, she had gone:** Lottman (1979), 323.

222 **Camus had to leave town:** Ibid., 324–25.

222 **His battalion reached:** http://www.marinettes-et-rochambelles.com/pages/JDM-1944-3trimestre.htm.

223 **After more than four years:** Jacob (1988), 106.

223 **"The soil of France":** Ibid., 106–7.

223 **"People of France":** Fondation Leclerc, http://www.fondation-leclerc.com/52/leclerc-et-ses-hommes/colonne-leclerc-2eme-db/2eme-france-allemagne.htm, trans. SBC.

224 **As the DB passed through:** Jacob (1988), 108; C-C Notin, *1061 Compagnons,* 686, trans. SBC.

224 **Despite the widespread destruction:** Patton and M. Blumenson (1974), 489.

225 **Patton's 3rd Army:** Blumenson (1993), 195.

225 **Leclerc seized the offer:** Fondation Leclerc, http://www.fondation-leclerc.com/52/leclerc-et-ses-hommes/colonne-leclerc-2eme-db/2eme-france-allemagne.htm.

225 **"an opportunity that comes":** Ripley (2003), 111.

225 **On the night of the eighth:** Jacob (1988), 110.

226 **He then heard the cries:** http://www.francaislibres.net/liste/fiche.php?index=54646.

226 **He and Jacob had become:** Jacob (1988), 109.

226 **Jacob looked once more:** Ibid., 110–11.

226 **Jacob, Benillouz, and several other:** http://www.marinettes-et-rochambelles.com/pages/JDM-1944-3trimestre.htm.

226 **As Jacob drifted:** Jacob (1988), 111.

226 **When Jacob woke up:** Ibid., 166–67.

Chapter 16: Les Jours de Gloire

227 **In order to encircle the Germans:** Jordan (2011), 377–80.
227 **Patton was furious:** Patton and Blumenson (1974), 510.
227 **Leclerc, too, was getting impatient:** Prados (2011), 254.
227 **In the meantime:** Patton and Blumenson (1974), 510.
228 **Leclerc, however, objected:** Ibid., 511.
228 **On Saturday, August 19:** (patriotic militia) Rol-Tanguy and Bourderon (1994), 185–86.
229 **"Gather yourselves by household":** Ibid., 180–81.
229 **A bicyclist going the other way:** Noufflard (unpublished), 110–11.
230 **Some of the trucks:** Bourderon (2004), 394; Ousby (2000), 291.
230 **His purpose that day:** Ajchenbaum (1994), 90–91.
230 **All fifty-six dailies:** Ibid., 92.
231 **Their authorization to begin:** Ibid., 98.
231 **In the meantime:** Ibid., 93–94.
231 **The police and the FFI:** Laub (2010), 287.
231 **"stamp out without pity":** Collins and Lappierre (1965), 36.
231 **To von Choltitz's surprise:** Ousby (2000), 291.
232 **Just before dark:** Noufflard (unpublished), 111.
232 **"The development of operations":** E.M.N. 3o Bureau à Region P1, August 20, 1944. Document courtesy of Geneviève Noufflard, trans. SBC.
233 **Monod turned to Noufflard:** Noufflard (unpublished), 111.
233 **Monod was wrong:** Bourderon (2004), 413; Rol-Tanguy and Bourderon (1994), 250.
234 **Noufflard bicycled over:** Noufflard (unpublished), 114.
234 **Before he hung up:** Collins and Lappierre (1965), 148; Schoenbrun, (1980), 455.
235 **Rol-Tanguy and his subordinates:** Bourderon (2004), 413.
235 **German vehicles were trapped:** Collins and Lappierre (1965), 165.
235 **and French sacrifices:** Schoenbrun (1980), 456.
235 **It was a momentous day:** Pétain would later be moved to Sigmaringen in southern Germany. See Rousso (1984), 78.
235 **Upon arrival, de Gaulle:** Collins and Lappierre (1965), 136–37.
235 **De Gaulle was promptly driven:** Ibid., 142–43.
235 **The next morning, Monod and Noufflard:** Noufflard (unpublished), 116–17; Bourderon (2004), 419–25.
237 **Monod and Noufflard would each return:** Noufflard (unpublished), 119; Monod identity card, courtesy of Olivier Monod, private archives, Monod family.
237 **For two days, the *Combat* staff:** Ajchenbaum (1994), 95.
237 **Since their July meeting:** *Combat,* August 21, 1944; Camus (2006), 13.
237 **Their answer was a France:** Ibid.
237 **It was a markedly different tone:** Ajchenbaum (1994), 99.
239 **"Today, August 21":** *Combat,* August 21, 1944; Camus (2006), 11–12.

238 **Indeed, the combat in the streets:** Bourderon (2004), 414; Collins and Lappierre (1965), 172–73.

238 **Tuesday, August 22:** Collins and Lappierre (1965), 184.

239 **But German tanks rolled:** Rol-Tanguy and Bourderon (1994), 271.

239 **The defenders' main weapon:** Pinault (2000), 260.

239 **The mixture combined:** Collins and Lappierre (1965), 109.

239 **Indeed, Joliot-Curie's lab:** Pinault (2000), 260.

239 **Rol directed the distribution:** Collins and Lappierre (1965), 184–85.

239 **To the west in Normandy:** Ibid., 165.

239 **"Information received today":** Prados (2011), 255.

240 **The situation was so critical:** Collins and Lappierre (1965), 165–66.

240 **While Eisenhower pondered:** Ibid., 180.

240 **The Falaise Gap:** Zuehlke (2007), 32.

240 **There was no reason to hold Leclerc:** *La 2e DB Général Leclerc: Combattants et Combats en France* (1945), 44–45.

240 **The next morning's issue:** *Combat,* August 23, 1944, author copy, translation from Camus (2006), 15.

241 **"Give through Swiss press":** Noufflard (unpublished), 118.

242 **"You know perfectly well":** Ibid., 127.

242 **"The Americans are coming":** Collins and Lappierre (1965), 53.

242 **"The defense of the Paris":** Ibid., 200.

243 **All of the bridges:** Ibid., 208–12.

243 **"We are coming":** Lottman (1979), 332.

243 **Fierce fighting continued:** Collins and Lappierre (1965), 234.

243 **Monod and Noufflard spent the day bicycling:** Noufflard (unpublished), 128.

243 **At 9:32, they heard:** Collins and Lappierre (1965), 156; 100 Ans de Radio, 24 août, 1944 http://100ansderadio.free.fr/HistoiredelaRadio/1944.html.

243 **"Parisians, rejoice":** Collins and Lappierre (1965), 256.

243 **"Awake! Be done with shame!":** 100 Ans de Radio, 24 août, 1944, http://100ansderadio.free.fr/HistoiredelaRadio/1944.html.

244 **Monod and Noufflard opened:** Noufflard (unpublished), 128.

244 **Noufflard and Monod wanted:** Ibid., 129.

244 **"You are coming with me":** The walk to the Ministry of War and the scene within are based on Noufflard (unpublished), 130–33.

246 **"AFTER FOUR YEARS":** *Combat,* August 25, 1944, author's collection.

246 **"As freedom's bullets":** Ibid.; Camus (2006), 17–18.

247 **Dawn broke to a perfect:** Account of the events at the War Ministry from Noufflard (unpublished), 134–38, and Horace Freeland Judson Papers, interview with Geneviève Noufflard, November 16, 1976.

248 **Noufflard went back:** Noufflard (unpublished), 137.

248 **At last, Monod and Noufflard:** Ibid., 138.

248 **Then, suddenly, the city lit up:** Collins and Lappierre (1965), 323–24, 285; Schoenbrun (1980), 475.

Chapter 17: The Talk of the Nation

253 **"The first thing for a writer":** Camus (1965), 37.
253 **"All Paris in the Street":** *Combat,* August 27, 1944, author's collection, trans. SBC.
253 **"Albert CAMUS, Henri FREDERIC":** Ibid. Henri Frederic was a pseudonym for Henri Cauquelin.
253 **"a unique opportunity":** *Combat,* September 1, 1944; Camus (2006), 25.
253 **"The Paris that is fighting":** *Combat,* August 24, 1944; Camus (2006), 17.
253 **It was, Camus would say:** *Combat,* September 30, 1944; Camus (2006), 54.
254 **In his signed editorial:** *Combat,* August 31, 1944; Camus (2006), 21–23.
254 **Camus hoped to influence:** *Combat,* September 1, 1944; Camus (2006), 25.
255 **"seek to inform":** *Combat,* September 8, 1944; Camus (2006), 32.
255 **"The truth is not the beneficiary":** Ibid.
255 **"To ensure that life":** *Combat,* September 8, 1944; Camus (2006), 31–32.
255 **"The affairs of this country":** *Combat,* September 4, 1944; Camus (2006), 28.
255 **"If our American friends":** *Combat,* September 30, 1944; Camus (2006), 53–54.
256 **Days after his fourth:** "A Guide to the United States' History of Recognition, Diplomatic, and Consular Relations, by Country, since 1776: France," U.S. Department of State, Office of the Historian, http://history.state.gov/countries/france.
256 **"with a language":** *Combat,* September 8, 1944; Camus (2006), 34.
256 **formed the habit:** Aron (1983), 208.
256 ***Combat* often sold out:** Lottman (1979), 341.
256 **After the liberation:** Ibid.
257 **When one or the other boy:** Letter, Odette Monod to Lucien and Charlotte Monod, October 29, 1944, private archives, Monod Family.
257 **"We have enjoyed":** *Combat,* September 29, 1944; Camus (2006), 51.
257 **"It is essential":** Yeide and Stout (2007), 181.
257 **By that time:** "The Siegfried Line Campaign," U.S. Army Center of Military History. http://www.history.army.mil/books/wwii/Siegfried/Siegfried%20Line/siegfried-ch01.htm#ch1.
258 **"To assist . . . in the study":** Ordre de Mission, November 13, 1944, signed by Gen. Joinville, MON. Bio. 02, SAIP.
258 **He took Toulon and Marseille:** Yeide and Stout (2007), 24–26; Davidson (1988), 99–100.
258 **"win over this vibrant":** Lattre (1952), 179.
259 **"most precious *auxiliaire*":** Letter, Jacques Monod to Odette Monod, December 8, 1944, private archives, Monod family.
259 **By February, an impressive:** Lattre (1952), 173.
259 **"My dear Odette":** Letter, Jacques Monod to Odette Monod, May 1, 1945, private archives, Monod family, trans. H. Dufour.
259 **"History is full":** *Combat,* May 9, 1944; Camus (2006), 195–96.

260 **"Many of our comrades":** *Combat*, September 16, 1944; Camus (2006), 39–40.

260 **"Those of us who are still waiting":** *Combat*, May 9, 1945; Camus (2006), 195–96.

260 **"Given the terrifying":** *Combat*, August 8, 1945; Camus (2006), 237.

261 **By the end of the war:** Aronson (2004), 46–47.

261 **"the French editorialist":** Ibid., 79.

Chapter 18: Secrets of Life

262 **Camus and *Combat* celebrated:** *Combat*, April 17, 1945; Camus (2006), 194–95.

262 **"*Combat* is waiting for you":** Ajchenbaum (1994), 207–8.

262 **Marcel Prenant, Monod's former FTP chief:** Prenant (1980), 245–71.

262 **While he had hoped:** Letter, Jacques Monod to Odette Monod, May 1, 1944, private archives, Monod family.

262 **his papers would not be signed:** Letter, Jacques Monod to Odette Monod, July 2, 1944, MON. Bio. 02, Fonds Monod, SAIP.

262 **"drew a curtain":** Judson (1979), 368.

263 **That fall, Lwoff invited Monod:** Ullmann (2003), 5.

263 **In the November 1943 issue:** Luria and Delbrück (1943), 491–511; Monod (1965), 190.

263 **The paper confirmed:** See chapter 13.

263 **"They have no genes":** Huxley (1963), 131–32.

264 **Indeed, so little was known:** Mayr (1997), 1–21.

264 **"How can the events":** Schrödinger (1992), 3.

264 **"The obvious inability:** Ibid., 4.

264 **Schrödinger speculated:** Ibid., 61.

265 **"strict and severe-looking":** Ullmann (2003), 37.

265 **"At any rate":** Ibid.

267 **"We are living in nihilism":** Lottman (1979), 374.

267 **"Revolt gives life its value":** Camus (1991a), 55.

267 **"feelings and images multiply":** Camus (1963), 210.

267 **"From now on":** Camus (1991b), 67.

267 **"the true embodiment":** Ibid., 134.

268 **"the flail of God":** Ibid., 95.

268 **"Calamity has come on you":** Ibid., 94.

268 **In the evenings, he would socialize:** Lottman (1979), 369; Aronson (2004), 50.

268 **"Man is nothing else":** J.-P. Sartre, "Existentialism Is a Humanism," http://www.marxists.org/reference/archive/sartre/works/exist/sartre.htm.

268 **"places the entire responsibility":** Ibid.

269 **It was a perk for prominent:** Lottman (1979), 376.

269 **"The Boldest Writer in France ":** *New York Times*, April 7, 1946.

269 **"After two wars have shattered":** Ibid.
270 **He told his audience:** Camus (1946–1947), 19–33.
270 **Camus explained that:** Ibid., 20.
270 **In order to illustrate:** Ibid., 21.
271 **"the death or torture":** Ibid., 22.
271 **Among the many contributors:** Ibid., 22–24.
271 **"the civilization of death":** Ibid., 27.
271 **"The great lesson":** Ibid., 27–28.
271 **"It took the war":** Ibid., 30.
271 **From this painful experience:** Ibid., 28–29.
272 **"man and woman":** Ibid., 31.
272 **"Those who met":** Lottman (1979), 381.
272 **That included nineteen-year-old:** Ibid., 388.
272 **"I, for one, am practically certain":** *Combat,* November 30, 1946; Camus (2006), 274.
273 **"whose first article would":** *Combat,* November 29, 1946; Camus (2006), 273.
273 **"civilization based on dialogue":** *Combat,* November 29, 1946; Camus (2006), 273.
273 **"Across five continents":** *Combat,* November 30, 1946; Camus (2006), 275–76.
273 **A few weeks after his series:** Lottman (1979), 407.
273 **"There's something lacking":** Camus (1991b), 254–55.
273 **"resolved to compile this chronicle":** Ibid., 308.
274 **"Nonetheless, he knew that the tale":** Ibid.
274 **The novel would be a bestseller:** Lottman (1979), 427, 431.

Chapter 19: Bourgeois Genetics

275 **"there is always something":** Ullmann, (2003), 39.
275 **Monod and Audureau's report:** Monod and Audureau (1946).
276 **During the war:** Avery et al. (1944).
276 **In early 1946, they reported:** McCarty and Avery (1946).
276 **"Twelve hours above the water":** Letter, Jacques Monod to Odette Monod, June 25, 1946, private archives, Monod family.
277 **"the difficulty in being current:** Letter, Jacques Monod to Odette Monod, July 2, 1946, private archives, Monod family.
277 **"of understanding how cells":** Monod (1947), 224.
277 **"may help in understanding":** Ibid.
278 **"Its significance appeared so profound":** Judson (1979), 370.
278 **In addition to scientific ideas:** Ullmann, (2003), 56.
278 **"absolute contempt and hatred":** Judson (1979), 368.
278 **By the end of 1945:** Ibid.
278 **"Heredity Is Not Commanded":** *Les Lettres Françaises,* August 26, 1948, MON Pol. 1.6, Fonds Monod, SAIP, trans. SBC.

279 **"mysterious and unforeseeable fashion":** Ibid.

279 **"Human intervention makes":** Ibid.

280 **"As in our world":** Ibid.

280 **Indeed, two days later in *Le Monde*:** *Le Monde,* August 28, 1948, MON. Pol. 1.7, Fonds Monod, SAIP.

281 **"A shadow has fallen":** "Winston Churchill's Iron Curtain Speech," The History Guide: Lectures on Twentieth Century Europe, http://www.historyguide.org/europe/churchill.html.

282 **In March 1947:** "Truman Doctrine (1947)," Our Documents, http://ourdocuments.gov/doc.php?flash=true&doc=81.

282 **"Our policy is directed":** "Marshall Plan Speech," George C. Marshall Foundation, http://www.marshallfoundation.org/library/MarshallPlanSpeechfromRecordedAddress_000.html.

282 **The Communist Information Bureau:** Tiersky (1974), 162–63.

282 **At the meeting:** Boterbloem (2004), 312.

282 **There was pressure:** Medvedev (2006), 196.

283 **Under Zhdanov:** Boterbloem (2004), 253–57, 305.

283 **The sealing off:** Ibid., 293.

283 **In the realm of science:** Medvedev (2006), 196–97.

283 **Well before the war:** Carroll (2006), 219–25. See also Z. Medvedev (1969), *The Rise and Fall of T.D. Lysenko,* and V. Sooyfer (1994), *Lysenko and the Tragedy of Soviet Science.*

283 **When Lysenko came under criticism:** Medvedev (2006), 190–95.

283 **"The Michurin position":** Ibid., 202.

284 **"is a return to the Middle Ages":** Molenaar (1981), 67–100.

284 **Camus and most of its editorial leadership:** Ajchenbaum (1994), 295.

284 **Its stated aim:** Tirard (1997), 98.

284 **"The recent Moscow debates":** Ibid.

284 **"According to Professor Marcel Prenant":** *Combat,* September 14, 1948, MON. Pol. 1, Fonds Monod, SAIP.

285 **"The really new point":** Ibid.

285 **"The important thing":** *Combat,* September 15, 1948, MON. Pol. 01, Fonds Monod, SAIP.

285 **"hardly be suspected":** Ibid.

285 **"Lysenko's claim":** Davies (1947), 344.

285 **"To every scientist":** Ibid., 343.

286 **"The system by which he":** Fyfe (1947), 348.

286 **"These judgments":** *Combat,* September 15, 1948, MON. Pol. 01, Fonds Monod, SAIP.

287 **"What emerges most clearly":** Ibid.

287 **The PCF continued:** Europe (1948), 31–68.

287 **Poet Louis Aragon:** Ibid., 3–30.

288 **The "debate" raged on:** Unidentified newspaper clippings, MON. Pol. 1.6, Fonds Monod, SAIP.

288 **"a purely theological affair":** Judson (1979), 372.

288 **pivotal because it started:** Ibid.

288 **For a time, he spent one Thursday:** Cohn (1978), 2.

288 **Cofounded by Camus:** Lottman (1979), 459–62; G. Walusinski, "Manifeste des Groupes de liaison internationale," A Contretemps: Bulletin de Critique Bibliographique, 22–24, www.acontretemps.org/spip.php?article238.

288 **He brought the scientist:** The exact timing of Camus and Monod's first meeting is not known, but is most likely between September 1948 and June 1949. Because of a newly discovered letter from Camus to Monod, it is certain they met prior to December 1949. Monod's recollection was that they met after he had published in *Combat* in September 1948 (H. F. Judson, "Transcript of Interview with Jacques Monod, December 1975," Judson Collection, American Philosophical Society, Philadelphia). Also, Lottman reports that Camus went to South America in June 1949 and was sick for a time thereafter, so a meeting prior to June, when the Lysenko affair was still fresh, would be most consistent. It is not correct that they met when Camus was at *Combat*, nor that Camus was editor when Monod published his article (Debré [1996], 200); Camus had been gone from *Combat* for over a year at the time of publication.

289 **"We are a group of men":** Walusinski, 22–24.

Chapter 20: On the Same Path

290 **The two former resistants:** Debré (1996), 200. La Closerie des Lilas was one of Monod's favorite brasseries: Melvin Cohn, e-mail to author, January 5, 2010.

290 **A Communist activist:** Scammell (2009), 162.

290 **From 1936 to 1938:** Aronson (2004), 70.

290 **The book was a sensation:** Scammell (2009), 286.

290 **Koestler visited Paris:** Ibid., 292.

291 **"It must be said":** Camus (1965), 145.

291 **Camus had met Louis Aragon:** Todd (1997), 165.

291 **and Camus's critic Emmanuel d'Astier:** See chapter 12.

291 **In 1948, d'Astier:** "Emmanuel d'Astier de la Vigerie," Ordre de la Libération. http://www.ordredelaliberation.fr/fr_compagnon/36.html.

292 **He asked Camus:** Lottman (1979), 436–37, 447–48.

292 **"The camps were part":** Camus Actuelles I. Ecrit Politique, second response to Emmanuel d'Astier de la Vigerie.

292 **"The majority among":** Ibid.

292 **"My role is not to transform":** Ibid.

293 **Upon encountering Merleau-Ponty:** Lottman (1979), 405.

293 **More than two years later:** Todd (1997), 248.

293 **"Whatever the nature":** Aronson (2004), 110–11.

293 **But with the benefit:** Todd (1997), 249.

293 **For Monod, who admired:** H. F. Judson, "Transcript of Interview with Jacques Monod, December 1975," Judson Collection, American Philosophical Society, Philadelphia.

293 **The other guests who enjoyed:** Melvin Cohn, e-mail to author, January 5, 2010. The date of the dinner is not known.

294 **"My dear Monod":** Letter, A. Camus to J. Monod, private archives, December 21, 1949. Monod family.

295 **Oleksandr Bogomoletz was a Ukrainian scientist:** Lawrence and Weisz (1998), 266.

295 **Stalin named Bogomoletz:** *Time*, January 17, 1944.

295 **The June 1949 issue:** Lawrence and Weisz (1998), 268.

296 **A Jacques Monod:** Jacques Monod's copy of *Actuelles I*, private archives, Monod family, inscription courtesy of Olivier Monod.

Chapter 21: A New Beginning

297 **Jacob had been named:** "Présentation de l'Ordre de la Libération," Ordre de la Libération, http://www.ordredelaliberation.fr/fr_doc/1_1_presentation.html.

298 **At one meeting:** Jacob (1988), 212.

298 **Jacob thought it was incredible:** Ibid., 209.

298 **He was astonished:** Ibid., 32.

298 **He then went to the director:** Ibid., 210–11.

298 **He went to see Monod:** Ibid., 211–12; Judson (1979), 385.

299 **Lwoff pondered Jacob:** Judson (1979), 385.

299 **"We have just found":** Jacob (1988), 213.

299 **The attic was cramped:** Ibid., 214–33.

300 **"like a cross between":** Ibid., 227.

300 **"Call me what you like":** Ibid., 229.

301 **"You should go":** Ibid., 244.

302 **He gave himself five years:** Ibid., 244–51.

302 **"chatting across the death mask":** Cohn, as quoted in Ullmann (2003), 102.

303 **Cohn and Monod developed:** Monod, Cohen-Bazire, and Cohn (1951).

303 **Monod's team found:** Cohn, as quoted in Ullmann (2003), 96–97.

304 **"had been answered with experimental":** Ibid., 97.

304 **The phenomenon was renamed:** Cohn et al. (1953), 1096.

305 **Among many measures:** "1950s: International Security Act of 1950," Documents of American History II, http://tucnak.fsv.cuni.cz/~calda/Documents/1950s/Inter_Security_50.html.

305 **"In view especially":** Letter, J. Monod to American Consul R. Clyde Larkin, June 4, 1951, MON. Pol. 06, Fonds Monod, SAIP.

306 **Monod's letter was subsequently published:** Monod (1952).

Chapter 22: Rebels with a Cause

307 **One evening, the two met:** J. Daniel, *Le Nouvel Observateur*, January 30, 1987, 22. The date of the dinner is not known, but Daniel's reference to Monod's visa problems and the date of Monod's letter to the American consul allow the inference that the date was May–June 1951.

308 **"What is a rebel":** "Remarque sur la Révolte," in Camus (1945), 9–23.
308 **"transcends the individual":** Ibid.
308 **"the solidarity of man":** Camus (1956), 18.
309 **"All modern revolutions":** Ibid., 177.
309 **"The greatest revolution":** Ibid., 240.
309 **"the dictatorship of":** Ibid., 232.
309 **"contrives the acceptance":** Ibid., 233.
309 **"the concentration camp system":** Ibid., 238.
309 **"dialogue and personal relations":** Ibid., 239–40.
309 **"the ration coupon":** Ibid., 240.
310 **"The answer is easy":** Ibid.
310 **"There remains of Marx's":** Ibid., 222–23.
310 **"Prophecy functions on":** Ibid., 189.
310 **"postpones to a point":** Ibid., 303.
310 **"real generosity to the future":** Ibid., 304.
310 **Well after his *Combat* article:** J. Monod, "Les Positions Scientifiques," unpublished manuscript, MON. Pol. 01, item 2, Fonds Monod, SAIP. This manuscript, in examining the role of chance in nature, undoubtedly planted seeds for what was to emerge two decades later in Monod's *Chance and Necessity*.
311 **"In ridding our science":** Ibid.
311 **"not only that *all*":** Ibid.
311 **And indeed, following:** Graham (1964).
311 **Monod attributed the denial:** Ibid., 30.
311 **"To make Marxism scientific ":** Camus (1956), 221–22.
312 **"has also had to rewrite":** Ibid., 236–37.
312 **"Rebellion indefatigably confronts evil":** Ibid., 303.
313 **rebellion "in moderation":** Ibid., 301.
313 **"When revolution in the name":** Ibid., 305.
313 **"complicity so intense":** J. Daniel, *Le Nouvel Observateur*, January 30, 1987.
313 **The two rebels:** Note language of letter from Camus to Monod in November 1957. See prologue, page 2.
313 **"à Jacques Monod cette réponse":** J. Monod, personal copy of A. Camus *L'Homme révolté*, private archives, Monod family, inscription courtesy of Olivier Monod [underscore in original].

Chapter 23: Taking Sides

314 **"Let's shake hands":** Todd (1997), 305.
314 **In *Le Figaro Littéraire*:** Lottman (1979), 496.
314 ***Le Monde* agreed:** Ibid.
314 **"More than the coming":** Ibid., 497.
315 **The book sold:** Todd (1997), 305.
315 **Jeanson, like Sartre:** Lottman (1979), 500–501; Todd (1997), 307; Aronson (2004), 135–36.
315 **A short time later:** Lottman (1979), 501.

315 **Jeanson mockingly questioned:** Translation in Sartre and Camus (2004), 79–80.

316 **"incoherent"—a "pseudophilosophical":** Translation in ibid., 101.

316 **"the privileges of":** Translation in ibid., 202.

316 **Sartre's secretary let Camus know:** Todd (1997), 307.

316 **"Dear Editor, I will":** A. Camus, "A Letter to the Editor of Les Temps Modernes," in Sartre and Camus (2004), 107.

317 **"A loyal and wise critic":** Ibid., 116.

317 **"In it I have found":** Ibid., 118.

317 **"I am beginning":** Ibid., 126.

317 **"the revolutionary living":** Sartre (1969), 10.

317 **"the USSR wants peace":** Ibid., 13.

318 **"My Dear Camus":** J.-P. Sartre, "Reply to Albert Camus," in Sartre and Camus (2004), 131–32.

318 **"The mixture of dreary":** Ibid., 132.

318 **"You do us the honor":** Ibid., 133.

319 **"Perhaps the Republic":** Ibid., 137.

319 **"Suppose you were wrong":** Ibid., 139.

319 **"You have been for us":** Ibid., 147–48.

319 **In 1944, Camus had been:** Ibid., 155.

319 **"I have said what you meant":** Ibid., 158.

319 **The newspaper headlines heralded:** Todd (1997), 312.

319 **"The Sartre-Camus Break":** Lottman (1979), 506.

319 **"Sartre, the man":** Camus (2008), 50.

320 **"I am anguished by Paris":** Todd (1997), 311.

320 **"At this point, the least sentence":** Ibid., 313.

320 **"Immediately after the attack":** Ibid., 314.

320 **"Leftist intellectuals":** Ibid.

320 **"The core of the problem":** Lottman (1979), 512.

321 **"Nuptials at Tipasa":** All passages quoted here are from Albert Camus, "Return to Tipasa," in Camus (1991a). The essay was completed and published initially in 1953 in the first issue of the Algerian review *Terrasses,* edited by Jean Sénac (H. Nacer-Khodja, A. Camus, and J. Sénac, *Albert Camus, Jean Sénac, ou Le Fils rebelle* 50), then published again in a compilation of Camus's essays entitled *L'Été,* published in 1954 (this quote 196).

321 **"And under the glorious":** "Return to Tipasa," in Camus (1991a), 200–201.

322 **"one by one the imperceptible":** Ibid., 201.

322 ***"Je redecouvrais à Tipaza":*** Ibid., 202.

322 **"bucked up and calmed":** Todd (1997), 314.

322 **"I have returned to Europe":** "Return to Tipasa," in Camus (1991a), 202–3.

322 **"à Jacques Monod":** J. Monod, personal copy of A. Camus *L'Été*, private archives, Monod family, inscription courtesy of Olivier Monod.

Chapter 24: The Attic

324 **"There is an element":** Camus (1965), 37.

324 **Arriving before most:** Jacob (1988), 241–42.

324 **The observation dated back decades:** Holmes (2006), 123–24.

325 **He first studied thirty strains:** Jacob (1988), 234–35.

326 **"The French don't live like this":** S. Benzer, interview by Heidi Aspaturian, Pasadena, California, September 1990–February 1991, Oral History Project, California Institute of Technology Archives, http://resolver.caltech.edu/CaltechOH:OH_Benzer_S, 32–33.

326 **"Did you ever try *tétine de vache*":** Benzer, interview, 42.

327 **His parents, Eugène and Elisabeth:** "Elie Wollman (1917–2008) Notice biographique," Institut Pasteur, http://www.pasteur.fr/infosci/archives/wol0.html.

327 **His godfather and namesake:** "Ilya Mechnikov—Biography," Nobel Prize, http://www.nobelprize.org/nobel_prizes/medicine/laureates/1908/mechnikov.html; Jacob (1988), 243.

327 **Eugène and Elisabeth Wollman were in fact:** A. Lwoff, in Comité à la mémoire des savants français (1959), 133–45; Jacques Tréfouël, statement of March 28, 1945, testimony for La Cour de Justice, Fonds Tréfouël, SAIP.

327 **After the war:** A. Ullmann, e-mail to author, April 13, 2012.

327 **"as much as had being part":** Jacob (1988), 274.

327 **"a revenge on the war":** Ibid., 271–72.

328 **"No sir, I am not":** Benzer, Interivew, 40.

329 **Convinced by Oswald Avery's original evidence:** Watson (1969), 73–74.

329 **The results indicated:** Jacob (1988), 264; Judson (1979), 131–32.

330 **What was the nature:** Jacob (1988), 253, 254.

330 **All of the top phage scientists:** Ibid., 263–65.

330 **The truth was:** Watson (1969), 89–90.

330 **Watson prudently borrowed:** Ibid., 90.

331 **"membership card to the club":** Jacob (1988), 263.

332 **After struggling for many months:** Watson (1969), 123–26.

332 **Although Watson did not:** James D. Watson, phone interview, March 27, 2012.

333 **It was not until six weeks later:** Jacob (1988), 269.

333 **"All this could not be false":** Ibid., 271.

333 **"one of the oldest problems":** Ibid.

334 **Before returning to France:** Ibid., 279.

Chapter 25: The Blood of the Hungarians

337 **"Rise Magyar!":** S. Petöfi, "Talpra Magyar," in E. Tappan, ed., *The World's Story: A History of the World in Story, Song, and Art* (Boston: Houghton Mifflin, 1914), Vol. 6: *Russia, Austria-Hungary, The Balkan States, and Turkey*, 408–10;

"Alexander Petofi: The National Song of Hungary, 1848," Internet Modern History Sourcebook, http://www.fordham.edu/Halsall/mod/1848hungary-natsong.asp.

337 **"not a tactical move":** Khrushchev (1956), 38.

338 **"certainty of the victory of communism":** Ibid., 40.

338 **"the ending of the arms race":** Ibid., 33.

338 **"to strengthen in every way":** Ibid., 13.

338 **Khrushchev again took:** "Speech to 20th Congress of the C.P.S.U.," Nikita Khruschchev Reference Archive, http://www.marxists.org/archive/khrushchev/1956/02/24.htm.

338 **Recounting the purges:** Ibid.

338 **"Our Party, armed":** Ibid.

340 **"looked like a Hitlerjugend":** Agnes Ullmann, interview, Paris, August 19, 2010.

340 **"The teaching was":** Ibid.

340 **"It was absolutely unbelievable":** Ibid.

340 **"You know, you told me once":** Ibid.

340 **"It was a fabulous discovery":** Ullmann (2003), 199.

341 **Adám's trial:** Lendvai (1998), 67.

341 **"It was absolutely awful":** Agnes Ullmann, interview, Paris, August 19, 2010.

341 **Between 1949 and 1953, an estimated 150,000:** Pryce-Jones (1969), 43.

342 **"Intellectuals must be esteemed":** Stillman (1958), 13.

343 **"In sleepless nights":** Lendvai (1998), 81.

344 **"anti-Party plot":** Pryce-Jones (1969), 57.

344 **"The hundreds of thousands":** Lendvai (1998), 123.

346 **"Poland Shows Us the Way":** Sebestyen (2006), 110.

347 **"those who seek to instill":** Ibid., 117.

348 **"Fascist and reactionary elements":** Pryce-Jones (1969), 71.

349 **"without difficulty in a few hours":** Sebestyen (2006), 125.

349 **"The soviet soldiers are risking their lives":** Ibid., 127.

349 **80 freedom fighters were killed:** Ibid., 136–37.

349 **"The army, the state security forces":** Ibid., 138.

350 **"there were people who did not get up":** Agnes Ullmann, interview, Paris, August 19, 2010.

351 **"Every street was smashed":** Pryce-Jones (1969), 81.

352 **"In consultation with the entire people":** UN General Assembly, *Report of the Special Committee on the Problem of Hungary,* Official Records: Eleventh Session, Supplement No. 18 (A/3592), 83; "Sixteen Political, Economic, and Ideological Points, Budapest, October 22, 1956," Modern History Sourcebook, http://www.fordham.edu/halsall/mod/1956hungary-16points.asp.

353 **"The Hungarian Government is initiating":** Pryce-Jones (1969), 86.

353 **"the tremendous force":** Sebestyen (2006), 208.

353 **"beginning a new chapter":** Ibid.

353 **"It is nothing short of a miracle":** Judt (2005), 322.

353 **"heroic and earth-shaking insurrection":** Lottman (1979), 589.
353 **"glimmering fires of joy":** *Le Figaro*, October 31, 1956.
354 **"violations and mistakes":** Sebestyen (2006), 199.

Chapter 26: Repression and Reaction

355 **"If ten or so Hungarian writers":** Sebestyen (2006), 81.
355 **"We should re-examine our assessment":** Mark Kramer, "The 'Malin Notes' on the Crisis in Hungary and Poland, 1956." *Cold War International History Project Bulletin* (1957), 394.
355 **"We'll do it then":** Sebestyen (2006), 219.
357 **"This is Imre Nagy speaking":** Pryce-Jones (1969), 105.
358 **"This is the Hungarian Writers Association":** Ibid.
358 **"SOS SOS SOS":** *Time,* November 12, 1956.
359 **"If ever there was a time":** Ibid.
359 **"interference with the internal affairs":** Ibid.
359 **"desist forthwith from all attack":** "Resolution 1004 (ES-II) adopted by the United Nations General Assembly on the Situation in Hungary (4 November 1956)," in *General Assembly Official Records*, 6, accessed at European Navigator, http://www.ena.lu/resolution_1004_es_ii_adopted_united_nations_general_assembly_november_1956-2-1260.
360 **"The Russians have given their answer":** *The Times,* London, November 5, 1956.
360 **In their statement:** *L'Humanité*, November 5, 1956.
360 **"Liberate Budapest":** Bernard (1991), 73, trans. SBC.
361 **thirty were wounded:** Ibid., trans. SBC.
361 **"fascist arsonists and vandals":** Ibid., 70, trans. SBC.
361 **"You are the only communists":** Ibid., 71, trans. SBC.
362 **"POETS, WRITERS, SCHOLARS":** Le Sueur (2001), 278.
362 **"Our Hungarian brothers":** Camus, Quilliot, and Faucon (1965), 1778, trans. SBC.
362 **"the genocide of which Hungary":** Ibid., trans. SBC.
363 **"to demonstrate to the world":** Ibid., 1780, trans SBC.
363 **"our Soviet friends":** Sartre (1974), 323; Aronson (2004), 200.
363 **"the Party has manifested":** Macridis (1958), 630.
363 **"I condemn absolutely":** Sartre (1957), 16.
363 **"the intervention was a crime":** Ibid., 6.
363 **"was made possible":** Ibid., 7.
363 **"Regretfully but completely":** Aronson (2004), 201–2.
364 **"gravest fault was probably Khrushchev's report":** Sartre in *L'Express*, November 9, 1956. Translation based on Sartre (1957), 14, and Birchall (2004), 163.
364 **"to reveal the truth to the masses":** Sartre (1957), 15.
364 **On November 28:** *Le Monde*, November 29, 1956. Note: Many secondary

sources in print incorrectly report the date of this event as November 23. However, primary sources reveal the event occurred on November 28.

364 **"The only thing that I can publicly":** Camus, Quilliot, and Faucon (1965), 1780–82, trans. SBC.

365 **More than 2,500 Hungarians:** Sebastyen (2006), 277.

366 **"Hungarian mothers":** Juhász (1999), 28.

366 **Ullmann was in the third row:** Agnes Ullmann, interview, Paris, August 19, 2010.

366 **Obersovszky and Gáli were charged:** "Hungarian Writers' Resistance to the Government," January 22, 1957, from the Evaluation and Research Section: Background Report, Hungarian Research, Open Society Archives, http://www.osaarchivum.org/files/holdings/300/8/3/text/29-4-179.shtml.

366 **On December 11:** Lomax (1982), 85.

366 **Workers continued to rebel at factories:** Ibid., 86.

367 **In the spring of 1957:** "Hungary–People Who Should Be Freed," March 28, 1963, PTO, Open Society Archives, http://www.osaarchivum.org/files/holdings/300/8/3/text_da/32-3-168.shtml.

367 **Other friends also were arrested:** Agnes Ullmann, interview, Paris, August 19, 2010.

367 **Russian soldiers in armored personnel carriers:** *The Times*, London, March 16, 1957.

368 **The men arrested Erdös anyway:** Agnes Ullmann, e-mail to author, December 14, 2010.

368 **Sartre wrote a 120-page exposition:** *Les Temps Modernes*, 129–31 (November–December 1956–January 1957): 577–97.

368 **"None of the evils that totalitarianism":** Camus in *Demain*, February 21–27, 1957, translated in Camus (1974), 171.

368 **"Communism appears to us":** Aronson (2004), 202.

369 **"Foreign tanks, police":** Camus (1974), 158.

369 **"all men of the left":** Birchall (2004), 165.

369 **"There is no possible evolution":** Camus (1974), 161.

369 **"the USSR is not imperialist":** Birchall (2004), 165.

369 **"The defects of the West":** Ibid., 163.

370 **"Our faith is that throughout the world":** *Franc-Tireur*, April 18, 1957; Camus (1974), 164.

Chapter 27: A Voice of Reason

371 **"in order not to add to its unhappiness":** Todd (1996), 339.

371 **"Algeria is the cause of my suffering":** Camus (1974), 126.

371 **"the no man's land between two armies":** Ibid., 128.

372 **"civilian truce" . . . "for the duration":** Ibid., 134.

372 **"duty, to come":** Ibid., 131–32.

372 **"My only qualifications":** Ibid., 132.

372 **"two Algerian populations":** Ibid., 135.
372 **"condemned to die together":** Ibid., 136.
372 **"on a single spot of the globe":** Ibid., 142.
373 **"such stupid and brutal initiatives":** Lottman (1979), 582.
373 **Camus suggested that if de Maisonseul deserved to be arrested:** Todd (1996), 348.
373 **"I owe you":** Camus (2008), 217–18.
374 **"not only useless":** Camus (1974), 178.
374 **"one of the last countries":** Camus (1974), 177.
375 **"let us recognize it":** Ibid., 197–98.
375 **"This is an emotion":** Ibid., 198.
375 **"is in the good it does":** Ibid., 178.
375 **"upholders cannot reasonably defend it":** Ibid., 179.
376 **"no less repulsive than the crime":** Ibid., 176. Note: This very passage (in a different translation) was explicitly cited in Justice William Brennan's dissenting opinion in the 1976 landmark United States Supreme Court case *Gregg v. Georgia,* which upheld the use of the death penalty in certain circumstances.
376 **"astronomical proportions":** Ibid., 227.
376 **"bloodthirsty laws":** Ibid., 227–28.
376 **"Without the death penalty":** Ibid., 228–29.
376 **"we must call a spectacular halt":** Ibid., 229.
376 **"in the unified Europe":** Ibid., 230, 234. Note: Membership in the EU today does indeed require abolition of the death penalty. See "EU Policy on Death Penalty," European Union External Action, http://eeas.europa.eu/human_rights/adp/index_en.htm.
377 **By the end of January:** Matthews (2007), 536.
377 **"intellectuals, students, and ne'er-do-wells":** *Time,* April 22, 1957.
377 **Tóth shared her political concerns:** Agnes Ullmann, e-mail to author, December 15, 2010.
378 **Fearing that the clerk would inform:** Eörsi (2006), 99–142.
378 **They would be charged with publishing:** *Time,* April 22, 1957.
378 **"I thought I had to do everything":** Ibid.
378 **"I want to be a free man":** Ibid.
378 **On June 20:** Eörsi (2006), 117; "Biographies of Condemned Writers," OSA Archivum, http://193.6.218.36/files/holdings/300/8/3/text_da/30-1-4.shtml.
379 **The consul immediately agreed:** Agnes Ullmann, e-mail to author, December 27, 2010.
379 **On July 4:** *Time,* July 8, 1957; "Biographies of Condemned Writers," OSA Archivum, http://www.osaarchivum.org/files/holdings/300/8/3/text/30-2-257.shtml. Note: Gáli was released in 1960–61, and Obersovszky was released in 1963.
379 **Another twenty or so writers:** "The Hungarian Writers After the Revolution," July 26, 1957, from RFE News and Information Service, Evaluation and Research Section: Background Report. Hungarian Research, Open Society Archives, http://www.osaarchivum.org/files/holdings/300/8/3/text/30-2-257.shtml.

380 **"Seldom in any country":** *New York Times*, February 17, 1957.

380 **"The balance between creation":** Todd (1996), 687, trans. SBC.

380 **He'd had the plan:** Camus interview with Dominique Aury, *New York Times*, February 24, 1957.

380 **"waiting for inspiration's wing":** Todd (1997), 366.

380 **"Goethe teaches courage":** Emerson (1996), 166.

380 **"I am resigned to failing":** Letter from Camus to Grenier, September 12, 1957, in Camus and Grenier (2003) 178–79.

381 **"As a French-Algerian":** Le Sueur (2001), 106; Camus to President Coty, September 26, 1957. Coty was president per Todd (1996), 684.

382 **"his important literary production":** *New York Times*, October 18, 1957.

382 **"a genuine moral pathos":** Ibid.

382 **"strange feeling of overwhelming":** Camus (2008), 197.

382 **The successful nomination:** E-mail from M. Holmstrom, archivist at Swedish Academy, to author, October 11, 2011. Sylvère Monod and Jacques Monod were descendants of the same great-great-grandfather Jean Monod (1765–1836).

382 **"I thought that the Nobel Prize":** Camus (2008), 197; Lottman (1979), 602.

382 **"Your reply honors both of us":** Lottman (1979), 603.

383 **Rebatet was spared:** Camus (2006), xv–xvi.

383 **"This prize which falls most often":** Todd (1997), 373.

383 **"My dear Camus":** Letter, J. Monod to A. Camus, Nobel 1957, October 18, 1957, copyright Catherine and Jean Camus, Fond Albert Camus, Bibliothèque Méjanes, Aix-en-Provence, rights reserved.

383 **"Remember that shit thou art":** Todd (1997), 373.

383 **Camus was deeply touched:** See Camus reply to Monod, prologue, page 2.

383 **"This young man":** *Le Monde*, October 19, 1957.

384 **"Frightened by what happens to me":** Camus (2008), 197.

384 **A week later, he exercised some of the prestige:** *The Times*, London, October 31, 1957.

384 **"a man almost young" . . . "with what feelings" . . . "two tasks that constitute the greatness":** Camus Nobel Banquet speech, December 10, 1957, available at Nobelprize.org, http://nobelprize.org/nobel_prizes/literature/laureates/1957/camus-speech-e.html.

385 **"For more than twenty years":** Ibid.

385 **"resound by means of his art":** Ibid.

385 **"The nobility of our craft":** Ibid.

385 **"as an homage":** Ibid.

385 **"of the revered master":** Camus (1974), 251.

385 **"among the police forces":** Ibid., 251b.

386 **"Let us rejoice":** Ibid., 270–71.

Chapter 28: The Logic of Life

388 **This notion led Jacob and Wollman:** Jacob (1988), 276.

388 **One gene was detectable:** Ibid., 279.

388 **Monod referred to the method:** Ibid., 280.

389 **That achievement:** F. Sanger, "The Chemistry of Insulin," Nobel Lecture, December 11, 1958, http://www.nobelprize.org/nobel_prizes/chemistry/laureates/1958/sanger-lecture.pdf.

389 **In mid-September, Jacob and other biologists:** "The Replication of Macromolecules," *Symposia of the Society for Experimental Biology* 12, Company of Biologists, Society for Experimental Biology (New York: Academic Press, 1958).

390 **"On Protein Synthesis":** Crick (1958), 138–63.

390 **"in biology proteins are":** Ibid., 138.

390 **"the main function":** Ibid., 138–39.

390 **"The direct evidence":** Ibid., 152.

390 **"assumes that the specificity":** Ibid.

391 **"passed into protein":** Ibid., 153.

392 **"it is remarkable":** Ibid., 160–61.

392 **"In comparison to the confusion":** F. Jacob, "47–Francis Crick," Web of Stories, http://www.webofstories.com/play/14627.

392 **"My God, they are":** Agnes Ullmann, interview, Paris, August 20, 2010.

392 **"The British were awful" . . . "They said that maybe":** Ibid.

393 **He presented Ullmann's work:** Straub (1958), 176–84.

393 **The shortcomings of Ullmann's:** Crick (1958), 146.

393 **"There was nothing new":** Ullmann, e-mail to author, December 8, 2010.

396 **Monod had constructed:** Judson (1979), 406.

396 **Pardee had to devise:** Pardee (2002), 585–86.

396 **Pardee did not speak French:** Ibid., 586.

396 **Pardee tried the experiment:** Ullmann (2003), 136.

397 **Prior to arriving in Paris:** Kay (2000), 213.

397 **"Sign there":** Jacob (1988), 293.

397 **Szilárd was given an office:** Kay (2000), 213.

398 **"repulsive" at first:** Monod (1972), 281.

398 **Monod painted a new picture:** Letter, Monod to Pardee, February 28, 1957, MON. Cor. 03, Fonds Monod, SAIP; translation Judson (1979), 410–11.

398 **Pardee, Jacob, and Monod wrote a paper:** A. Pardee, F. Jacob, and J. Monod (1958).

399 **Getting from Budapest to Paris:** Agnes Ullmann, interview, Paris, August 20, 2010.

399 **"They won't let you leave":** Ibid.

400 **When the Hungarian military attaché:** Agnes Ullmann, e-mail to author, January 11, 2011.

400 **Ullmann had not tried to write:** Agnes Ullmann, interview, Paris, August 20, 2010.

400 **She then heard a man:** Perrin, as quoted in Ullmann (2003), 165.

400 **Ullmann introduced herself:** Ullmann (2003), 200.

400 **"What are you doing in Paris":** Ibid.; Agnes Ullmann, interview, Paris, August 20, 2010.

400 **"If you would allow me":** Ullmann (2003), 200.

400 **"What would you like to do?":** Ibid.

400 **"If Monsieur Gros":** Ibid.

401 **"I want to leave Hungary":** Ibid.

401 **He encouraged Ullmann to discuss her situation:** Ibid.; Agnes Ullmann, interview, Paris, August 20, 2010.

401 **"What will happen to you" . . . "My poor child":** Agnes Ullmann, interview, Paris, August 20, 2010; Agnes Ullmann, e-mail to author, January 11, 2011.

402 **"I didn't sleep":** Agnes Ullmann, interview, Paris, August 20, 2010.

Chapter 29: Making Connections

404 **The note appeared:** A. Pardee, F. Jacob, and J. Monod (1958).

404 **Other members of the group:** Wall (2001).

404 **Jacob, however, did not get caught up:** Jacob (1988), 295, 309.

405 **By late July, Monod was sailing:** Debré (1996), 217.

405 **"with no taste for work":** Jacob (1988), 297.

405 **"invaded by a sudden excitement":** Ibid., 297–98.

405 **"In both cases":** Ibid., 298.

406 **"With the phage":** Ibid., 297–98.

406 **"had climbed a mountain":** Ibid.

406 **"You've had enough?":** Ibid., 298.

406 **"I think I've just thought up":** Ibid., 298.

407 **silly, even "childish":** Ibid., 300.

407 **To Jacob, the analogy seemed so strong:** Jacob, in Ullmann (2003), 121.

407 **"as if it closed a single lock":** Jacob (1988), 301.

408 **"like a switch":** Ibid.

408 **"Actually, there is no direct evidence":** Jacob, as quoted in Ullmann (2003), 122.

409 **"How stupid I had been":** Ibid., 125.

409 **The repartee was more than scientific:** Jacob (1988), 307–8.

410 **"may lead to a generalizable picture":** A. Pardee, F. Jacob, and J. Monod (1959), 177.

410 **As a child during the Nazi occupation:** The Gábor Sztehlo Foundation for the Help of Children and Adolescents, http://www.sztehlo-gabor-alapitvany.hu/bemutatke.htm.

411 **"Last year prices are":** Kövesi to Monod, April 18, 1959, MON. Ser. 04, Fonds Monod, SAIP.

411 **"We have been busy here":** Monod to Kövesi, April 21, 1959, MON. Ser. 04, Fonds Monod, SAIP.

412 **Kövesi would have $4,700:** Monod to Kövesi, April 28, 1959, MON. Ser. 04, Fonds Monod, SAIP.

412 **"I have entire confidence":** Ibid.

412 **"My interview with the captain":** Kövesi to Monod, May 20, 1959, MON. Ser. 04, Fonds Monod, SAIP.

413 **"TOM TO COME TO ESZTERGOM":** Kövesi to Monod and Ullmann, June 3, 1959, MON. Ser. 04, Fonds Monod, SAIP.

414 **They underscored how mutations:** Jacob and Monod (1959), 1282–84.

414 **"Ah, Jacob. Pleased to see you again":** Jacob (1988), 308–9.

Chapter 30: The Possible and the Actual

417 **The former resistant loved the idea:** Agnes Ullmann, interview, Paris, August 20, 2010.

417 **In Monod's "Code Agnes":** Note in J. Monod's handwriting, MON. Ser. 04, Fonds Monod, SAIP.

417 **"This is like a game of cards":** Letter, Ullmann and Kövesi to Monod, September 3, 1959, MON. Ser. 04, Fonds Monod, SAIP, trans. SBC.

418 **She made one request of Monod:** Ibid.

419 **There was no reaction from the audience:** Account of meeting based on Jacob (1988), 310–11.

419 **"at the same time I sent you a chromatogram":** Letter, Ullmann to Monod, September 17, 1959, MON. Ser. 04, Fonds Monod, SAIP.

419 **Kövesi had sent Monod a summary:** Letter, Kövesi to Monod, September 14, 1959, MON. Ser. 04, Fonds Monod, SAIP.

420 **"I am not able to do any more waiting":** Letter, Kövesi to Monod, October 3, 1959, MON. Ser. 04, Fonds Monod, SAIP.

420 **"Thank you so kindly":** Letter, Monod to Ullmann, September 28, 1959, MON. Ser. 04, Fonds Monod, SAIP, trans. SBC.

421 **"I am very confident":** Letter, Monod to Kövesi, October 6, 1959, MON. Ser. 04, Fonds Monod, SAIP.

421 **"Thank you for your good faith":** Letter, Kövesi to Monod, October 9, 1959, MON. Ser. 04, Fonds Monod, SAIP.

421 **"If you believe that this trip":** Message in letter from Monod to Kövesi, October 16, 1959, MON. Ser. 04, Fonds Monod, SAIP.

422 **"I adore Mozart":** Letter, Brunerie to Ullmann, October 19, 1959, MON. Ser. 04, Fonds Monod, SAIP, trans. SBC.

422 **"They are still working in the same field":** Letter, Monod to Kövesi, October 16, 1959, and October 23, 1959, MON. Ser. 04, Fonds Monod, SAIP.

423 **"I've met somebody":** Letter, Kövesi to Monod, October 22, 1959, MON. Ser. 04, Fonds Monod, SAIP.

423 **"This new plan":** Letter, Monod to Kövesi, October 26, 1959, MON. Ser. 04, Fonds Monod, SAIP.

424 **"IT IS ABOUT A CAR":** Letter, Kövesi to Monod, October 29, 1959, MON. Ser. 04, Fonds Monod, SAIP.

424 **Kövesi explained to Monod:** Ibid.

424 **One of the financial supporters:** Letter, Csapo to Monod, November 6, 1959, MON. Ser. 04, Fonds Monod, SAIP.

425 **"the main difficulty":** Letter, Monod to Csapo, October 26, 1959, MON. Ser. 04, Fonds Monod, SAIP.

425 **"I refuse to believe":** Ibid.

425 **"I must therefore ask you":** Letter, Csapo to Monod, October 28, 1959, MON. Ser. 04, Fonds Monod, SAIP.

425 **"Dear Dr. Csapo":** Letter, Monod to Csapo, November 2, 1959, MON. Ser. 04, Fonds Monod, SAIP.

426 **Kövesi wrote to Monod:** Letter, Kövesi to Monod, November 6, 1959, MON. Ser. 04, Fonds Monod, SAIP.

426 **"I am fantastically busy":** Letter, Monod to Kövesi, November 10, 1959, MON. Ser. 04, Fonds Monod, SAIP.

428 **the writing of the paper that would introduce the new concept:** F. Jacob, D. Perrin, C. Sanchez, and J. Monod (1960).

428 **"the process of science":** Jacob (1988), 288.

Chapter 31: Unfinished

429 **"fundamental misunderstanding":** Letter, Csapo to Monod, November 6, 1959, item 53, MON. Ser. 04, Fonds Monod, SAIP.

430 **"If the previous attempt has failed":** Letter, Monod to Csapo, November 19, 1959, item 56–58, MON. Ser. 04, Fonds Monod, SAIP.

430 **Monod suggested that Csapo request an extension:** Letter, Csapo to Monod, November 24, 1959, item 59, MON. Ser. 04, Fonds Monod, SAIP; Letter, Monod to Csapo, December 8, 1959, item 60, MON. Ser. 04, Fonds Monod, SAIP; Letter, Monod to Csapo, December 8, 1959, item 61, MON. Ser. 04, Fonds Monod, SAIP.

430 **"Must I point out":** Letter, Monod to Csapo, December 29, 1959, item 64, MON. Ser. 04, Fonds Monod, SAIP.

431 **He wrote appeals:** Lottman (1979), 625.

431 **Camus wrote to ask:** Ibid., 637–38.

431 **"To keep quiet":** *New York Times*, April 22, 1958.

431 **He also wrote the preface:** Secker and Warburg, *The Truth About the Nagy Affair: Facts, Documents, Comments* (New York: F. A. Praeger, 1959).

432 **"qui nous persecutent encore":** J. Monod, personal copy of A. Camus, *Les Possédés*, inscription courtesy of Olivier Monod.

432 **"Enough of you must":** Camus (2008), 203–4.

432 **"My job is to make":** Ibid., 205.

432 **It was purchased:** E-mail from Olivier Monod (Jacques Monod's son) to author, August 30, 2012. Olivier Monod of Lourmarin and Jacques Monod were descendants of the same great-great-grandfather, Jean Monod (1765–1836).

433 **He had conceived of the title:** Ibid., 86.

433 **"a 'direct' novel":** Letter from Camus to Grenier, August 24, 1955, Camus and Grenier (2003), 168.

433 **"fresco of the contemporary world":** Todd (1997), 405.

433 **"He wrote *War and Peace*":** Camus (1965), 230.

433 **As early as February 1957:** Camus interview with Dominique Aury, *New York Times*, February 24, 1957.

433 **"It's also the novel of my maturity":** Lottman (1979), 615.
433 **Camus's plan was:** Camus (2008), 172.
434 **"I am finding a little peace":** Letter from Camus to Grenier, May 8, 1959, in Camus and Grenier (2003), 191.
434 **"Frère Albert, O.D.":** Lottman (1979), 644.
434 **"Nietzsche is here":** Letter from Camus to Grenier, May 8, 1959, in Camus and Grenier (2003), 191.
434 **"The Wisdom of Lourmarin":** Grenier (1936).
434 **"I put my footsteps in yours":** Lottman (1979), 636.
434 **"The greatest works":** Camus (2008), 244.
434 **"I think it's all over":** Ibid., 646.
435 **"ailing, tense, stubborn":** Camus (1995), 27.
435 **"four women at the same time":** Ibid., 296.
435 **"I have never worked with such dense material":** Letter from Camus to Mi, November 22, 1959, in Todd (1997), 407.
435 **"the best in the world":** Letter from Camus to his mother, December 21, 1959, in Todd (1997), 411.
436 **"To you who will never":** Camus (1995), 3.
436 **"I must finish the first draft":** Todd (1997), 407.
436 **"By the time you read this":** Ibid., 411.
436 **"Alright, this is a last letter":** Ibid., 411–12.
436 **"This is my last letter, my tender one":** Letter from Camus to Catherine Sellers, in Todd (1997), 412.
436 **The next day they all lunched:** The account of the car trip back to Paris is from Lottman (1979), 660–64.
437 **"You shouldn't have bought it":** Lottman (1979), 660.
438 **inside his black leather briefcase:** Brée (1961), vii; Lottman (1979), 664–65.
438 **"For over twenty years":** Lottman (1979), 669.
438 **The director of the Théâtre de France:** O'Brien, *New York Times*, January 10, 1960.
438 ***"Absurde"*:** *Time*, January 18, 1960.
438 **"A Conscience Against Chaos" and "The Best of Us":** *Combat,* January 5, 1960.
438 **"a creed which calls on men":** *New York Times*, January 5, 1960.
438 **"is one of the greatest losses":** Ibid.
439 **"We shall recognize":** J.-P. Sartre, *France-Observateur,* January 7, 1960, published in Brée (1962), 173–75.
439 **"an irreparable catastrophe":** G. Brée, *New York Times*, January 24, 1960.
439 **"We are not just weeping":** *New York Times*, January 10, 1960.
439 ***"Je suis triste pour toi"*:** Letter, Kövesi to Monod, January 9, 1960, item 66, MON. Ser. 04, Fonds Monod, SAIP.
439 **"The death of Camus":** Letter, Monod to Kövesi, January 14, 1960, item 67, MON. Ser. 04, Fonds Monod, SAIP.

CHAPTER 32: MESSENGERS

440 **"Dear Jacques, There are good hopes":** Letter, Kövesi to Monod, February 15, 1960, MON. Ser. 04, Fonds Monod, SAIP.

440 **"What would happen if the lion":** Brunerie (2008), 73, and Ullmann e-mail to author, March 7, 2011.

441 **"A. and T. . . . now have experimental proof":** Letter, Monod to Kövesi, March 1, 1960, MON. Ser. 04, Fonds Monod SAIP.

441 **"Would you let me know":** Letter, Monod to Kövesi, March 7, 1960, MON. Ser. 04, Fonds Monod, SAIP.

442 **The results meant that:** Riley et al. (1960).

443 **To Jacob, it felt more like:** Jacob (1988), 311.

443 **"That's when the penny dropped":** Judson (1979), 431.

443 **Several minutes passed:** Jacob (1988), 312.

443 **In their frenzied reaction:** Volkin and Astrachan (1956).

444 **"once again, a creature":** Jacob (1988), 313.

445 **"A possible model":** Judson (1979), 434.

445 **"You know exactly":** Ibid., 435.

445 **"I believe that your visit":** Letter, Ullmann to Monod, April 19, 1960, MON. Ser. 04, Fonds Monod, SAIP.

446 **"I talked to one of my new friends":** Letter, Kövesi to Monod, April 29, 1960, MON. Ser. 04, Fonds Monod, SAIP.

446 **"so that we can discuss":** Letter, Monod to Ullmann, May 5, 1960, MON. Ser. 04, Fonds Monod, SAIP.

446 **In order to understand:** Telegram, Monod to Kövesi, May 11, 1960; trip details in letter from Monod to Csapo, June 3, 1960, MON. Ser. 04, Fonds Monod, SAIP.

447 **The collaboration entailed:** Riley et al. (1960).

447 **"Oh, let's go and have a walk":** Agnes Ullmann, interview, Paris, August 20, 2010.

447 **Monod and Erdös had never met:** Tamás Erdös, interview by phone, January 14, 2011.

448 **At the time, Hungary's borders:** Letter, Monod to Csapo, June 3, 1960, MON. Ser. 04, Fonds Monod, SAIP.

448 **"Oh, I should like to have a look":** Agnes Ullmann, interview, Paris, August 20, 2010.

448 **Along the shores of the Danube:** Agnes Ullmann, e-mail to author, March 18, 2011.

449 **In his lecture:** Agnes Ullmann, interview, Paris, August 20, 2010.

449 **"Our extractions, in particular":** Letter, Monod to Ullmann, June 3, 1960, MON. Ser. 04, Fonds Monod, SAIP.

449 **"I hardly need to tell you":** Letter, Monod to Kövesi, June 3, 1960, MON. Ser. 04, Fonds Monod, SAIP.

450 **"Also, may I mention":** Ibid.

450 **In early June, Helm made a preliminary trip:** Agnes Ullmann, e-mail to author, March 24, 2011.

451 **Helm and his wife:** Tamás Erdös, phone interview, January 14, 2011.

451 **"I will not get in there":** Agnes Ullmann, interview, Paris, August 20, 2010.

452 **"Don't move":** Brunerie (2008), 83.

452 **Helm closed the box:** Agnes Ullmann, e-mail to author, March 24, 2011.

452 **At the border, two customs officers:** The journey across the border is reconstructed from Agnes Ullmann, interview, Paris, August 20, 2010; Tamás Erdös, phone interview, January 14, 2011; and Brunerie (2008), 78 and 82.

454 **"You have organized something":** Brunerie (2008), 79.

Chapter 33: Synthesis

455 **"I don't believe it":** Jacob (1988), 314; Judson (1979), 436.

456 **It also did not help:** Jacob (1988), 315.

456 **"It's the magnesium!":** Ibid., 317; and S. Brenner, "100—Using Magnesium to Compete with Caesium: Radioactive Coca-Cola," Web of Stories, http://www.webofstories.com/play/13313?o=MS.

456 **He and Jacob hurried back:** S. Brenner, "100—Using Magnesium to Compete with Caesium: Radioactive Coca-Cola," Web of Stories, http://www.webofstories.com/play/13313?o=MS.

456 **They were rushing:** Ibid.; Judson (1979), 439.

457 **At the end of the run:** S. Brenner, "101—Using Magnesium to Compete with Caesium: The Experiment," Web of Stories, http://www.webofstories.com/play/13314?o=MS.

457 **Nervously but skillfully:** Jacob (1988), 317; Judson (1979), 440; S. Brenner, "101—Using Magnesium to Compete with Caesium: The Experiment," Web of Stories, http://www.webofstories.com/play/13314?o=MS.

457 **His task of obtaining French visas:** Brunerie to Monod, June 27, 1960, MON. Ser. 04, Fonds Monod, SAIP.

458 **"croissant of liberty":** Brunerie (2008), 82.

458 **a succession of papers had appeared:** At the same time Jacob and Brenner were conducting their experiments in Pasadena, Francois Gros was working in Jim Watson's laboratory at Harvard also in pursuit of the RNA intermediate. The two groups eventually decided to publish their papers simultaneously in May 1961. See Brenner, Jacob, and Meselson (1961); Gros et al. (1961); and Kay (2000), 229.

459 **The challenge of writing in English:** Jacob (1988), 318.

460 **The next morning, Monod asked Madeleine:** Story and quotes of exchange concerning 1960 Nobel Prize is from Brunerie (2008), 62–63, 84–85.

461 **"messenger RNA":** Jacob and Monod (1961), 350.

461 **"to understand why":** Ibid., 354.

462 **"anything found to be true":** Monod and Jacob (1961), 393.

CHAPTER 34: CAMUS IN A LAB COAT

467 **"for their discoveries":** Brunerie (2008), 204.

467 **"The beautiful and hard adventure":** *Paris-Presse*, October 16, 1965, trans. SBC.

467 **It showed pictures:** *France-Soir*, October 15, 1965.

468 **Of the three, no one:** *Le Nouvel Observateur*, October 20–26, 1965; MON. Bio. 08, item 2, Fonds Monod, SAIP. Note that sources regarding the date of the interview are in conflict; Jacob (1988) indicates the interview was the evening of October 14; Debré (1996) reports that the interview took place on October 20.

468 **"Camus's existentialism, in the widest sense":** J. Monod, *Lire,* November 1975, 243, trans. SBC.

468 **"the proper response":** *New York Times*, January 6, 1960.

469 **"Communists condemn":** Duchen (1994), 181.

469 **an estimated 400,000 to 800,000 procedures:** *Time,* July 21, 1961.

470 **"Because of scientific and technical":** Letter, Monod, Jacob, and Lwoff to Dr. Lagroua Weill-Hallé, November 18, 1965, MON. Pol. 02, Fonds Monod, SAIP, trans. SBC.

470 **"Many of our fellow citizens":** J. Monod in "La Contraception. Problèmes biologiques et psychologiques" by J. Dalsace and R. Palmer (1966), MON. Pol. 02, Fonds Monod, SAIP, trans. SBC.

471 **"The Country of the rights of man":** J. Monod, speech to Le Comité Parisien Martin Luther King, Palais des Sports, Paris, March 28, 1966, MON. Mss. 02, Fonds Monod, SAIP, trans. SBC; H. Dufour; *L'Express* April 4–10, 1966, 78–79.

473 **"Let us listen to the lesson":** J. Monod, speech at Cirque d'Hiver, April 9, 1968, in homage to Dr. Martin Luther King, MON. Mss. 02, Fonds Monod, SAIP, trans. SBC.

474 **More than 20,000 protestors:** "La Contestation: La Terrible Semaine Qu'a Vécue le Quartier Latin," Ina.fr, http://www.ina.fr/economie-et-societe/education-et-enseignement/video/AFE86001191/la-contestation-la-terrible-semaine-qu-a-vecue-le-quartier-latin.fr.html.

475 **In his inaugural interview:** *Le Nouvel Observateur*, October 20–26, 1965, 2, MON. Bio. 08, item 2, Fonds Monod, SAIP.

475 **"These so-called representatives":** *Le Nouvel Observateur*, May 15–21, 1968, 31, trans. SBC.

476 **Monod telephoned Alfred Kastler:** Kastler, "Les 07 et 08 mai," Come4News, http://www.come4news.com/mai-68-suite-29-9995.

476 **General de Gaulle:** Telegram to de Gaulle, Fonds Monod, SAIP, trans. SBC.

476 **The rector could not:** *Le Nouvel Observateur*, May 15–21, 1968, 31.

476 **Ullmann told Monod:** Ullmann interview, Paris, August 20, 2010.

477 **"Mr. Minister":** *Le Nouvel Observateur*, May 15–21, 1968, 31.

477 **"Are you in agreement":** Kastler, "Les 07 et 08 mai," Come4News, http://www.come4news.com/mai-68-suite-29-9995.

477 **"In the present circumstances":** *France Soir*, May 12–13, 1968.

478 **Early Saturday morning:** G. Kopelowicz, bhikku, http://www.bhikku.net/lugworm/com-bhikku-ments.php?429.

478 **Four hours of battle:** *Le Monde*, May 12–13, 1968.

478 **"The undersigned professors declare":** *Le Nouvel Observateur*, May 15–21, 1968, 31.

479 **In the provinces:** "Résumé Chronologieque Des Principaux Faits Touchant a L'Ordre Public Survenus Au Cours Des Mois de Mai-Juin 1968," http://polices.mobiles.free.fr/documents/mai 68.htm.

479 **"prodigious and exceeded the imagination":** *Le Nouvel Observateur*, May 15–21, 1968, 30.

479 **The student crisis:** *Time*, May 24, 1968.

480 **"Otherwise we will tumble":** *Time*, May 31, 1968.

480 **"There comes a moment":** *Le Monde*, June 14, 1968; MON. Bio. 09, Fonds Monod, SAIP.

481 **Monod and a few colleagues:** *Le Monde*, May 15, 1968.

481 **Sweeping reforms:** "La loi d'orientation de l'enseignement supérieur," November 12, 1968, http://guilde.jeunes-chercheurs.org/Textes/Txtfond/L68-978.html.

481 **"They are crazy!":** Ullmann interview, Paris, August 20, 2010.

Chapter 35: Chance and Necessity: Sisyphus Returns

482 **"Whenever objectivity":** Monod (1953), 319–20.

482 **"There's always the tendency":** J. Monod, interview with Gerald Leach in Paris, January 3, 1967, broadcast on the BBC February 1, 1967, transcript MON. Bio. 09, Fonds Monod, SAIP, 11.

483 **"The most unnecessary science":** Ibid., 7.

483 **"If we still thought":** Ibid., 12.

483 **"Science has molded":** Ibid.

484 **"The mind's deepest desire":** Camus (1991a), 17.

484 **"The urge, the anguish":** Monod (1969), 19.

484 **"At that subtle moment":** Camus (1991a), 123.

485 **"philosophical, religious, and political":** Monod (1971), xi.

486 **"remained as if suspended":** Ibid.

486 **"Thirty years ago":** Ibid.

486 **"The 'secret of life'":** Ibid., xii.

486 **"everything existing in the universe":** Credited to Democritus, although scholars have not been able to locate such a phrase in any particular work by the ancient Greek philosopher.

486 **"Drawn out of the realm":** Monod (1971), 118.

487 **"man was the product":** Monod (1968), 27, trans. SBC.

487 **"And, in any case":** Monod (1969), 24.

487 **"this fundamental scientific result":** Monod (1968), 27, trans. SBC.

487 **"Should he despair":** Ibid., 28, trans. SBC.

487 **"We must despair":** Camus (1991a), 18.

488 **"Each conquest of Science is a victory of the absurd":** Monod (1968), 27.

488 **"In virtually all the mythic":** Monod (1969), 22.

488 **"between Man and the Universe":** Ibid.

488 **"the scientific approach":** Ibid., 24.

488 **"It remained for modern Biology":** Ibid., 23.

488 **"none of the gracious":** Ibid., 24.

489 **"Society by definition":** *Le Nouvel Observateur,* October 20, 1965, 2, trans. SBC.

489 **"Modern societies had accepted":** Monod (1971), 170.

489 **"our societies are still":** Ibid., 171.

490 **"a nauseating mixture":** Monod (1969), 24.

490 **"They all lie":** Ibid., 25.

490 **"Man must wake":** Monod (1971), 172–73. Alternative translation of parts of the original French by SBC.

490 **"A world that can":** Camus (1991a), 6.

490 **"heart open to":** Camus (1946), 32.

490 **"this universe without":** Camus (1991a), 123.

491 **"that he alone creates":** Monod (1969), 25.

491 **"Of all the schools":** Camus (1991a), 115.

491 **"the absurd joy":** Ibid., 93.

491 **"authentic creation is a gift":** Camus (1991a), 212.

491 **"And what other ultimate values":** Monod (1969), 26.

491 **"A society that would":** Ibid., 27.

491 **"A utopia. Perhaps":** Monod (1971), 180.

492 ***Le Hasard et La Necessité* appeared:** This date has been reported differently elsewhere, but no reviews appear before this date; the date cited here is from *Le Figaro,* November 3, 1970, 10, in MON. Bio. 12, Fonds Monod, SAIP.

492 **Almost 200,000 copies:** *Atlantic Monthly,* November 1970, 126, MON. Bio. 12, Fonds Monod, SAIP.

492 **Foreign translations quickly followed:** *Monod Notes en bas de la page,* MON. Bio. 12, Fonds Monod, SAIP.

492 **The book garnered scores of reviews:** *The Economist,* May 13, 1972; *Atlantic Monthly,* November 1971, 125–30; H. Kenner, *LIFE,* November 5, 1971, *Newsweek,* April 26, 1971, all MON. Bio. 12, Fonds Monod, SAIP.

492 **Monod received two reviews in the *New York Times:*** C. Lehmann-Haupt, *New York Times,* October 15, 1971; G. Steiner, *New York Times,* November 21, 1971, MON. Bio. 12, Fonds Monod, SAIP.

492 **also published two interviews:** J. C. Hess, *New York Times,* March 13, 1971; *New York Times,* November 8, 1971, MON. Bio. 12, Fonds Monod, SAIP.

492 **"I read Jacques Monod":** *Le Figaro,* November 3, 1970; MON. Bio. 12, Fonds Monod, SAIP.

493 **"Even the most cautious":** *The Economist,* May 13, 1972, MON. Bio. 12, Fonds Monod, SAIP.

493 **"One could ask oneself":** Interview with Emile Noël, "Notre Temps,"

November 25, 1970, MON. Bio. 09, item 1, Fonds Monod, SAIP, transcript 50, trans. SBC.

493 **"Myself, I do believe":** Ibid.

494 **he became the eighth general director:** "Direction Institut Pasteur 1887–1940," Institut Pasteur, http://www.pasteur.fr/infosci/archives/dr1.html.

494 **"Where does genius come from":** "Louis Pasteur: A Great Benefactor of Humanity," Vigyan Prasar Science Portal, http://www.vigyanprasar.gov.in/scientists/PLouis.htm.

494 **Monod hurled himself:** Stanier (1977), 1–12.

495 **He told Agnes Ullmann:** Ullmann (2003), 204.

495 **"Sir, I am a thirteen year-old boy":** Letter, Bruno to Monod, January 7, 1976, MON. Cor. 02, Fonds Monod, SAIP.

495 **"My Dear Bruno, Thank you very much:** Letter, Reply, J. Monod to Bruno, February 13, 1976, MON. Cor. 02, Fonds Monod, SAIP.

496 **He enjoyed drinking:** Kosiński (1986), 81–89.

496 **"See you Monday":** Ullmann (2003), 204.

497 **Philo heard his brother say:** Brunerie (2008), 264; Ullmann, e-mail to author, June 15, 2011; Olivier Monod, interview, Paris August 17, 2010.

497 **After his funeral:** Brunerie, M. "Once Upon a Time," in Ullmann (2003), 52.

Bibliography

Ajchenbaum, Y. M. (1994) *A la Vie, à la Mort: L'histoire du Journal Combat, 1941–1974*. Paris: Monde Éditions.

Ambrose, S. (1994) *D-Day, June 6, 1944: The Climactic Battle of World War II*. New York: Simon & Schuster.

Aron, R. (1983) *Mémoires*. Paris: Julliard.

Aronson, R. (2004) *Camus & Sartre: The Story of a Friendship and the Quarrel That Ended It*. Chicago: University of Chicago Press.

Asprey, R. B. (1975) *War in the Shadows: The Guerrilla in History*. Garden City, NY: Doubleday.

Avery, O., et al. (1944) "Studies on the Chemical Nature of the Substance Inducing Transformation of Pneumococcal Types." *Journal of Experimental Medicine* 79.2: 137–58.

Barber, N. (1976) *The Week France Fell*. London: Macmillan.

Battistelli, P. P., and A. Hook. (2011) *Heinz Guderian: Leadership, Strategy, Conflict*. Oxford; Long Island City, NY: Osprey.

Beavan, C. (2006) *Operation Jedburgh: D-Day and America's First Shadow War*. New York: Viking.

Belot, R., and G. Karpman. (2009) *L'Affaire Suisse: La Résistance, a-t-elle trahi de Gaulle (1943–1944)*. Paris: Armand Colin.

Bernard, J. (1967) "The Background of *The Plague*: Albert Camus' Experience in the French Resistance." *Kentucky Romance Quarterly* 14.2: 165–73.

Birchall, I. (2004) *Sartre against Stalinism*. New York: Berghahn Books.

Blaisdell, R. (2011) *Great Speeches of the Twentieth Century*. Mineola, NY: Dover.

Blatt, J. (2000) *The French Defeat of 1940: Reassessments*. Oxford: Berghahn.

Blumenson, M. (1977) *The Vildé Affair: Beginnings of the French Resistance*. Boston: Houghton Mifflin.

———. (1993) *The Battle of the Generals: The Untold Story of the Falaise Pocket: The Campaign That Should Have Won World War II*. New York: Morrow.

Bodson, H. (2005) *Downed Allied Airmen and Evasion of Capture: The Role of Local Resistance Networks in World War II*. Jefferson, NC: McFarland.

Boterbloem, K. (2004) *The Life and Times of Andrei Zhdanov, 1896–1948*. Montréal; Ithaca: McGill-Queen's University Press.

Bourderon, R. (2004) *Rol-Tanguy*. Paris: Tallandier.

Bourdet, C. (1975) *L'aventure Incertaine: De La Résistance à la Restauration*. Paris: Stock.

Brassaï. (1999) *Conversations with Picasso*. Trans. Jane Marie Todd. Chicago: University of Chicago Press.

Brée, G. (1961) *Camus*. New Brunswick: Rutgers University Press.

———. (1962) *Camus: A Collection of Critical Essays*. Edited by Germaine Brée. Prentice Hall.

Brown, A. C. (1975) *Bodyguard of Lies*. Guilford, CT: The Lyons Press.

Brunerie, M. (2008) *Cinquante-huit ans a l'institut Pasteur, vingt-deux ans pres de Jacques Monod*. www.pasteur.fr/infosci/archives/mado_bio.pdf.

Buell, T., et al. (1978) *The Second World War: Europe and the Mediterranean*. West Point, NY: Department of History, US Military Academy.

Burleigh, M. (2011) *Moral Combat: A History of World War II*. New York: Harper.

Camus, A. (1945) *L'Existence*. Paris: Gallimard.

———. (1946) *The Stranger*. Translated by Stuart Gilbert. New York: A. A. Knopf.

———. (1946–1947) "The Human Crisis." Translated by Lionel Abel. *Twice a Year*, Fall–Winter 1946–1947.

———. (1956) *The Rebel: An Essay on Man in Revolt*. Translated by Anthony Bower. New York: Knopf.

———. (1962) *Théâtre, Récits, Nouvelles*. Préface par Jean Grenier. Textes établis et annotés par Roger Quilliot. Paris: Gallimard.

———. (1963) *Notebooks, 1935–1942*. Translated by Philip Thody. New York: Knopf.

———. (1965) *Notebooks, 1942–1951*. Translated by Justin O'Brien. New York: Knopf.

———. (1974) *Resistance, Rebellion, and Death*. Translated by Justin O'Brien. First Vintage Books Edition. New York: Random House.

———. (1991a) *The Myth of Sisyphus and Other Essays*. Translated by Justin O'Brien, *Le Mythe de Sisyphe*, 1942. First Vintage International Edition, 1991. New York: Random House.

———. (1991b) *The Plague*. Translated by Stuart Gilbert. First Vintage International Edition. New York: Vintage Books, Random House.

———. (1995) *The First Man*. Translated by David Hapgood. New York: Alfred A. Knopf.

———. (2006) *Camus at Combat: Writing 1944–1947*. Edited by Jacqueline Lévi-Valensi, translated by Arthur Goldhammer. Princeton, NJ: Princeton University Press.

———. (2008) *Notebooks, 1951–1959*. Translated by Ryan Bloom. Chicago: Ivan R. Dee.

Camus, A., R. Quilliot, and L. Faucon. (1965) *Essais*. Paris: Gallimard.

Camus, A., and J. Grenier. (2003) *Correspondence, 1932–1960*. Translated by Jan F. Rigaud. Lincoln, NE: University of Nebraska Press.

Camus, A., and R. Char. (2007) *Correspondance, 1946–1959*. Paris: Gallimard.

Caracciolo, P., and A. Camus. (1957) "Letters: M. Camus and Algeria." Edited by Stephen Spender and Irving Kristol. *Encounter* 8.6: 68.

Carroll, S. B. (2006) *The Making of the Fittest: DNA and the Ultimate Forensic Record of Evolution*. New York: W. W. Norton & Co.

Champoux, R. J. (1975) "The Massilia Affair." *Journal of Contemporary History*, 10.2: 283–300.

Chevassus-Au-Louis, N. (2004) *Savants Sous L'Occupation*. Paris: Seuil.

Churchill, W. (1948) *The Second World War: The Gathering Storm*. Cambridge, MA: Houghton Mifflin.

———. (1949) *The Second World War: Their Finest Hour*. Cambridge, MA: Houghton Mifflin.

———. (1964) *The Second World War: Tide of Victory*. London: Cassell.

Cohn, M. (1978) "In Memoriam." *The Operon* 7: 1–9.

Cohn, M., et al. (1953) "Terminology of Enzyme Formation." *Nature*,172: 1096.

Collins, L., and D. Lappierre. (1965) *Is Paris Burning?* New York: Simon & Schuster.

Comité à la mémoire des savants français. (1959) *À la mémoire de quinze savants français lauréats de l'Institut assassinés par les allemands, 1940–1945*. Paris: Comité à la mémoire des savants français.

Cornick, M. (2000) *France at War in the Twentieth Century: Propaganda, Myth, and Metaphor*. Edited by Valerie Holman and Deborah Kelly. New York: Berghahn Books.

Cox, T. (1997) *Hungary 1956—Forty Years On*. London; Portland, OR: Frank Cass.

Crémieux-Brillhac, J. L. (1975) *Ici Londres, 1940–1944: Les Voix De La Liberté*. Paris: Documentation Française.

Crick, F. (1958) "On Protein Synthesis." *Symposia of the Society for Experimental Biology* 12: 138–63.

Dainton, F. (1989) "François Jacob—In Search of Meaning." *New Scientist*, January 21, 326.

Dallas, G. (2005) *1945: The War That Never Ended*. New Haven: Yale University Press.

Dantzer, R., and K. Kelley. (2009) *Elie Wollman 1917–2008: A Biographical Memoir*. Washington, DC: National Academy of Sciences. Available online at books.nap .edu/html/biomems/ewollman.pdf (accessed January 25, 2013).

Davidson, P. B. (1988) *Vietnam at War: The History, 1946–1975*. Novato, CA: Presidio Press.

Davies, R. G. (1947) "Genetics in the U.S.S.R." *Modern Quarterly* 2: 336–46.

Debré, P. (1996) *Jacques Monod*. Paris: Flammarion.

De Gaulle, C. (1964) *The Complete War Memoirs of Charles de Gaulle*. Translated by Jonathan Griffin and Richard Howard. New York: Simon & Schuster.

Delpla, François. (1997) "Montoire: Du Nouveau?" *Guerres Mondiales et Conflits Contemporains* 186: 81–94.

Duchen, C. (1994) *Women's Rights and Women's Lives in France, 1944–1968*. London, New York: Routledge.

Dulles, A., et al. (1996) *From Hitler's Doorstep: The Wartime Intelligence Reports of Allen Dulles*. University Park, PA: Pennsylvania State University Press.

Eisenhower, J. S. D. (1982) *Allies, Pearl Harbor to D-Day*. Garden City, NY: Doubleday.

Emerson, R. W. (1996) *Representative Men: Seven Lectures*. Harvard University Press.

Eörsi, L. (2006) *The Hungarian Revolution of 1956: Myths and Realities*. Translated by Mario D. Fenyo. New York: Columbia University Press.

Europe. (1948a) "De la Libre Discussion des Idées, par Aragon." *Europe. Numéro Spécial*. 26 Année No. 33–34: 3–30.

———. (1948b) "T. D. Lysenko: Rapport à l'Académie d'Agronomie de l'U.R.S.S. sur l'État de la Science Biologique." *Europe. Numéro Spécial*. 26 Année No. 33–34: 31–68.

Feliciano, H. (1997) *The Lost Museum: The Nazi Conspiracy to Steal the World's Greatest Works of Art*. New York: Basic Books.

Frenay, H. (1976) *The Night Will End*. New York: McGraw-Hill.

Fyfe, J. L. (1947) "The Soviet Genetics Controversy." *Modern Quarterly* 2: 347–56.

Germain, M. (2008) *Glières: Mars 1944: "Un Grande et Simple Histoire."* Montmélian: La Fontaine de Siloé.

Gilbert, M. (2004) *D-Day*. Hoboken, NJ: J. Wiley & Sons.

Gildea, R., et al. (2006), *Surviving Hitler and Mussolini: Daily Life in Occupied Europe*. Oxford; New York: Berg.

Gros, F. et al. (1961) "Unstable ribonucleic acid revealed by pulse labeling of Escherichia coli." *Nature*, 190: 581–585.

Goebbels, J. (1939) "The New Year 1939/1940," from "Jahreswechsel 1939/1940. Sylvesteransprache an das deutsche Volk." *Die Zeit Ohne Beispiel*. Munich: Zentralverlag der NSDAP, 1941: 229–39. Available online at the German Propaganda Archive, http://www.calvin.edu/academic/cas/gpa/goeb21.htm (accessed January 2, 2010).

Graham, L. R. (1964) "A Soviet Marxist View of Structural Chemistry: The Theory of Resonance Controversy." *ISIS* 55: 20–31.

Granville, J. (2003) "Reactions to the Events of 1956: New Findings from the Budapest and Warsaw Archives." *Journal of Contemporary History* 38.2: 261–90.

Grenier, J. (1936) "Sagesse de Lourmarin." *Cahiers du Sud*, May 1936, 183: 390–97.

Guérin, A. (2010) *Chronique de la Résistance*. Paris: Omnibus.

Guillain de Bénouville, P. (1949) *Unknown Warriors, a Personal Account of the French Resistance*." Translated by Lawrence G. Blochman. New York: Simon & Schuster.

Haarr, G. H. (2009) *The German Invasion of Norway: April 1940*. Great Britain: Seaforth Publishing.

Hardré, J. (1964) "Camus Dans La Résistance." *The French Review* 37.6: 646–50.

Harrison, G. (1951) *Cross-Channel Attack*. Washington, DC: Office of the Chief of Military History, Department of the Army.

Henry, P. (2007) *We Only Know Men: The Rescue of Jews in France During the Holocaust*. Washington, DC: Catholic University of America Press.

Hitler, A. (1939) *Mein Kampf: An Unexpurgated Digest*. Translated by B. D. Shaw. New York: Political Digest Press.

Hitler, A., and M. Domanus. (1990) *Hitler, Speeches and Proclamations: 1932–1945*. Volume 4. Translated by Mary Fran Gilbert. Wauconda, IL: Bolchazy-Carducci.

Holmes, F. L. (2006) *Reconceiving the Gene: Seymour Benzer's Adventures in Phage Genetics*. Edited by W. C. Summers. New Haven and London: Yale University Press.

Humbert, A. (2008) *Résistance: A Woman's Journal of Struggle and Defiance in Occupied France*. Translated by Barbara Mellor. New York: Bloomsbury.

Huxley, J. (1963) *Evolution: The Modern Synthesis*. London: George Allen & Unwin.

Jackson, J. (2003a) *The Fall of France: The Nazi Invasion of 1940*. Oxford: Oxford University Press.

———. (2003b) *France: The Dark Years, 1940–1944*. Oxford; New York: Oxford University Press.

Jacob, F. (1976) *The Logic of Life: A History of Heredity*. Translated by Betty E. Spillmann. New York: Random House.

———. (1988) *The Statue Within: An Autobiography/François Jacob*. Translated by Franklin Philip. New York: Basic Books.

———. (1998) *Of Flies, Mice, and Men*. Cambridge: Harvard University Press.

Jacob, F., D. Perrin, C. Sanchez, and J. Monod. (1960) "L'Opéron: Groupe de Gènes à expression coordonnée par un opérateur." *Comptes Rendus Hebdomadaires des Séances de l'Académie des Sciences*. 250: 1727–29.

Jacob, F., and J. Monod. (1959) "Genes of Structure and Genes of Regulation in the Biosynthesis of Proteins." *Comptes Rendus Hebdomadaires des Séances de l'Académie des Sciences*. 249: 1282–84.

———. (1961) "Genetic Regulatory Mechanisms in the Synthesis of Proteins." *Journal of Molecular Biology* 3: 318–56.

Jordan, J. W. (2011) *Brothers, Rivals, Victors: Eisenhower, Patton, Bradley, and the Partnership That Drove the Allied Conquest in Europe*. New York: NAL Caliber/New American Library.

Judson, H. F. (1979) *The Eighth Day of Creation: The Makers of the Revolution in Biology*. First Touchstone Edition, 1980. New York: Simon & Schuster.

Judt, T. (2005) *Postwar: A History of Europe Since 1945*. New York: Penguin Press.

Juhász, B. (1999) "Women in the Hungarian Revolution of 1956. The Women's Demonstration of December 4th." In *Construction. Reconstruction. Wieder. Aufbau.* Edited by Andrea Pet and Béla Rásky. Budapest: Central European University, the Program on Gender and Culture; New York: Open Society Institute, Network Women's Program.

Kay, L. (2000) *Who Wrote the Book of Life? A History of the Genetic Code*. Palo Alto, CA: Stanford University Press.

Kersaudy, F. (1991) *Norway 1940*. New York: St. Martin's Press.

Kershaw, I. (2001) *Hitler, 1936–45: Nemesis*. New York: W. W. Norton.

Khrushchev, N. S. (1956) *Report of the Central Committee of the Communist Party of the Soviet Union to the 20th Party Congress*. Moscow: Foreign Languages Publishing House.

Kosiński, J. (1986) "Death in Cannes." *Esquire*, March, 81–89.

Kupferman, F. (2006), *Le Procès de Vichy: Pucheu, Pétain, Laval, 1944–1945*. Bruxelles: Complexe.

Lattre, J. (1952) *The History of the French First Army*. Translated by Malcolm Barnes. London: Allen & Unwin.

La 2e DB Général Leclerc: Combattants et Combats en France. (1945) Paris: Arts et Métiers Graphiques.

Laub, T. (2010) *After the Fall: German Policy in Occupied France 1940–1944*. Oxford: Oxford University Press.

Lawrence, C., and G. Weisz. (1998) *Greater Than the Parts: Holism in Biomedicine 1920–1950*. Oxford: Oxford University Press.

Lazareff, P. (1942) *Deadline: The Behind-the-Scenes Story of the Last Decade in France*. Translated by David Partridge. New York: Random House.

Le Sueur, J. (2001) *Uncivil War: Intellectuals and Identity Politics During the Decolonization of Algeria*. Philadelphia: University of Pennsylvania Press.

Lendvai, P. (1998) *Blacklisted: A Journalist's Life in Central Europe*. London; New York: I. B. Tauris.

Liebling, A. J. (2008) *World War II Writings*. New York: Library of America.

Lomax, B. (1982) "Twenty-Five Years After 1956: The Heritage of the Hungarian Revolution." *Socialist Register* 19: 79–104.

Lottman, H. R. (1979) *Albert Camus: A Biography*. Garden City, NY: Doubleday.

———. (1992) *The Fall of Paris: June 1940*. New York: HarperCollins.

Luneau, A. (2005) *Radio Londres: Les Voix de la Liberté (1940–1944)*. Paris: Perrin.

Luria, S. E., and M. Delbrück. (1943) "Mutations of Bacteria from Virus Sensitivity to Virus Resistance. *Genetics* 28: 491–511.

Macridis, R. (1958) "The Immobility of the French Communist Party." *Journal of Politics* 20.4: 613–34.

Matthews, J. (2007) *Explosion: The Hungarian Revolution of 1956*. New York: Hippocrene Books.

May, E. R. (2000) *Strange Victory: Hitler's Conquest of France*. New York: Hill and Wang.

Mayr, E. (1997) *This Is Biology: The Science of the Living World*. Cambridge, MA: Belknap Press of Harvard University Press.

McCarty, M., and O. T. Avery. (1946) "Studies on the Chemical Nature of the Substance Inducing Transformation of Pneumococcal Types." *Journal of Experimental Medicine* 83(2): 89–96.

McCullough, D. (1999) *The Art of Biography II. Paris Review* 151, Fall.

McIntire, S., and W. E. Burns. (2008) *Speeches in World History*. New York: Facts on File.

Medvedev, J. (1971) *Grandeur et Chute de Lyssenko*. Traduit de [l'édition anglaise] par Pierre Martory. Préface de Jacques Monod. Paris: Gallimard.

Medvedev, Z. A., and R. A. Medvedev. (2006) *The Unknown Stalin*. London: Tauris.

Mitchell, A. (2008) *Nazi Paris: The History of an Occupation, 1940–1944*. New York: Berghahn Books.

Molenaar, L. (1981) "The Lysenko Affair (1927–1981)." *Komma*, December, 67–100.

Monod, J. (1941a) "Croissance des populations bactèriennes en function de la concentration de l'aliment hydrocarboné." *Comptes Rendus Hebdomadaires des Séances de l'Académie des Sciences. Série* D, *Sciences Naturelles*. Paris, France: Gauthier-Villars, 771–74.

———. (1941b) "Sur un Phénomène Nouveau de Croissance Complexe Dans Les Cultures Bactériennes." *Comptes Rendus Hebdomadaires des Séances de l'Académie des Sciences. Série* D, *Sciences Naturelles*. Paris, France: Gauthier-Villars, 934–36.

———. (1942) "Recherches Sur La Croissance Des Cultures Bactériennes, par

Jacques Monod." *Actualités Scientifiques et Industrielles; 911. Microbiologie, 1.* Paris: Hermann & Cie.

———. (1947) "The Phenomenon of Enzymatic Adaptation and Its Bearings on Problems of Genetics and Cellular Differentiation." Seventh Symposium of the Society for the Study of Development and Growth. Held at the Univeristy of Connecticut, Storrs, CT, August 26–29, 1947. *Growth* 11 (4): 223–89.

———. (1952) "Passports and Visas." *Science,* 116: 178–79.

———. (1953) "Letter to the Editor." *Bulletin of the Atomic Scientists* 9.8: 319–20.

———. (1965) "From Enzymatic Adaptation to Allosteric Transitions." *Nobel Lecture, December 11, 1965.* In *Nobel Lectures, Physiology or Medicine 1963–1970* (1972). Amsterdam: Elsevier. Available at http://www.nobelprize.org/nobel_prizes/medicine/laureates/1965/monod-lecture.html

———. (1968) *Lecon Inaugurale Faite Le Vendredi, 3 Novembre 1967.* Nogent-le-Rotrou: Daupeley-Gouverneur.

———. (1969) "On Values in the Age of Science." In *The Place of Value in a World of Facts: Proceedings of the Fourteenth Nobel Symposium Stockholm, September 15–20, 1969.* Edited by Arne Tiselius and Sam Nilsson, 19–27. New York: Wiley Interscience Division.

———. (1971) *Chance and Necessity.* Vintage Books Edition, October 1972. New York: Alfred A. Knopf.

———. (1972) "The Man Who Didn't Find Time to Write His Autobiography." *New Scientist* 56.818: 280–81.

Monod, J., and A. Audureau. (1946) "Mutation et Adaptation Enzymatique Chez Escherichia Coli-Mutabile." *Annales de l'Institut Pasteur* 72: 868–77.

Monod, J., G. Cohen-Bazire, and M. Cohn. (1951) "Sur La Biosynthese de la Galactosidase (Lactase) Chez Escherichia Coli. La Specificite de l'Induction." *Biochimica et Biophysica Acta* 7: 585–99.

Monod, J., and F. Jacob (1961) "General Conclusions: Teleonomic Mechanisms in Cellular Metabolism, Growth, and Differentiation." *Cold Spring Harbor Symposium of Quantitative Biology* 26: 389–401.

Müller-Hill, B. (1996) *The Lac Operon: A Short History of a Genetic Paradigm.* Berlin; New York: Walter Gruyter.

Murphy, R., et al. (1943) *National Socialism; Basic Principles, Their Application by the Nazi Party's Foreign Organization, and the Use of Germans Abroad for Nazi Aims.* Washington, DC: US Government Printing Office.

Nicholas, L. (1994) *The Rape of Europa: The Fate of Europe's Treasures in the Third Reich and the Second World War.* New York: Knopf.

Notin, J. C. (2000) *1061 Compagnons: Histoire des Compagnons de la Libération.* Paris: Perrin.

Noufflard, G. (unpublished) *Second World War Years.* Unpublished memoir, used by permission.

Orme, J. (1998) "Dismounting the Tiger: Lessons from Four Liberalizations." *Political Science Quarterly* 103.2: 245–65.

Ousby, I. (2000) *Occupation: The Ordeal of France, 1940–1944.* New York: Cooper Square Press.

Pardee, A. (2002) "PaJaMas in Paris." *TRENDS in Genetics* 18.11: 585–87.

Pardee, A., F. Jacob, and J. Monod. (1958) "The Role of the Inducible Alleles and the Constitutive Alleles in the Synthesis of Beta-Galactosidase in Zygotes of Escherichia Coli." *Comptes Rendus Hebdomadaires des Séances de l'Académie des Sciences* 246 (21): 3125–58.

———. (1959) "The Genetic Control and Cytoplasmic Expression of 'Inducibility' in the Synthesis of ß-galactosidase by *E. Coli*." *Journal of Molecular Biology* 1: 165–78.

Patton, G. S., and M. Blumenson. (1974) *The Patton Papers*. Boston: Houghton Mifflin.

Pinault, M. (2000) *Frédéric Joliot-Curie*. Paris: O. Jacob.

Ponty, J., et al. (1996). *La Résistance et les Français: Lutte Armée et Maquis: Colloque International de Besançon 15–17 Juin*. Paris: Les Belles Lettres.

Prados, J. (2011) *Normandy Crucible: The Decisive Battle That Shaped World War II in Europe*. New York: NAL Caliber.

Prenant, M. (1980). *Toute Une Vie a Gauche*. Paris: Editions Encre.

Pryce-Jones, D. (1969) *The Hungarian Revolution*. London: Benn.

———. (1981) *Paris in the Third Reich: A History of the German Occupation, 1940–1944*. New York: Holt, Rinehart, and Winston.

Riley, M., et al. (1960) "On the Expression of a Structural Gene." *Journal of Molecular Biology* 2: 216–25.

Ripley, T. (2003) *Patton Unleashed: Patton's Third Army and the Breakout from Normandy, August–September, 1944*. St. Paul, MN: MBI.

Rol-Tanguy, H., and R. Bourderon. (1994) *Libération de Paris: Les Cent Documents*. Paris: Hachette.

Rottman, G., and P. Dennis. (2010). *World War II Allied Sabotage Devices and Booby Traps*. Botley, Oxford: Osprey Publishing.

Rousso, H. (1984) *Pétain et la Fin de la Collaboration: Sigmarigen, 1944–1945*. Bruxelles: Editions Complexe.

Ryan, C. (1959) *The Longest Day: June 6, 1944*. New York: Simon & Schuster.

Saint-Exupéry, A. (1942) *Flight to Arras*. New York: Reynal & Hitchcock.

Sartre, J. (1957) "After Budapest." *Evergreen Review* 1.1: 5–23.

———. (1969) *The Communists and Peace: With an Answer to Claude Lefort*. Translated by Irene Celphane. London: Hamish Hamilton.

———. (1974) *Select Writings of Jean-Paul Sartre,* Volume 1: *A Bibliographical Life*. Edited by Michael Contat and Michel Rybalka. Translated by Richard C. McCleary. Evanston, IL: Northwestern University Press.

Sartre, J., and A. Camus (2004) *Sartre and Camus: A Historic Confrontation*. Edited and translated by David Sprintzen and Adrian van den Hoven. Amherst, NY: Humanity Books.

Scammell, M. (2009) *Koestler: The Literary and Political Odyssey of a Twentieth-Century Skeptic*. New York: Random House.

Schoenbrun, D. (1980) *Soldiers of the Night: The Story of the French Resistance*. New York: Dutton.

Schrödinger, E. (1992) *What Is Life? With Mind and Matter and Autobiographical Sketches*. Cambridge: Press Syndicate of the University of Cambridge.

Schwartz, L. (1997) *Un mathèmaticien aux prises avec le siècle*. Paris: O. Jacob.

Sebestyen, V. (2006) *Twelve Days: Revolution 1956: How the Hungarians Tried to Topple Their Soviet Masters*. London: Weidenfeld & Nicolson.

Shamir, H. (1976) "The Drôle de Guerre and French Public Opinion." *Journal of Contemporary History* 11: 129–43.

Shirer, W. L. (1969) *The Collapse of the Third Republic; an Inquiry into the Fall of France in 1940*. New York: Simon & Schuster.

———. (1990) *The Rise and Fall of the Third Reich: A History of Nazi Germany*. New York: Simon & Schuster.

Stanier, R. Y. (1977) "Obituary: Jaques Monod, 1910–1976." *Journal of General Microbiology* 101: 1–12.

Stillman, E. (1958) "The Beginning of the 'Thaw,' 1953–1955." *Annals of the American Academy of Political and Social Science*, 317, The Satellites in Eastern Europe: 12–21.

Straub, F. B. (1958) "Formation of Amylase in the Pancreas." *Symposia of the Society for Experimental Biology* 12: 176–84.

Tiersky, R. (1974) *French Communism, 1920–1972*. New York: Columbia University Press.

Tirard, S. (1997) "Les Biologistes Francais et L'Affaire Lyssenko, a L'automne 1948." *Historiens & Géographes* 358: 95–105.

Todd, O. (1996) *Albert Camus: Une Vie*. Paris: Gallimard.

———. (1997) *Albert Camus: A Life*. New York: Alfred A. Knopf.

Toulmond, A. (2005) "Un Biologiste Engagé Dans Son Siècle: Georges Teissier (1900–1972)." Available at http://www.sb-roscoff.fr/histoire-et-patrimoine-sbr/1176-un-biologiste-engage-dans-son-siecle-georges-teissier-1900-1972.html.

Ullmann, A. (2003) *Origins of Molecular Biology: A Tribute to Jacques Monod*. Revised edition. Washington, DC: ASM Press.

Volkin, E., and L. Astrachan. (1956) "Phosphorus Incorporation in *Escherichia coli* Ribonucleic Acid after Infection with Bacteriophage T2." *Virology* 2: 149–61.

Wall, I. (2001) *France, the United States, and the Algerian War*. Berkeley: University of California Press.

Watson, J. D. (1969) *The Double Helix: A Personal Account of the Discovery of the Structure of DNA*. New York: Mentor.

Yeide, H., and M. Stout. (2007) *First to the Rhine: the 6th Army Group in World War II*. St. Paul, MN: MBI.

Zinner, P. (1959) "Revolution in Hungary: Reflections on the Vicissitudes of a Totalitarian System." *Journal of Politics* 21.1: 3–36.

Zuccotti, S. (1993) *The Holocaust, the French, and the Jews*. New York: Basic Books.

Zuehlke, M. (2007) *Terrible Victory: First Canadian Army and the Scheldt Estuary Campaign, September 13–November 6, 1944*. Vancouver: Douglas & McIntyre.

Index

Permissions

Grateful acknowledgment is made to the following for permission to reprint previously published material:

Alfred A. Knopf: Excerpt from THE FIRST MAN by Albert Camus, translated by David Hapgood, translation copyright © 1995 by Alfred A. Knopf, a division of Random House LLC. Excerpt from THE MYTH OF SISYPHUS by Albert Camus, translated by Justin O'Brien, translation copyright © 1955, copyright renewed 1983 by Alfred A. Knopf, a division of Random House LLC. Excerpt from THE PLAGUE by Albert Camus, translated by Stuart Gilbert, translation copyright © 1948, copyright renewed 1975 by Stuart Gilbert. Excerpt from THE REBEL by Albert Camus, translated by Anthony Bower, translation copyright ©1956 by Alfred A. Knopf, a division of Random House LLC. Excerpt from RESISTANCE, REBELLION AND DEATH by Albert Camus, translated by Justin O'Brien, translation copyright © 1960, copyright renewed 1988 by Alfred A. Knopf, a division of Random House LLC. Excerpt from CHANCE AND NECESSITY by Jacques Monod, translated by Austryn Wainhouse, translation copyright © 1971 by Alfred A. Knopf, a division of Random House LLC. Excerpt from ALBERT CAMUS: A LIFE by Olivier Todd, translated by Benjamin Ivry, copyright ©1997 by Alfred A. Knopf LLC. Reprinted by permission of Alfred A. Knopf, a division of Random House LLC.

Don Congdon Associates Inc.: Excerpt from THE COLLAPSE OF THE THIRD REPUBLIC by William L. Shirer. Copyright © 1969 by William L. Shirer, renewed 1997 by Irina Lugovskaya, Linda Rae and Inga Dean. Reprinted by permission of Don Congdon Associates, Inc.

Perseus Books Group: Excerpt from THE STATUE WITHIN by Francois Jacob, translation copyright © 1988 by Basic Books, Inc. Reprinted by permission of the Perseus Books Group as administered by Copyright Clearance Center.

Princeton University Press: Excerpt from CAMUS AT COMBAT: Writing 1944-1947 by Albert Camus, edited by Jacqueline Levi-Valensi. Copyright © 2006 by Princeton University Press. Reprinted by permission of Princeton University Press.

Printed in the United States
by Baker & Taylor Publisher Services